Biotechnology Fundamentals

Biotechnology Fundamentals

Third Edition

Firdos Alam Khan

CRC Press
Taylor & Francis Group
Boca Raton London New York

CRC Press is an imprint of the
Taylor & Francis Group, an **informa** business

CRC Press
Taylor & Francis Group
6000 Broken Sound Parkway NW, Suite 300
Boca Raton, FL 33487-2742

First issued in paperback 2023

ISBN 13: 978-1-03-265345-7 (pbk)
ISBN 13: 978-1-138-61208-2 (hbk)
ISBN 13: 978-1-00-302475-0 (ebk)

DOI: 10.1201/9781003024750

Library of Congress Cataloging-in-Publication Data

Names: Khan, Firdos Alam, author.
Title: Biotechnology fundamentals / authored by Firdos Alam Khan.
Description: Third edition. | Boca Raton : CRC Press, 2020. | Includes bibliographical
   references and index. | Summary: "Biotechnology Fundamentals, Third Edition breaks
   down the basic fundamentals of this discipline, and highlights both conventional and
   modern approaches unique to the industry. The revised work presents new information
   on Forensic Science, Bioinformatics, Synthetic Biology, Biosimilars and Regenerative
   Medicine. In addition to recent advances and updates relevant to the previous edition,
   the revised work also covers ethics in biotechnology and discusses career possibilities
   in this growing field"—Provided by publisher.
Identifiers: LCCN 2019051191 | ISBN 9781138612082 (hardback acid-free paper) |
   ISBN 9781003024750 (ebook)
Subjects: LCSH: Biotechnology. | MESH: Biotechnology.
Classification: LCC TP248.2 .K43 2020 | DDC 660.6—dc23
LC record available at https://lccn.loc.gov/2019051191

Visit the Taylor & Francis Web site at
www.taylorandfrancis.com

and the CRC Press Web site at
www.crcpress.com

eResource material is available for this title at https://www.crcpress.com/9781138612082.

Contents

Preface

After successful launching of the first and second editions of *Biotechnology Fundamentals*, we thought let us find out the feedbacks from our esteemed readers, faculty members, and students about their experiences. After receiving their suggestions and recommendations we thought it would be great idea to write a third edition of the book. Being a teacher of biotechnology, I always wanted a book that covers all aspects of biotechnology, right from basics to applied and industrial levels. In our previous editions, we included all topics of biotechnology which are important and fundamentals for students learning. One of the important highlights of the book is that it has a dedicated chapter for the career aspects of biotechnology and you may agree that many students are eager to know what are the career prospects that they have in biotechnology. There are a great number of textbooks available that deal with molecular biotechnology, microbial biotechnology, industrial biotechnology, agricultural biotechnology, medical biotechnology, or animal biotechnology independently; however, there is not a single book available that deals with all aspects of biotechnology in one book. Today the field of biotechnology is moving with lightning speed. It becomes very important to keep track of all the new information that affects the biotechnology field directly or indirectly.

In this book, I have tried to include all the topics that are directly or indirectly related to the field of biotechnology. The book discusses both historical and modern aspects of biotechnology with suitable examples and gives the impression that the field of biotechnology has been there for ages with different names; you may call them plant breeding, cheese making, in vitro fertilization, or alcohol fermentation—they are all the fruits of biotechnology. The primary aim of this book is to help the students to learn biotechnology with classical and modern approaches and take them from basic information to complex topics. There is a total of 15 chapters in this textbook covering topics ranging from an introduction to biotechnology, genes to genomics, protein to proteomics, recombinant DNA technology, microbial biotechnology, plant biotechnology, animal biotechnology, environmental biotechnology, medical biotechnology, nanobiotechnology, product development in biotechnology, industrial biotechnology, and ethics in biotechnology. All chapters begin with a brief summary followed by text with suitable examples. Each chapter is illustrated by simple line diagrams, pictures, and tables. In addition, we have also included trends in biotechnology topics which help the readers to know major trends in biotechnology such as forensic science, regenerative medicine, biosimilars, synthetic biology, bioinformatics, pharmacogenomics, biorobotics, and biomimetics. Each chapter concludes with a question session, assignment, and field trip information. The brief answers to all questions are in Annexure-I. I have included laboratory tutorials as a separate chapter to expose the students to various laboratory techniques and laboratory protocols. This practical information would be an added advantage to the students while they learn the theoretical aspects of biotechnology.

No one walks alone on the journey of life, but when one is walking, it is where you start to thank those that joined you, walked beside you, and helped you along the way. First, I am grateful to Almighty Allah who gave me strength to write the third edition of the book on stipulated time and made me focused for the entire period of writing. I am thankful to many people who helped me in producing this book especially to Dr. T. Michael Slaughter (former Executive Editor) who encouraged me and supported me to write the first and second editions of the book. For the third edition, I am grateful to Dr. Marc Gutierrez, Editor–Electrical & Biomedical Engineering, CRC Press, Taylor & Francis Group for his support and encouragement. I am also thankful to the entire team of CRC Press/Taylor & Francis Group for making this edition of the book a reality.

I would like to thank the entire management team of Institute for Research & Medical Consultations (IRMC), Imam Abdulrahman Bin Faisal University, Dammam, Saudi Arabia for their support and encouragement, especially to Professor Ebtessam Al-Suhaimi, Dean, IRMC, Imam Abdulrahman Bin Faisal University, Dammam, Saudi Arabia for her constant support.

I am thankful to all my teachers and mentors especially to Professor Nishikant Subhedar PhD and Late Professor Obaid Siddiqi PhD FRS for their immense contributions in making me a true scientist. I am also thankful to all my friends, well-wishers, and colleagues for their support and cooperation.

I am grateful to my entire family, especially to my father Late Nayeemuddin Khan and mother Sarwari Begum, my brothers (Aftab Alam Khan, Javed Alam Khan, Intekhab Alam Khan, Sarfaraz Alam Khan), my sisters (Syeeda Khanum, Faheemida Khanum, Kahkashan Khanum, Aysha Khanum), my wife Samina Khan, and my sons (Zuhayr Ahmad Khan, Zaid Ahmad Khan, and Zahid Ahmad Khan) and my daughter (Azraa Khan). I am also grateful to my father in-law Abdul Qayyum Siddiqi and mother-in-law Uzma Siddiqi and brothers-in-law (Rehan Ahmed Siddiqi, Haroon Ahmed Siddiqi, Noman Ahmed Siddiqi). All of them, in their own ways supported me to complete this project.

I welcome your comments and suggestions to make this book error-free and more interesting in the future. You may send your comments or recommendations to the address below.

Enjoy reading!

Firdos Alam Khan PhD
Professor and Chairman
Department of Stem Cell Biology
Institute for Research and Medical Consultations
Imam Abdulrahman Bin Faisal University
Post Box No. 1982
Dammam 31441,
Saudi Arabia
Email: fakhan@iau.edu.sa

Author

Firdos Alam Khan is a professor and chairman at the Department of Stem Cell Biology, Institute for Research and Medical Consultations, Imam Abdulrahman Bin Faisal University, Dammam, Saudi Arabia. Before joining Imam Abdulrahman Bin Faisal University, Dammam, Saudi Arabia, Professor Khan was a Professor and Chairperson of School of Life Sciences (formerly known as Department of Biotechnology), Manipal University Dubai Campus, in the United Arab Emirates (UAE). He received his PhD degree from Nagpur University, India in 1997. He has more than 25 years of research and teaching experience in various domains of biotechnology and life sciences. He did his first postdoctoral research fellowship from the National Centre for Biological Sciences (NCBS) in Bangalore, India. In 1998, Professor Khan moved to the United States and did his second postdoctoral research fellowship at Department of Brain and Cognitive Sciences, Massachusetts Institute of Technology (MIT), and worked on the research project entitled "Axonal nerve regeneration in adult Syrian hamster." In 2001, he returned to India and joined Reliance Life Sciences, a Mumbai-based biotechnology company, where he associated with adult and embryonic stem cell research projects. During his association with Reliance Life Sciences, Professor Khan along with his colleagues showed that both adult and embryonic stem cells differentiate into neuronal cells. He also developed novel protocols to derive neuronal cells from both adult and embryonic stem cells. Over the past 10 years, he has taught IPR, business of biotechnology, cell biology and regenerative medicine courses to undergraduate and post-graduate students. His area of specialty in biotechnology includes stem cell technology and neuroscience. He has written many articles in various national and international journals in the areas of neuroscience, nanomaterials, and nanomedicine and stem cell biology. His current research interests are stem cell biology and nanomedicine. He has been working on stem cells especially differentiation of stem cells into neural cells (GABAergic and dopaminergic neurons). Professor Khan is also exploring the therapeutic potentials of nanoparticles, biomaterials, and nanocomposites. Professor Khan has been granted three patents from the US patent office on stem cell technology. In 2014, Professor Khan received an Academic Excellence Award from Manipal Global Education Services, Bangalore, India, for his contribution. Professor Khan has been an associate editor of three biotech journals published by Springer. He has been associated with various international scientific organizations like the International Brain Research Organization, France, and Society for Neuroscience, USA. He has presented at more than 25 different national and international conferences in India, Singapore, UAE, and the USA.

1 Introduction to Biotechnology

LEARNING OBJECTIVES
- Define biotechnology
- Discuss the historical perspectives of biotechnology
- Explain the classifications of biotechnology based on its applications
- Explain how biotechnology has revolutionized the healthcare, agricultural, and environmental sectors
- Explain how biotechnology became the science of integration of diverse fields
- Explain ethical issues in biotechnology

1.1 WHAT IS BIOTECHNOLOGY?

Let's quickly learn the difference between bioscience and biotechnology; because there is a difference between the disciplines. Bioscience is the science that studies the basics and fundamentals of living organisms (bacteria or viruses), which include their structure and functions, whereas biotechnology deals with the use of living organisms for making useful products, like bacteria can be used for making an antibiotic medicine and viruses can be used for making a vaccine. Here we can say bioscience teaches you about the internal organization of a living organism, whereas biotechnology teaches you how to use these living organisms for human benefit. Interestingly, the fruits of biotechnology are very evident in everyday life, but sometimes we do not realize that we are benefitting from it, such as when we eat yogurt or when we receive a vaccine. Everyone may not be aware of the formal definition of biotechnology, but one thing is certain, we all have benefitted from the products of biotechnology, such as cheese, detergents, biodegradable plastics, and antibiotics.

It is important to know how these useful products are developed and passed on to us for our own benefits. To really appreciate the benefits of the biotechnology; the best example is the making of a cheese. One may argue that cheese making is no big deal, as cheese can be found in almost every city in the world. In addition, what is the relationship between cheese making and biotechnology? Yes, there is a relationship if you know the different ingredients that are required for making cheese. Let us first learn how to make cheese at home, which will give us a fair idea about cheese making, and then we can learn how to make cheese at the industrial level.

CHEESE MAKING AT HOME:

- Place one cup of milk in a saucepan; bring the milk slowly to a boil while stirring constantly. It is very important to constantly stir the milk, or it will burn.
- Turn the burner off once the milk is boiling but leave the saucepan on the element or gas grate.
- Add vinegar to the boiling milk, at which point the milk should turn into curd and whey.
- Stir well with a spoon and let it sit on the element for 5–10 min.
- Pass the curd and whey through cheesecloth or a handkerchief to separate the curd from the whey.
- Press the cheese using the cloth to get as much of the moisture out as possible.
- Open the cloth and add a pinch of salt if desired.
- Mix the cheese and salt and then press again to remove any extra moisture.

- Put the cheese in a mold or just leave it in the form of a ball.
- Refrigerate for a while before eating.

It's so easy to make a cheese at home in a small quantity; in the same way, if you want to make the cheese in a large quantity (thousands of pounds/kilograms) then adding acid-like vinegar is necessary and sometimes bacteria are also used. These starter bacteria convert milk sugars into lactic acid. The same bacteria and the enzymes they produce also play a large role in the eventual flavor of aged cheeses. Most cheeses are made with starter bacteria from the *Lactococci*, *Lactobacilli*, or *Streptococci* families. Swiss starter cultures also include *Propionibacterium shermanii*, which produces carbon dioxide gas bubbles during aging, giving Swiss cheese (or Emmental) its holes (Figure 1.1). You may now know the application of microorganisms in industrial-level production of cheese, but the role of microorganisms is not limited to cheese making. It has multiple applications in other products as well, such as curd and antibiotic production.

1.1.1 DEFINITIONS OF BIOTECHNOLOGY

The general definition of *biotechnology* is a field that involves the use of biological systems or living organisms to manufacture products or develop processes that ultimately benefit humans. The following are some of the most commonly used definitions of biotechnology:

- The use of living organisms (especially microorganisms) in industrial, agricultural, medical, and other technological applications.
- The application of the principles and practices of engineering and technology to the life sciences.

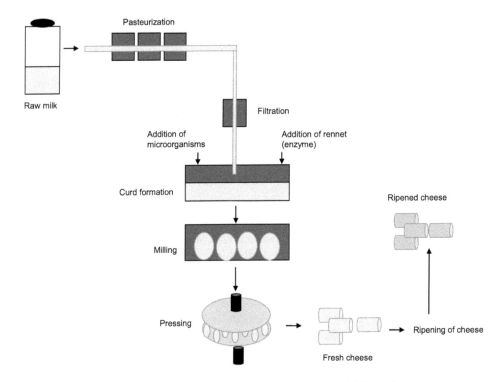

FIGURE 1.1 Schematic representation of cheese making in industry, which involves many steps such as pasteurization, filtration, curd formation, milling, pressing, and ripening of the cheese.

- The use of biological processes to make products.
- The production of genetically modified organisms or the manufacture of products from genetically modified organisms.
- The use of living organisms or their products to make or modify a substance. Biotechnology includes recombinant DNA (deoxyribonucleic acid) techniques (genetic engineering) and hybridoma technology.
- A set of biological techniques developed through basic research and applied to research and product development.
- The use of cellular and biomolecular processes to solve problems or make useful products.
- An industrial process that involves the use of biological systems to make monoclonal antibodies and genetically engineered recombinant proteins.
- Development of 3D organs or tissues under *in vitro* conditions

We should not debate on which of the given definitions is true because all of them are true in their respective ways. For example, if you ask a **farmer** about what biotechnology is, he or she may say, "Biotechnology is to produce high yield or pest-resistant crops." If you pose the same question to a **doctor**, he or she may say, "Biotechnology is about making new vaccines and antibiotics." If you ask the question to an **engineer**, he or she may say, "Biotechnology is about designing new diagnostic tools for better understanding of human diseases," and if you ask the question to a **patient** suffering from Parkinson's disease, he or she may say "Biotechnology is about stem-cell-based therapy and has tremendous capability to cure Parkinson's disease." All of these different definitions of biotechnology suggest that biotechnology has immensely impacted our daily life with arrays of products. As the field of biotechnology keeps expanding, efforts are being made to subclassify this field into various types. The field of biotechnology may be broadly subclassified into animal, plant, medical, industrial, and environmental biotechnology. Nonetheless there are other emerging fields of biotechnology, such as regenerative medicine (Figure 1.2), biosimilars, pharmacogenomics, bioinformatics, therapeutic proteins, forensic science, synthetic biology, bio-robotics, and biomimetics which we have separately discussed in Chapter 12. Now let's briefly go through some of the major fields of biotechnology to learn the diverse applications of biotechnology.

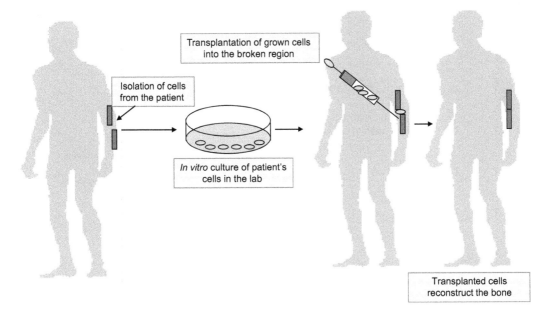

FIGURE 1.2   The application of regenerative medicine in reconstruction of bone by using the patient's own cells.

1.2 MICROBIAL BIOTECHNOLOGY

Microbes are small organisms but possess tremendous capacities in product development. For example, for thousands of years microorganisms have been used in the production of bread, cheese, yogurt, etc. Interestingly, traditional microbial biotechnology began during World War I and resulted in the development of the acetone-butanol and glycerol fermentations, followed by production of vitamins and antibiotics. With the advent of molecular biology and decoding of DNA, microorganisms were used in the development of biopharmaceutical products, such as recombinant insulin (Figure 1.3), recombinant erythropoietin, recombinant human growth hormone, and recombinant interferons. Today, microorganisms are a major contributor in global industry, especially in the dairy, pharmaceutical, biopharmaceutical, food, and chemical industries.

1.3 ANIMAL BIOTECHNOLOGY

Animal biotechnology is the application of scientific and engineering principles to the processing or production of materials from animals or aquatic species to provide research models and to make healthy products. Some examples of animal biotechnology are generation of transgenic animals (animals with one or more genes introduced by human intervention), use of gene knockout technology to generate animals with a specified gene inactivated, production of nearly identical animals by somatic cell nuclear transfer (also referred to as *clones*), and production of infertile aquatic species. Since the early 1980s, methods have been developed and refined to generate transgenic animals. For example, transgenic livestock and transgenic aquatic species have been generated with increased growth rates, enhanced lean muscle mass, enhanced resistance to disease, or improved use of

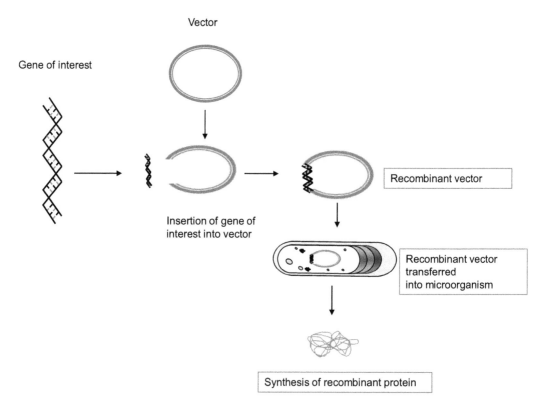

FIGURE 1.3 Therapeutic or recombinant protein is prepared by inserting the gene of interest into a vector and then into a microorganism resulting in synthesis of recombinant protein.

dietary phosphorous to lessen the environmental impacts of animal manure. Transgenic poultry, swine, goats, and cattle have also been produced to generate large quantities of human proteins in eggs, milk, blood, or urine with the goal of using these products as human pharmaceuticals. Some examples of human pharmaceutical proteins are enzymes, clotting factors, albumin, and antibodies. The major factor limiting the widespread use of transgenic animals in agricultural production systems is the relatively inefficient rate (success rate <10%) of production of transgenic animals.

With the help of genetic tools; it's possible to knock out or inactivate a specific gene in animals; this technology is commonly known as Knockout Technology. Knockout technology creates a possible source of replacement organs from animals which can be used for human benefit. The process of transplanting cells, tissues, or organs from one species (animal) to another (human) is referred to as *Xenotransplantation*. Currently, pigs are considered as a viable source for the xenotransplantation in humans. But due to non-matching of pig tissues with human cells, there is a rejection of pig tissues/organs by the human body. The cause of rejection is because pig cells express a carbohydrate epitope (alpha 1, 3 galactose) on their cell surface that is not normally found in human cells. Upon transplantation of pig cells, humans can generate antibodies to this epitope, which will result in acute rejection of the xenograft. Research work has been done to minimize the immune rejection and with the help of genetic engineering; it's now possible to either knock out or inactivate the pig gene (alpha1, 3 galactosyl transferase) that attaches this carbohydrate epitope on pig cells. Another example of knockout technology in animals is the inactivation of the prion-related peptide gene that

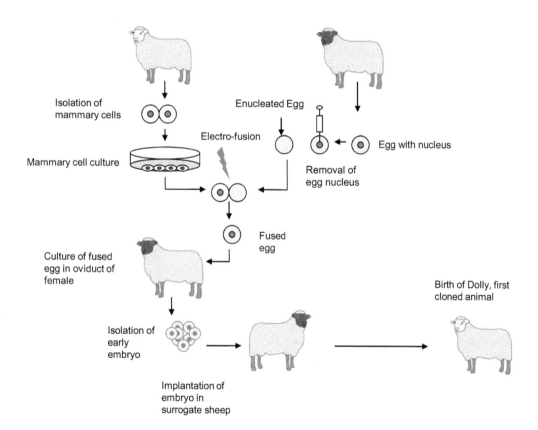

FIGURE 1.4 The method of animal cloning involves isolation of mammary cells, electro-fusion of the enucleated egg, transplantation of the fused egg, isolation of the early embryo, and implantation of the embryo in surrogate sheep.

may produce animals that are resistant to diseases associated with prions, such as bovine spongiform encephalopathy, Creutzfeldt—Jakob disease, etc.

Another application of animal biotechnology is the use of somatic cell nuclear transfer to produce genetically identical copies of an organism. This process has been referred to as *cloning*. To date, somatic cell nuclear transfer has been used to clone cattle, sheep, pigs, goats, horses, mules, cats, rats, and mice. The technique involves culturing somatic cells from an appropriate tissue (fibroblasts) from the animal to be cloned. Nuclei from the cultured somatic cells are then microinjected into an enucleated oocyte obtained from another individual of the same or a closely related species. Through a process that is not yet understood, the nucleus from the somatic cell is reprogrammed to a pattern of gene expression suitable for directing normal development of the embryo. The embryo is further cultured in an *in vitro* environment, and then it is transferred to a recipient female for normal fetal development (Figure 1.4). Another very important application of genetically modified animals is in the drug testing and toxicity evaluations. In Chapter 7, we have discussed various applications of animal biotechnology with great details supported by beautiful illustrations.

1.4 PLANT BIOTECHNOLOGY

Plant biotechnology is also known as *green biotechnology* and involves the use of environment-friendly solutions as an alternative to traditional industrial agriculture, horticulture, and animal breeding processes. The following are some examples of green biotechnology:

- The use of bacteria to facilitate the growth of plants
- Development of pest-resistant grains
- Engineering of plants to express pesticides
- Accelerated evolution of disease-resistant animals
- The use of bacteria to assure better crop yields (instead of using pesticides and herbicides)
- Production of superior plants by stimulating the early development of their root systems
- The use of plants to remove heavy metals such as lead, nickel, or silver, which can then be extracted (or "mined") from the plants
- Genetic manipulation to allow plant strains to be frost-resistant
- The use of genes from soil bacteria to genetically alter plants to promote tolerance to fungal pathogens
- The use of bacteria to have the plants grow faster, resist frost, and ripen earlier (Figure 1.5)

In Chapter 6, we have discussed various applications of plant biotechnology with great details supported by beautiful illustrations.

1.5 NANOBIOTECHNOLOGY

Nanobiotechnology is a discipline in which tools or techniques from nanotechnology are developed and applied to study the field of biology. For example, nanomaterials or nanoparticles can be used as probes, sensors for diagnostic purposes, or as vehicles for drug or biomolecule delivery or as drug molecules to treat various diseases such as cancer. In Chapter 10, we have discussed various applications of nanobiotechnology with great details supported by beautiful illustrations (Figure 1.6).

1.6 ENVIRONMENTAL BIOTECHNOLOGY

Environmental biotechnology is when biotechnology is applied to and used to study the natural environment. Environmental biotechnology could also imply that one tries to harness the biological process for commercial use and exploitation. The International Society for Environmental Biotechnology defines environmental biotechnology as "the development, use and regulation of biological systems for remediation of contaminated environments (land, air, water), and for

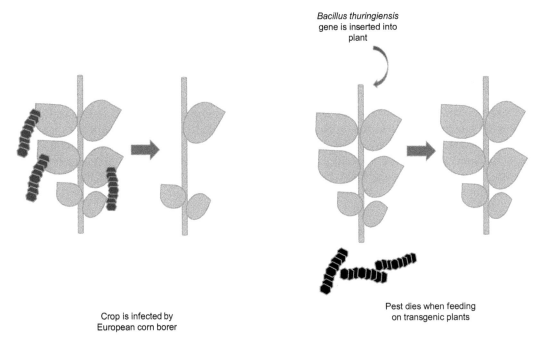

FIGURE 1.5 Bioengineered plant: Following the insertion of a gene from the bacteria *Bacillus thuringiensis*, corn becomes resistant to corn borer infection.

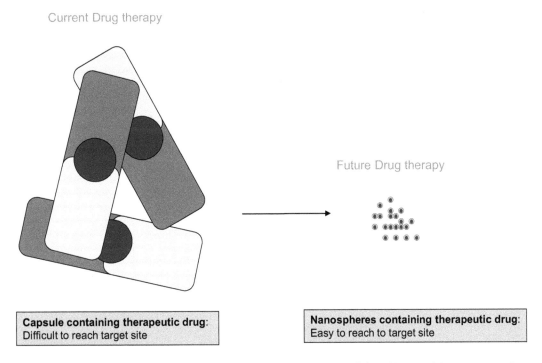

FIGURE 1.6 Difference between traditional medicine and nanomedicine: Nanoparticles constructed to carry a therapeutic payload.

environment-friendly processes (green manufacturing technologies and sustainable development)." In Chapter 8, we have discussed various applications of environmental biotechnology with great details supported by beautiful illustrations.

1.7 MEDICAL BIOTECHNOLOGY

Medical biotechnology deals with the development of therapy using cells or microorganisms by employing molecular engineering techniques. It includes the designing of organisms to manufacture pharmaceutical products like therapeutic proteins (growth hormones, insulin), antibiotics, vaccines, regenerative medicine, and gene therapy. The term medical biotechnology is also used in forensics through DNA Profiling (Figure 1.7). Chapter 9 is dedicated to medical biotechnology.

1.8 INDUSTRIAL BIOTECHNOLOGY

Industrial biotechnology, also known as *white biotechnology*, is the application of biotechnology for industrial purposes such as manufacturing of biomolecules, enzymes or chemicals, and biomaterials. It includes the practice of using cells or components of cells like enzymes to generate industrially useful products. It uses living cells from yeasts, molds, bacteria, plants, and enzymes to synthesize products that are easily degradable, require less energy, and create less waste during their production. Some examples include the designing of an organism to produce a useful chemical and the use of enzymes as industrial catalysts to either produce valuable chemicals or destroy hazardous/polluting chemicals. White biotechnology consumes fewer resources (compared to the traditional processes) to produce industrial goods (Figure 1.8). In Chapter 11, we have discussed various applications of industrial biotechnology with great details supported by beautiful illustrations.

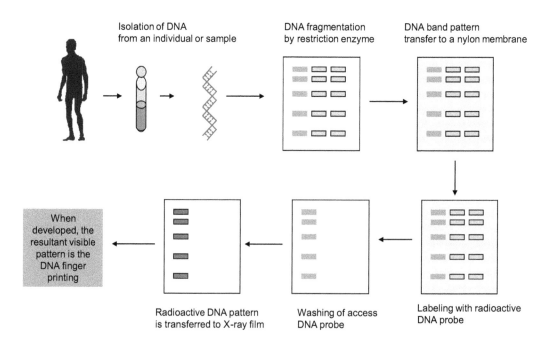

FIGURE 1.7 DNA fingerprinting technique shows the isolation of DNA, DNA fragmentation, DNA band pattern transfer, labeling with a radioactive probe, and transferring of the DNA on X-ray film.

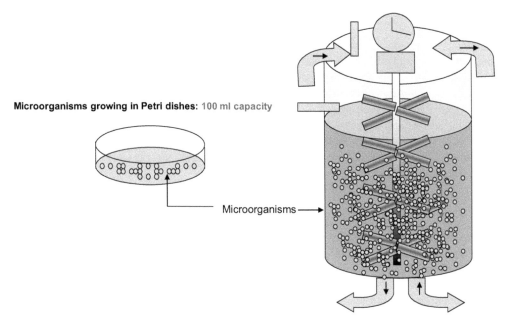

FIGURE 1.8   Difference between small-scale and large-scale production of microorganisms depicted in a Petri dish and a bioreactor, respectively.

1.9   HISTORY OF BIOTECHNOLOGY

After learning various definitions and applications of biotechnology, we will now go through the historical aspects of biotechnology. Most people still think that biotechnology is a relatively new discipline that is only recently getting a lot of attention. It may surprise you to know that, in many ways, this technology involves several ancient practices and methodologies. In fact, our ancestors have been using biotechnology for their benefit in many processes for many centuries. Although the word "biotechnology" was not yet in use at that time, the application itself already existed. One of the early applications of biotechnology was the use of microorganisms in making bread, cheese, yogurt, and alcoholic beverages.

We can view the development of biotechnology through the traditional as well as the modern window. One of the common links between traditional and modern biotechnology is that in both periods, man has been exploiting organisms to generate products. In modern biotechnology, human manipulation of the genes of organisms and their insertions into other organisms are used to acquire desired traits, whereas in traditional biotechnology, microorganisms were used in fermentation.

1.9.1   Ancient Biotechnology

An old saying states, "Necessity is the mother of all inventions." Over the centuries, our ancestors have been using breeding techniques based on phenotype characteristics to create animals and plants with desirable traits (high milk-producing cows) and benefits (high-yield crops). During this period, the best animals and plants have been bred together, and each successive generation has been more likely to carry the desirable traits of the parent animal or plant. A hundred years ago, an organism's DNA would have been scanned first to look for desirable traits, and then organisms with those traits would have been bred. Today this is no longer necessary since we can now genetically engineer animals.

Another form of biotechnology that has been around for thousands of years is the use of micro-organisms in food. Microorganisms are used to turn milk into cheese and yogurt and to ferment alcohol. Yeast is used in bread to make it rise. These are considered as biotechnology because they utilize microorganisms.

Biotechnology is as old as the ancient cultures of the Indians, Chinese, Greeks, Romans, Egyptians, Sumerians, and other ancient communities of the world. Some examples of ancient bio-technology are the use of microorganisms for fermentation, domesticating animals for livestock, alcohol in the form of wine and beer, herbal remedies, and plant balms for the treatment of wounds and ailments. The contribution of other scientific fields has greatly helped the development of bio-technology as it utilizes the sciences of biology, chemistry, physics, engineering, computers, and information technology to develop tools and products that hold great promise and hope for thou-sands of patients who are suffering from various incurable diseases such as cancer, diabetes, and Parkinson's disease.

1.9.2 Modern Biotechnology

Modern biotechnology deals more with the treatment of ailments and the alteration of organisms to better human life. Most breakthroughs in biotechnology have been relatively recent, with the earliest advancement about 170 years ago with the discovery of microbes. Proteins were discov-ered only in 1830, with the isolation of the first enzyme following closely 3 years later. In 1859, Charles Darwin published his revolutionary book *On the Origin of Species*. Six years later, Gregor Mendel, who is considered the father of modern genetics, discovered the laws of heredity and set the groundwork for genetic research. Near the turn of the century, Louis Pasteur and Robert Koch provided the basis for research in microbiology. These numerous advancements allowed modern biotechnology to rise. In the early twentieth century, the modern biotechnology movement started, particularly in immunology and genetics. Penicillin, computers, the discovery of DNA as the genetic basis, the use of bacteria to treat raw sewage (bioremediation project) are significant developments in this direction. Revolution in forensics and biomedical science took place with the new laboratory methods such as DNA sequencing, protein analysis, and polymerase chain reaction (PCR). The millennium ended with the introduction of the first cloned sheep (named Dolly), which started the debate over the ethical issues relating to biotechnology, stem cell research, genetic testing, and genetically modified organisms. Modern biotechnology received a big boost when Watson and Crick discovered the double helix of DNA structure, which allowed researchers to study the genetic code of life in detail and opened an era of genetic engineering, genetic mapping, or genetic manipulation. The twenty-first century started with the development of the rough draft of the human genome, or the map of human life. The milestones in the field of biotechnology are listed in Table 1.1.

TABLE 1.1
Milestones of the Human Genome Project

1986: The birth of the human genome project

1990: Project initiated by the US Department of Energy and the National Institutes of Health

1994: Genetic Privacy Act was proposed

1996: Welcome Trust joined the project

1998: Celera Genomics formed to sequence much of the human genome in 3 years

1999: Completion of the sequence of Chromosome 22

2000: Completion of the working draft of the entire human genome

2001: Analysis of the working draft was published

2003: Human Genome Project was completed

1.10   HUMAN GENOME PROJECT

In the quest to chart the innermost reaches of the human cell, scientists have set out biology's most important mapping expedition, the *Human Genome Project* (HGP). Its mission is to identify the full set of genetic instructions contained in human cells and to read the complete text written in the language of the hereditary chemical DNA. The project began with the culmination of several years of work supported and subsequently initiated by the United States Department of Energy. A 1987 report stated boldly, "The ultimate goal of this initiative was to understand the human genome" and "knowledge of the human is necessary to the continuing progress of medicine and other health sciences as knowledge of human anatomy has been for the present state of medicine." Candidate technologies were already being considered for the proposed undertaking at least as early as 1985. James D. Watson was the head of the National Center for Human Genome Research at the National Institutes of Health (NIH) in the United States starting from 1988. Largely due to his disagreement over the issue of patenting genes, Watson was replaced by Francis Collins in April 1993, and the name of the center was changed to the National Human Genome Research Institute (NHGRI) in 1997.

The $3 billion project was founded in 1990 by the United States Department of Energy and the US NIH and took almost 15 years to complete. In addition to the United States, the international consortium comprised geneticists from the United Kingdom, France, Germany, Japan, China, and India. Due to widespread international cooperation and advances in the field of genomics, the study of genomes of organisms, as well as major advances in computing technology, a "rough draft" of the genome was finished in 2000. Ongoing sequencing led to the announcement of the essentially complete genome in April 2003, 2 years earlier than planned. In May 2006, another milestone was passed on the way to completion of the project when the sequence of the last chromosome was published in the journal *Nature*.

Inherited diseases are rare, but there are more than 3000 disorders which are known to be caused by alteration of a single gene. Current knowledge of genetic tools are not sufficient to provide a successful cure for most of these disorders. However, having a gene in hand allows scientists to study its structure and characterize the molecular alterations or mutations that result in disease. Progress in understanding the causes of cancer, for example, has taken a leap forward with the recent discovery of cancer genes. The goal of the HGP was to provide scientists with powerful new tools to help them clear the research hurdles that now keep them from understanding the molecular essence of other tragic and devastating illnesses, such as schizophrenia, alcoholism, Alzheimer's disease, and manic depression.

Gene mutations probably play a role in many of today's most common diseases, such as heart disease, diabetes, immune system disorders, and birth defects. These diseases are believed to result from complex interactions between genes and environmental factors. When genes for diseases have been identified, scientists can study how specific environmental factors, such as food, drugs, or pollutants, interact with those genes. Once a gene is located on a chromosome and its DNA sequence worked out, scientists can then determine which protein in the gene is responsible for the disease and then find out its function in the body. This is the first step in understanding the mechanism of a genetic disease and eventually conquering it. One day it may be possible to treat genetic diseases by correcting errors in the gene itself, replacing its abnormal protein with a normal one, or by switching the faulty gene off. Finally, the HGP research will help solve one of the greatest mysteries of life: How does one fertilized egg "know" how to generate so many different specialized cells, such as those making up muscles, brain, heart, eyes, skin, blood, and so on? For a human being or any organism to develop normally, a specific gene or set of genes must be switched on in the right place in the body at exactly the right moment in development. Information generated by the HGP will shed light on how this intimate dance of gene activity is choreographed into the wide variety of organs and tissues that make up a human being.

1.11 MAJOR SCIENTIFIC DISCOVERIES IN BIOTECHNOLOGY

To understand the significance of biotechnology, it is very important to know the various constituents of science and technology that greatly helped the field of biotechnology. In this section, we will go through various milestones that greatly impacted biotechnology. Among these discoveries, the application of selective breeding of plants and animals has produced a great impact on improving crop and livestock productions for human consumption. In selective breeding, organisms with desired features are purposely mated to produce offspring with the same desirable characteristics. For example, mating plants that produce the largest, sweetest, and most tender ears of corn is a good way for farmers to maximize their land to produce the most desirable crops. Selective breeding in plants has been done primarily based on the phenotype information on plants and recent advancement in the field of genetic engineering. This made it possible to create a unique individual plant, called a *transgenic plant*. The selective breeding technique was not confined to plants alone. It had also been extensively used in animals such as cows, chickens, goats, and pigs to improve the population and quality of farm animals. Like plants, selective breeding in animals has been done primarily based on the phenotype information of animals and recent advancement in the field of genetic engineering. It also made it possible to create transgenic animals by manipulating specific genes of interest.

Another discovery which revolutionized the treatment procedure for microbial infections in humans is the use of antibiotics in treating microbial infections. In 1918, Sir Alexander Fleming discovered that the mold *Penicillium* inhibited the growth of the human skin disease-causing bacteria called *Staphylococcus aureus*. He was then able to successfully isolate and purify the antibiotic substance of the mold to use it for human purposes. *Antibiotics* are substances produced by microorganisms that normally inhibit the growth of other microorganisms. In the1940s, penicillin became a widely available drug for treating microbial infections in human beings. Currently, a wide variety of microorganisms have been used to generate thousands of liters of antibiotic drugs by using advanced biotechnology tools.

Between 1950 and 1960, a series of discoveries about the human genetic code unfolded. This started in 1953 with the publication of a research article by James Watson and Francis Crick, who had discovered the structure of the human DNA. Nine years later, in 1962, they shared the Nobel Prize in Physiology or Medicine with Maurice Wilkins for solving one of the most important biological riddles. This discovery has led to the birth of genetic manipulation and genetic engineering. With the help of genetic engineering, the genes of interest can be identified and manipulated to develop the desired product. This process of genetic manipulation is called *recombinant DNA technology*. The recombinant DNA technique was first proposed by Peter Lobban, a graduate student with A. Dale Kaiser at the Stanford University Department of Biochemistry. The technique was then realized by a group of researchers and the finding was published in reputed international journals. Recombinant DNA technology was made possible by the discovery, isolation, and application of restriction endonucleases by Werner Arber, Daniel Nathans, and Hamilton Smith, for which they received the 1978 Nobel Prize in Medicine. Cohen and Boyer applied for a patent on the process for producing biologically functional molecular chimeras which could not exist in nature in 1974.

In recent years, recombinant DNA technology has been extensively employed to generate therapeutic products for treating human diseases. For example, with a rapid increase in diabetic patients, there was an urgent need for a drug which could help this condition. Through their efforts, scientists were able to successfully synthesize insulin molecules by using recombinant DNA technology. The application of recombinant DNA technology has produced arrays of therapeutic products and disease-resistant crops that produce a greater harvest. Scientists were also able to generate a better quality of rice, such as golden rice. Recombinant DNA technology has also been applied to create engineered bacteria that are capable of degrading environmental pollutants.

1.12 BIOTECHNOLOGY AS THE SCIENCE OF INTEGRATION

We have learned about various applications of biotechnology to create diverse products, and that this could not have been achieved without the contributions of various fields of science and technology. In this section, we will discuss the fundamentals of biotechnology.

The foundation of biotechnology is based on biology and chemistry, supported by mathematics, physics, engineering, and computer and information technology. The integration of various disciplines in biotechnology is well described in Figure 1.9. We know that biotechnology is initially known for microbial-related products. So, to make a microbial product, microbiology is integrated with *Bioreactor technology*, a machine used to manufacture microbes in great number, to develop a new technology known as microbial biotechnology.

The discovery of the DNA double helix and its recombinant capabilities has given a completely new dimension to the field of biotechnology by integrating molecular biology, genetics, human pathology, biochemistry, and microbiology and resulted in a recombinant insulin product. The integration is not limited only to the fields of biology and engineering. We also find integration in the field of information technology. One such example is Bioinformatics, which studies and analyzes genetic data from a large databank using specialized tools and software (Figure 1.10). The integration of the sciences of pharmacology, genetics, and information technology gave birth to a new technology called Pharmacogenomics. Pharmacogenomics is an emerging field of biotechnology that makes genetically tailored medicines. All of these examples suggest that biotechnology is a highly diversified field with enormous applications. Some new integration is also happening in other fields such as nanobiotechnology, biometrics, and biomedical equipment. The field of biotechnology keeps expanding with the emergence of new technology and knowledge, so we expect to see more integration in the future as well.

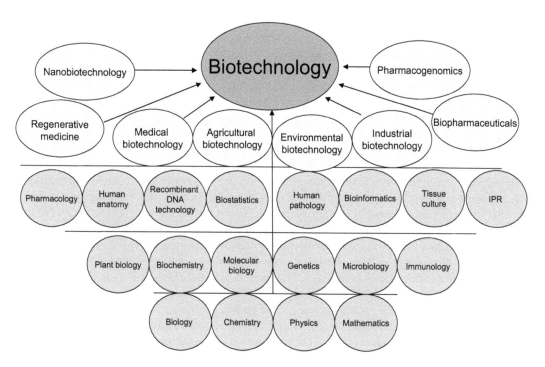

FIGURE 1.9 Biotechnology is an integration of diverse fields.

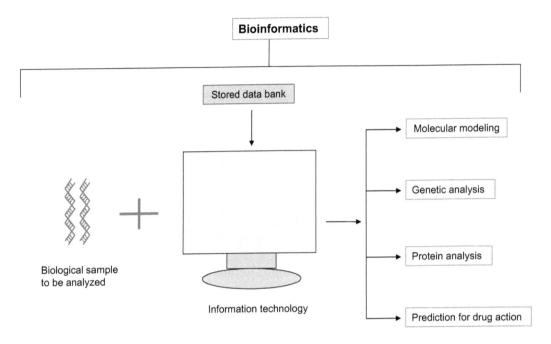

FIGURE 1.10   Applications of bioinformatics in molecular modeling, genetic analysis, protein analysis, and prediction of drug action.

1.13   BIO-REVOLUTION

The bio-revolution resulting from advances in molecular biosciences and biotechnology had already outstripped the advances of the "Green Revolution." In the early 1960s, the pioneering studies of Nobel Prize winner Norman Borlaug, using crossbreeding techniques based on classical genetics, offered for the first time a weapon against hunger in the countries of Latin America, Asia, and Africa. As a direct result of the comprehensive studies of Borlaug and his contemporaries, new wheat hybrids began to transform the harvests of India and China, although they had a relatively minor influence on agriculture in more temperate climates. There is little doubt that genetic manipulation will open more new doors in this field and will dramatically alter farming worldwide. It does not require a crystal ball to imagine the potential of the immediate biotechnological future. From the advances in recent years, it is possible to extrapolate the number of likely developments based on the researches which are now in progress.

In the plant world, the 1978 development of the "pomato," a laboratory-generated combination of two members of the Solanaceae family (the potato and the tomato), was very significant. The Flavr Savr tomato was reviewed by the US Food and Drug Administration (FDA) in the spring of 1994 and found to be as safe as conventionally produced tomatoes. This was the first time the FDA had evaluated a whole food produced by biotechnology. Exciting prospects are likely to result from industrial-scale plant tissue culture. This may soon obviate the need for rearing whole plants to generate valuable commodities such as dyes, flavorings, drugs, and chemicals. Cloning techniques could prove to be the way to tackle some of the acute problems of reforesting in semidesert areas. Seedlings grown from the cells of mature trees could greatly speed up this process. In the summer of 1987, a Belgian team introduced into crop plants a group of genes encoding for insect resistance and resistance to widely used herbicides. This combination of advantageous genes brought about a new era in plant protection. The crop can be treated safely with more effective doses of weed killer, and it is also engineered to be less susceptible to insect damage.

Dairy farming is also benefiting from advances in biotechnology. Bovine somatotropin (growth hormone) enhances milk yields with no increase in feed costs. Embryo duplication methods mean that cows will bear more calves than in the past, and embryo transfer techniques are enabling cattle of indifferent quality to rear good quality stock, a potentially important development for nations with less advanced agriculture. Genetic manipulation of other stock, such as sheep and pigs, appears to be feasible, and work is in progress on new growth factors for poultry.

The outcome of this intense activity will be improvements in the texture, quality, variety, and availability of traditional farm products, as well as the emergence of newly engineered food sources. Such bioengineered superfoods will be welcomed, and will offer new varieties, and hence find new markets in the quality-conscious advanced countries. Despite the enormous potential gains, the economic consequences of possible overproduction in certain areas must also be faced. It will be essential for those concerned with making agricultural policies to keep abreast of the pace of modern biotechnology. Short-term benefits to the consumer of lower agricultural prices must be weighed against a long-term assessment of the impact of new discoveries on the farming industry.

In the medical field, considerable efforts will be devoted to the development of vaccines for killer diseases such as AIDS. Monoclonal antibodies will be used to boost the body's defenses and guide anticancer drugs to their target sites. This technology may also help to rid the human and animal world of a range of parasitic diseases by producing specific antibodies to parasites. Synthesis of drugs, hormones, and animal health products, together with drug-delivery mechanisms, are all advancing rapidly. Enzyme replacement and gene replacement therapy are other areas where progress is anticipated. The next decade will see significant advances in medicine, agriculture, and animal health directly attributable to biotechnology. The impact of new technology will not, however, be confined to bio-based industries. Genetically engineered microbes may become more widely used to extract oil from the ground and valuable metals from factory wastes. Although most biotechnology companies are primarily based in North America and Europe, a tremendous growth has been witnessed in biotechnology-related activities in Singapore, Australia, China, and India.

1.14   FUTURE OF BIOTECHNOLOGY

The recent emphasis on environmental awareness has challenged scientists to find solutions for better and safer living conditions. The added threat of deadly diseases such as AIDS and resistant strains of tuberculosis, gonorrhea, bird flu, and swine flu have forced scientists to look for new therapies within the field of biotechnology. The structure of DNA was deciphered by James Watson, a geneticist, and Francis Crick, a physicist, thus marking the beginning of molecular biology in the twentieth century. Their determination of the physical structure of the DNA molecule became the foundation for modern biotechnology, enabling scientists to develop new tools to improve the future of mankind. The HGP was a major biotechnological endeavor, the aim of which was to make a detailed map of the human DNA. The hereditary instructions inscribed in the DNA guide the development of the human being from the fertilized egg cell to his or her death. In this project, which took 15 years to complete, chromosome maps were developed in various laboratories worldwide through a coordinated effort guided by the NIH. The genetic markers for over 4000 diseases caused by single mutant genes have been mapped. To get an idea of the magnitude of this project, imagine a stack of 25,000 books. If each book is 2 cm thick, the stack would measure 500 m, the height of a 15-story building. Consider locating a word within one of the books in the stack. For a molecular biologist, this would be analogous to finding one gene in the human genome. Up to this point, molecular biologists have mapped only a tiny fraction of the genome. The 23 pairs of human chromosomes are estimated to contain between 50,000 and 100,000 genes, of which apparently only about 5% have so far been transcribed.

Recombinant DNA biotechnology has aroused public interest and concern and has influenced medicine, industry, agriculture, and environmental problem solving in the past 20 years since its inception. In medicine, faster and more efficient diagnosis and treatment of diseases such as cystic fibrosis, cancer, sickle cell anemia, and diabetes are soon to be developed. Recombinant organisms will be used in industry to produce new vaccines, solvents, and chemicals of all kinds. Biotechnology also has applications in both plant and animal breeding. Scientists are developing disease- and herbicide-resistant crops, disease-resistant animals, seedless fruits, and rapidly grow-ing chickens. Microbes are also being engineered to digest compounds that are currently polluting our environment.

Some of the most exciting frontiers of biotechnology include protein-based biochips, which may replace silicon chips. It is believed that the biochips would be faster and more energy-efficient and that these biochip implants in the body could deliver precise amounts of drugs to affect heart rate and hormone secretion or to control artificial limbs. Moreover, biosensors are monitors that use enzymes, monoclonal antibodies, or other proteins to test air and water quality, to detect hazardous substances, and to monitor blood components *in vivo*.

Gene therapy involves correction of defects in genetic material. In this process, a normal gene is introduced to replace a malfunctioning one. Gene therapy will be the "expression" of the medi-cal research branch of biotechnology. It may, in time, form the basis of its own industry or join the traditional pharmaceutical industry. New delivery systems, called *liposomes*, are being developed to get cytotoxic drugs to tumor sites with minimal damage to surrounding healthy tissues. New monoclonal antibodies will be isolated for use in cancer treatment, diagnostic testing, bone marrow transplantation, and other applications.

Progress in biotechnology is currently working on environmental-friendly biodegradation pro-cesses for a cleaner and healthier planet. This includes experimenting with the until now untapped energy sources, and devising useful consumer chemicals such as adhesives, detergents, dyes, fla-vors, perfumes, and plastics. With the progress made thus far in the fight against deadly diseases such as polio and smallpox, it is not unreasonable to expect biotechnology to hold the promise for effective treatments or even cures for, say, cancer and AIDS. Gene therapy may well become the method whereby we correct congenital diseases caused by faulty genes. Stem cell research may prove the panacea for Parkinson's disease, multiple sclerosis, and muscular dystrophy. Moreover, given the genetic improvements made with crop yield and nutritive value, world hunger and malnu-trition may witness their end with the continual advancement of biotechnology.

PROBLEMS

Section A: Descriptive Type

Q1. What are different definitions of biotechnology?
Q2. Differentiate ancient biotechnology from modern biotechnology.
Q3. Explain the significance of the HGP.
Q4. What is bio-revolution?
Q5. How does academic research differ from industrial research?

Section B: Multiple Choice

Q1. Most cheeses are made with starter bacteria from the *Lactococci, Lactobacilli,* or *Streptococci* families. True/False
Q2. Which of the following is not a product of recombinant DNA technology?
   a. Human insulin
   b. Human growth hormone
   c. Antibiotic
   d. Antibody
Q3. Which is the branch of pharmacology that deals with the influence of genetic variation on drug response?
   a. Gene therapy
   b. Bioinformatics
   c. Pharmacogenomics
   d. Medicine

Q4. Recombinant human proteins are manufactured using nonhuman mammalian cells that are engineered to express certain human genetic sequences to produce specific proteins. True/False

Q5. Which of the following is not a product of ancient biotechnology?

    a. Cheese
    b. Alcohol
    c. Human insulin
    d. Animal breeding based on traits

Q6. When was the HGP founded?

    a. 1989
    b. 1990
    c. 1991
    d. 1995

Q7. Sir Alexander Fleming discovered that the mold Penicillium promoted the growth of human skin disease-causing bacteria called *Staphylococcus aureus*. True/False

Q8. Recombinant DNA technology was made possible by the discovery, isolation, and application of . . .

    a. Exonucleases
    b. Restriction endonucleases
    c. RNases
    d. Telomerases

Q9. Flavr Savr tomato is the first genetically engineered fruit approved by the US FDA. True/False

Q10. Which of the following is the first genetically engineered rice?

    a. Silver rice
    b. Golden rice
    c. Brown rice
    d. Basmati rice

Section C: Critical Thinking

Q1. Why is biotechnology a field of diverse sciences? Explain.

Q2. More than 3000 diseases are known to result from genetic mutations, so how can one correct these genetic mutations to cure the disorders?

Q3. Do you believe that ethical issues are hindrances in the development and progress of the field of biotechnology? Why?

ASSIGNMENT

Prepare a poster based on various applications of biotechnology in medicine, agriculture, environment, and industry. Describe the application in each field using suitable examples. Posters can be displayed in the classroom or in the lobby to enhance general awareness.

ONLINE RESOURCES

Current information pertaining to biotechnology may be directly accessed from biotechnology journals, discussion panels, and societies.

- www.nature.com/articles/twas08.40a
- www.pbslearningmedia.org/collection/biot/#.WtcjuoVOL5o
- www.ncbi.nlm.nih.gov/pubmed/
- www.sciencedaily.com/news/plants_animals/biotechnology/

- www.usda.gov/topics/biotechnology
- www.bio.org/
- www.usda.gov/topics/biotechnology/biotechnology-frequently-asked-questions-faqs

REFERENCES AND FURTHER READING

Bains, W.E. *Biotechnology from A to Z*. Oxford University Press, Oxford, UK, 1993.

Barnhart, B.J. DOE human genome program. *Hum. Genome Quart*. 1: 1, 1989. www.ornl.gov/sci/techresources/ Human_Genome/publicat/hgn/v1n1/01doehgp.html (retrieved on February 3, 2005).

Benton, D. Bioinformatics-principles and potentials of a new multidisciplinary tool. *Trends Biotechnol*. 14: 261–272, 1996.

Cavalieri, D., McGovern, P.E., Hart, D.L., Mortimer, R., and Polsinelli, M. Evidence for *S. cerevisiae* fermentation in ancient wine. *J. Mol. Evol*. 57: S226–S232, 2003.

DeLisi, C. Genomes: 15 years later a perspective by Charles DeLisi, HGP pioneer. *Hum. Genome News*. 11: 3–4, 2001. http://genome.gsc.riken.go.jp/hgmis/publicat/hgn/v11n3/05delisi.html (retrieved on February 3, 2005).

Dirar, H. *The Indigenous Fermented Foods of the Sudan: A Study in African Food and Nutrition*. CAB International, Cambridge, UK, 1993.

Fermented fruits and vegetables: A global perspective. *FAO Agricultural Services Bulletins* 134, January 19, 2007. www.fao.org/docrep/x0560e/x0560e05.htm (retrieved on January 28, 2007).

Kreuzer, H., and Massay, A. *Recombinant DNA and Biotechnology: A Guide for Students*. ASM Press, Washington, DC, 2001.

Nath, I. Bio-revolution. *Nature*. 456: 40, 2008.

Pederson, R.A. Embryonic stem cells for medicine. *Sci. Am*. 280: 68–73, 1999.

Steinkraus, K.H. *Handbook of Indigenous Fermented Foods*. Marcel Dekker, Inc., New York, 1995. www.phppo.cdc.gov/phtn/botulism/alaska/alaska.asp (retrieved on January 28, 2007).

Sugihara, T.F. Microbiology of breadmaking. In: *Microbiology of Fermented Foods*, B.J.B. Wood (ed.). Elsevier Applied Science Publishers, London, UK, 1985.

2 Genes and Genomics

LEARNING OBJECTIVES

- Define the cell as the building block of the human body
- Discuss intracellular and extracellular organization of the cell
- Discuss the nucleus and its constituents
- Explain the structure and function of DNA
- Explain cell division by meiosis and mitosis
- Explain DNA replication in cell division
- Discuss molecular and genetic tools of biotechnology

2.1 INTRODUCTION

We learned various attributes of biotechnology in the previous chapter and how the field of biotechnology have revolutionized the agricultural, environmental, industrial, and healthcare sectors. This incredible phase of biotechnology is called modern biotechnology, where modern tools are employed in creating arrays of products which include antibiotics, vaccines, monoclonal antibodies, and recombinant insulin. In this chapter, we will learn about modern molecular biology tools and how these tools influence the development of new biotechnology products. We know the fact that the molecule *deoxyribonucleic acid* (DNA) plays a critical role in the development and functioning of all known living organisms and some viruses. The main role of DNA molecules is the long-term storage of information. It is often compared to a set of blueprints, a recipe, or a code because it contains the instructions needed to construct other components of cells, such as proteins and *ribonucleic acid* (RNA) molecules. It would be interesting to know the structure and function of DNA in various species and to understand how DNA controls the physiological functions in the human body, plants, animals, and microbes. In this chapter, we will compare the difference between animal cells and plant cells, or between plant cells and microbial cells. Let us first learn about the organization of cells in animals, plants, and microorganisms which contain genetic material.

2.2 CELL AS THE BUILDING BLOCK OF LIFE

The cell is the smallest unit of a living organism and is often called the building block of life. Humans have approximately 100 trillion or 10^{14} cells and are an example of a *multicellular organism*. On the other hand, a single-celled bacterium is called a *unicellular organism*. A typical cell size is 10 micrometers (μm) and a typical cell mass is 1 nanogram (ng).

Let's go back to history and learn how cells of plant and animal origin were being investigated by many researchers over many centuries. The concept that plants and animals are made of cells was originally coined by Aristotle (384–322 BC). In 1665, Robert Hooke observed for the first time the structure of a cell under a very primitive microscope. In 1674, Antonie van Leeuwenhoek discovered single-celled organisms and the structural organization within the cell. Earlier in the nineteenth century, scientists proposed theories about the origin of a cell and how a cell becomes a living organism. H.J. Dutrochet, a French scientist, presented the idea of cell theory in 1824, then the German botanist M.I. Schleiden and the German zoologist T. Schwann formulated and outlined the basic features of the theory in 1839. In 1858, R. Virchow extended the cell theory and suggested that all living cells arise from preexisting living cells.

To prove Virchow's hypothesis, Louis Pasteur performed some experiments and concluded that living things are composed of cells, and that all living cells arise from preexisting cells. The one exception that does not fit into the cell theory are viruses which may be defined as an infectious subcellular and ultramicroscopic organism. Viruses are simple as they lack the internal organization which is the main characteristic of a living cell. Because of this unique characteristic, viruses, mycoplasma, viroids, and prions do not easily fit in the definition of cell theory. In addition, there are other organisms such as protozoa and algae, which also do not fit in the definition of cell theory.

2.3 CLASSIFICATION OF CELLS

There are two dissimilar types of cells, *eukaryotic* and *prokaryotic* cells. Prokaryotic cells are usually independent, whereas eukaryotic cells are often found in multicellular organisms. Let us first learn to understand the basic similarities and differences between them.

2.3.1 Prokaryotic Cell

The prokaryote cell lacks a nucleus and most of the other organelles of eukaryotes. There are two types of prokaryotes, *bacteria* and *archaea,* and these prokaryotes share a similar overall structure. A prokaryotic cell has three regions: (1) *flagella* and *pili* project from the cell's surface. These are structures (not present in all prokaryotes) made of proteins that facilitate movement and communication between cells. (2) The *cell envelope* generally consisting of a cell wall covering a plasma membrane though some bacteria also have a further covering layer called a *capsule.* The envelope gives rigidity to the cell and separates the interior of the cell from its environment, serving as a protective filter. Though most prokaryotes have a cell wall, there are exceptions, such as *Mycoplasma* (bacteria) and *Thermoplasma* (archaea). The cell wall consists of *peptidoglycan* in bacteria, and acts as an additional barrier against exterior forces. It also prevents the cell from expanding and finally bursting, called *cytolysis,* from osmotic pressure against a hypotonic environment. Some eukaryote cells (plant cells and fungi cells) also have a cell wall. (3) The *prokaryotic chromosome* is usually circular in shape. An exception to this is the bacterium *Borrelia burgdorferi,* which causes Lyme disease, whose DNA is linear in shape. One of the distinct features of a prokaryote is the absence of a nucleus in the cytoplasm, and DNA is usually condensed in a *nucleoid.* Prokaryotes can carry extra-chromosomal DNA elements called *plasmids,* which are usually circular. Plasmids enable additional functions, such as antibiotic resistance. The presence of plasmid DNA in bacteria not only makes bacteria distinct from animals and plants, but also makes it an important genetic engineering tool (Figure 2.1).

2.3.2 Eukaryotic Cell

Eukaryotic cells are about 10 times the size of a typical prokaryote and the major difference between prokaryotes and eukaryotes is that eukaryotic cells contain membrane-bound compartments in which specific metabolic activities take place. Most important among these is the presence of a cell nucleus, a membrane-delineated compartment that houses the DNA. It is this nucleus that gives the eukaryote its name, which means "true nucleus". The plasma membrane resembles that of prokaryotes in function, with minor differences in the arrangement. The eukaryotic DNA is organized in one or more linear molecules, called *chromosomes,* which relate to histone proteins. All chromosomal DNA is stored in the cell nucleus, separated from the cytoplasm by a membrane. Some eukaryotic organelles such as mitochondria also contain some DNA. Eukaryotes can move using *cilia* or *flagella* and the flagella of a eukaryote are more complex than those of prokaryotes.

All cells (prokaryotic or eukaryotic) have a membrane that envelops the cell, separates its interior from its environment, regulates what moves in and out (selectively permeable), and maintains the electric potential of the cell. Inside the membrane, a salty cytoplasm takes up most of the cell volume. All cells

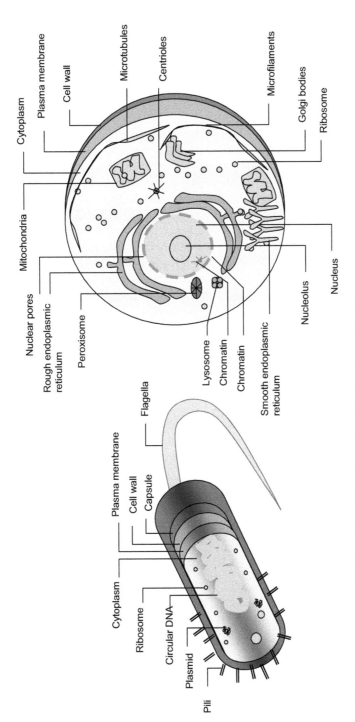

FIGURE 2.1 Diagrammatic illustrations of a prokaryotic cell and a eukaryotic cell.

have DNA and RNA to produce the various proteins such as enzymes, the cell's primary machinery. In the next section, we discuss all the constituents present in bacteria, animal, or plant cells (Figure 2.1).

2.4 EXTRACELLULAR COMPONENTS

The cell consists of various components which play a critical role in protecting the cell from external invasion, to make cell—cell contact, and to help the cell to perform regulated physiological functions.

2.4.1 Cell Membrane

The cytoplasm of a cell is surrounded by a *cell membrane* or *plasma membrane*. The plasma membrane in plants and prokaryotes is usually covered by a cell wall. This membrane serves to separate and protect a cell from its surrounding environment and is made mostly from a double layer of lipids (hydrophobic fat-like molecules) and hydrophilic phosphorus molecules. Hence, the layer is called a *phospholipid bilayer.* It may also be called a *fluid mosaic membrane.* A variety of protein molecules are embedded in the membrane and act as channels and pumps that transport different molecules into and out of the cell in a regulated manner. The membrane is said to be "semipermeable" in that it can either let a substance (molecules or ions) pass through freely, pass through to a limited extent, or not pass through at all. Cell surface membranes also contain receptor proteins that allow cells to detect external signaling molecules such as hormones and ligands (Figure 2.2).

2.4.2 Cell Capsule

The *cell capsule* is a very large organelle of some prokaryotic cells, such as bacterial cells. It is a layer that lies outside the cell wall of bacteria. It is usually composed of polysaccharides, but could

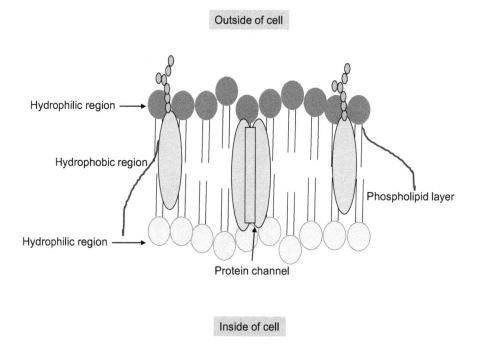

FIGURE 2.2 Diagrammatic illustration of a cell membrane showing the phospholipid layer which is embedded with protein channels

be composed of other materials (e.g., polypeptide in *B. anthracis*). It is a well-organized layer, not easily washed off, and it can be the cause of various diseases. Because most capsules are water soluble, they are difficult to stain using standard stains because most stains do not adhere to the capsule. Capsules also contain water, which protects bacteria against desiccation. For examination under the microscope, the bacteria and their background are stained darker than the capsule, which does not stain. Since the capsule protects the bacteria against phagocytosis, it is considered a virulence factor. A capsule-specific antibody may be required for phagocytosis to occur. Capsules also exclude bacterial viruses and most hydrophobic toxic materials such as detergents. Further than that, bacterial capsules allow bacteria to adhere to surfaces and other cells.

2.4.3 Flagella

A *flagellum* is a tail-like structure that projects from the cell body of certain prokaryotic and eukaryotic cells and is mostly involved in locomotion and movement. There are some notable differences between prokaryotic and eukaryotic flagella, such as protein composition, structure, and mechanism of propulsion. An example of a flagellated bacterium is the ulcer-causing *Helicobacter pylori*, which uses multiple flagella to propel itself through the mucus lining to reach the stomach epithelium. An example of a eukaryotic flagellated cell is the sperm cell, which uses its flagellum to propel itself through the female reproductive tract. Eukaryotic flagella are structurally identical to eukaryotic cilia, although distinctions are sometimes made according to function and structure.

2.5 INTRACELLULAR COMPONENTS

2.5.1 Cytoplasmic Constituents

There are several types of organelles within an animal cell and some, such as the nucleus and Golgi apparatus, are present in a solitary state, whereas others, such as mitochondria, peroxisomes, and lysosomes, can be found in numerous numbers (hundreds to thousands) in the cytoplasm. The *cytosol* is the gelatinous fluid that fills the cell and surrounds the organelles.

2.5.1.1 Mitochondria and Chloroplasts

Mitochondria are self-replicating organelles that occur in various numbers, shapes, and sizes in the cytoplasm of all eukaryotic cells. Mitochondria play a critical role in generating energy in the eukaryotic cell. Mitochondria generate the cell's energy by the process of oxidative phosphorylation, utilizing oxygen to release energy stored in cellular nutrients (typically pertaining to glucose) to generate adenosine triphosphate (ATP). Mitochondria multiply by splitting in two. In plant cells, organelles that are modified chloroplasts are broadly called *plastids* and are involved in energy storage through the process of photosynthesis, which utilizes solar energy to generate carbohydrates and oxygen from carbon dioxide and water. Mitochondria and chloroplasts each contain their own genome, which is separate and distinct from the nuclear genome of a cell. Both organelles contain this DNA in circular plasmids, much like prokaryotic cells. Since these organelles contain their own genomes and have other similarities to prokaryotes, they are thought to have developed through a symbiotic relationship after being engulfed by a primitive cell.

2.5.1.2 Ribosomes

The *ribosome* is a large complex of RNA and is called a protein molecule. This enzyme produces proteins from mRNA. These ribosomes can be found either floating freely or bound to the membrane rough endoplasmic reticulum in eukaryotes, or the cell membrane in prokaryotes.

2.5.1.3 Endoplasmic Reticulum

The *endoplasmic reticulum* (ER) is only present in eukaryotic cells and is the transport network for molecules targeted for certain modifications and specific destinations, as compared to molecules that will float freely in the cytoplasm. The ER has two forms: the rough ER, which has ribosome on its surface and secretes proteins into the cytoplasm, and the smooth ER, which lacks them. Smooth ER plays a role in calcium sequestration and release.

2.5.1.4   Golgi Apparatus

The primary function of the *Golgi apparatus* is to process and package the macromolecules, such as proteins and lipids, that are synthesized by the cell. It is particularly important in the processing of proteins for secretion. The Golgi apparatus forms a part of the endomembrane system of eukaryotic cells. Vesicles that enter the Golgi apparatus are processed in a *cis* to *trans* direction, meaning they coalesce on the *cis* side of the apparatus and after processing pinch off on the opposite (*trans*) side to form a new vesicle in the animal cell.

2.5.1.5   Lysosomes and Peroxisomes

The *lysosomes* contain digestive enzymes (acid hydrolases) and they digest excess or worn-out organelles, food particles, and engulfed viruses or bacteria. *Peroxisomes* have enzymes that remove the toxic peroxides from the cells and these destructive enzymes are contained in a membrane-bound system. These organelles are often called a "suicide bag" because of their ability to detonate and destroy the cell.

2.5.1.6   Centrosome

The *centrosome* produces the microtubules of a cell, a key component of the cytoskeleton. It directs the transport through the ER and the Golgi apparatus. Centrosomes are composed of two centrioles, which separate during cell division and help in the formation of the mitotic spindle. A single centrosome is present in animal cells. They are also found in some fungi and algae cells.

2.5.1.7   Vacuoles

Vacuoles store food and waste. Some vacuoles store extra water. They are often described as a liquid-filled space and are surrounded by a membrane. Some cells, such as *Amoebae*, have contractile vacuoles, which can pump excess water out of the cell.

2.5.1.8   Cytoskeleton

The *cytoskeleton* organizes and maintains the shape of a cell, anchors organelles in place, helps during *endocytosis* (the uptake of external materials by a cell) and *cytokinesis* (the separation of daughter cells after cell division), and moves parts of the cell in processes of growth and mobility. The eukaryotic cytoskeleton is composed of microfilaments, intermediate filaments, and microtubules. There are a great number of proteins associated with them, each controlling a cell's structure by directing, bundling, and aligning filaments. The prokaryotic cytoskeleton is less studied, but it plays an essential role in the maintenance of a cell's shape, polarity, and cytokinesis.

2.5.2   NUCLEAR CONSTITUENTS

2.5.2.1   Deoxyribonucleic Acid

The cell nucleus in a eukaryotic cell contains the genetic information and is where almost all *deoxyribonucleic acid* (DNA) replications and RNA syntheses occur. The nucleus is spherical in shape and separated from the cytoplasm by a double membrane called the nuclear envelope. The nuclear envelope isolates and protects a cell's DNA from various molecules that could accidentally damage its structure or interfere with its processing. In most living organisms (except for viruses), genetic information is stored in the DNA, which resides in the nucleus of the living cells. It gets its name from the sugar molecule contained in its backbone (deoxyribose). However, it gets its significance from its unique structure. Four different nucleotide bases occur in DNA: adenine (A), cytosine (C), guanine (G), and thymine (T). The versatility of DNA comes from the fact that the molecule is double-stranded. The nucleotide bases of the DNA molecule form complementary pairs. The nucleotides hydrogen bond to another nucleotide base in a strand of DNA opposite the original. This bonding is specific, and adenine always bonds to thymine (and vice versa) and guanine always bonds to cytosine (and vice versa). This bonding occurs across the molecule, leading to a double-stranded system as shown in Figure 2.3. The human DNA consists of about 3 billion bases, and more than 99% of those bases are the same in all people. DNA bases pair up with each other, A with T and C with G, to form units called *base pairs* (Figure 2.5). Each base is also attached to a sugar molecule and a phosphate molecule together making up a nucleotide. Nucleotides are arranged in two long strands forming a *double helix*. The structure of the double helix is somewhat like a ladder, with the base pairs forming the ladder's rungs and the sugar and phosphate molecules forming the vertical sidepieces of the ladder. An important property of DNA is that it can replicate or make copies of itself. Each strand of DNA in the double helix can serve as a pattern for duplicating the sequence of bases. This is critical when cells divide because each new cell needs to have an exact copy of the DNA present in the old cell.

The DNA chain is 22–26 Å wide (or 2.2–2.6 nano meter), and one nucleotide unit is 3.3 Å (or 0.33 nano meter) long. Although each individual repeating unit is very small, DNA polymers can be very large molecules containing millions of nucleotides. For instance, the largest human chromosome, chromosome number 1, is approximately 220 million base pairs long. These two strands run in opposite directions to each other and are therefore anti-parallel. Attached to each sugar is one of four types of molecules called bases. It is the sequence of these four bases along the backbone that encodes information. This information is read using the genetic code, which specifies the sequence of the amino acids within proteins.

In all living organisms, DNA usually exists as a pair of molecules that are held tightly together in the shape of a double helix. These two long strands entwine like vines, in the shape of a double

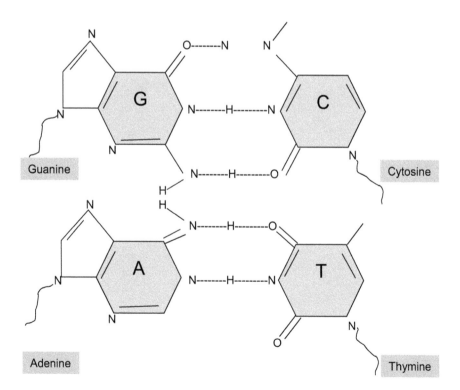

FIGURE 2.3 DNA structure shows the association of guanine with cytosine and adenine with thymine by hydrogen bonds.

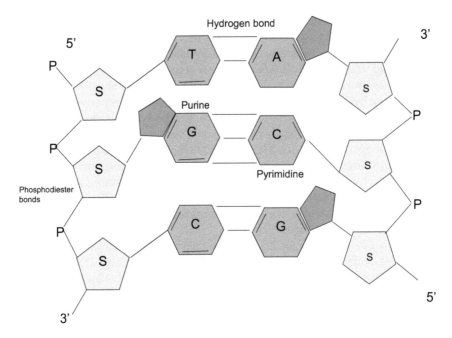

FIGURE 2.4 DNA base pairing shows the pairing of purine and pyrimidine with sugars by phosphodiester bonds.

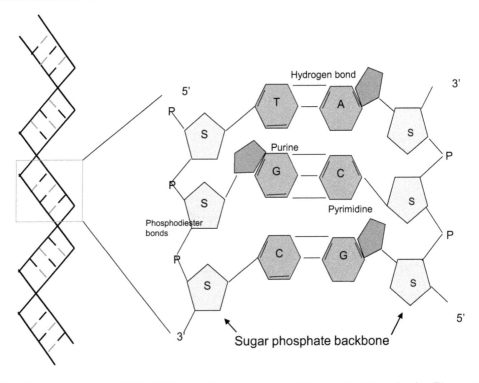

FIGURE 2.5 The structure of DNA: DNA strands are composed of four nucleotides subunits. These are adenine (A), thymine (T), cytosine (C), and guanine (G).

helix. The nucleotide repeats contain both the segment of the backbone of the molecule, which holds the chain together, and a base, which interacts with the other DNA strand in the helix. In general, a base linked to a sugar is called a *nucleoside*, and a base linked to a sugar and one or more phosphate groups is called a *nucleotide*. If multiple nucleotides are linked together, as in DNA, this polymer is called a *polynucleotide*. The backbone of the DNA strand is made from alternating phosphate and sugar residues. The sugar in DNA is 2-deoxyribose, which is a pentose (five-carbon) sugar. The sugars are joined together by phosphate groups that form phosphodiester bonds between the third and fifth carbon atoms of adjacent sugar rings (Figure 2.4). These asymmetric bonds mean a strand of DNA has a direction. In a double helix, the direction of the nucleotides in one strand is opposite to their direction in the other strand. This arrangement of DNA strands is called *antiparallel*. The asymmetric ends of DNA strands are referred to as the 5′ (*five prime*) and 3′ (*three prime*) ends, with the 5′ end being that with a terminal phosphate group and the 3′ end that with a terminal hydroxyl group. One of the major differences between DNA and RNA is the sugar, with 2-deoxyribose being replaced by the alternative pentose sugar ribose in RNA.

One of the interesting characteristics of DNA is its ability to form a coil-like structure. This process of coiling is called *DNA supercoiling* (Figure 2.6). With DNA in its "relaxed" state, a strand usually circles the axis of the double helix once every 10.4 base pairs, but if the DNA is twisted, the strands become tighter. When the DNA is twisted in the direction of the helix, it is called *positive supercoiling*, and the bases are held more tightly together. When they are twisted in the opposite direction, it is called *negative supercoiling*, and the bases come apart more easily. In nature, most DNA has slight negative supercoiling that is introduced by enzymes called *topoisomerases*. These enzymes are also needed to relieve the twisting stresses introduced into DNA strands during transcription and DNA replication.

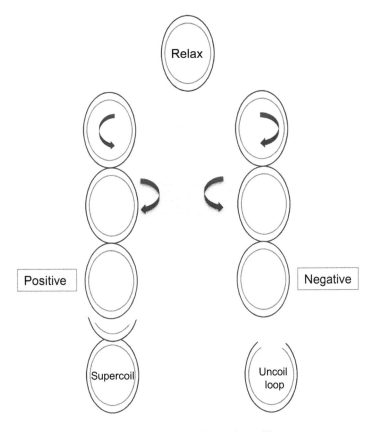

FIGURE 2.6   DNA supercoiling shows both positive and negative coiling.

2.5.2.2 Ribonucleic Acid

Ribonucleic acid (RNA) is a biologically important type of molecule that consists of a long chain of nucleotide units. RNA and DNA are both nucleic acids but differ in three main ways. First, unlike DNA, which is double-stranded, RNA is a single-stranded molecule in most of its biological roles and has a much shorter chain of nucleotides. Second, although DNA contains *deoxyribose*, RNA contains *ribose*, (there is no hydroxyl group attached to the pentose ring in the 2′ position in DNA). These hydroxyl groups make RNA less stable than DNA because it is more prone to hydrolysis. Third, the complementary base to adenine is not thymine, as it is in DNA, but rather uracil, which is an unmethylated form of thymine (Figure 2.7). Like DNA, most biologically active RNAs, including messenger ribonucleic acid (mRNA), transfer RNA (tRNA), ribosomal RNA (rRNA), small nuclear RNA (snRNA) and other non-coding RNAs, contain self-complementary sequences that allow parts of the RNA to fold and pair with itself to form double helices. Structural analysis of these RNAs has revealed that they are highly structured. Unlike DNA, their structures do not consist of long double helices, but rather collections of short helices packed together into structures akin to proteins. In this fashion, RNAs can achieve chemical catalysis, like enzymes. For instance, the determination of the structure of the ribosome, an enzyme that catalyzes peptide bond formation, revealed that its active site is composed entirely of RNA.

2.5.2.2.1 mRNA

mRNA is involved in protein synthesis by carrying coded information to the sites of protein synthesis: the ribosomes. Here, the nucleic acid polymer is translated into a polymer of amino acids: a protein. In mRNA, as in DNA, genetic information is encoded in the sequence of nucleotides arranged into codons consisting of three bases each. Each codon encodes for a specific amino acid, except the stop codons that terminate protein synthesis. This process requires two other types of RNA: *transfer RNA* (tRNA), which mediates recognition of the codon and provides the corresponding amino acid, and *ribosomal RNA* (rRNA), which is the central component of the ribosome's protein manufacturing machinery.

2.5.2.2.2 Transfer RNA

tRNA is a small RNA molecule (usually about 74–95 nucleotides) that transfers a specific active amino acid to a growing polypeptide chain at the ribosomal site of protein synthesis during translation. It has a 3′ terminal site for amino acid attachment. This covalent linkage is catalyzed by an aminoacyl tRNA synthetase. It also contains a three-base region called the *anticodon* that can base

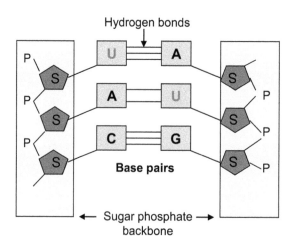

FIGURE 2.7   RNA structure shows base pairing with sugar phosphate.

pair to the corresponding three-base codon regions on mRNA. Each type of tRNA molecule can be attached to only one type of amino acid, but because the genetic code contains multiple codons that specify the same amino acid, tRNA molecules bearing different anticodons may also carry the same amino acid (Figure 2.8).

2.5.2.2.3 rRNA

rRNA is the central component of the ribosome, the protein manufacturing machinery of all living cells. The function of the rRNA is to provide a mechanism for decoding mRNA into amino acids and to interact with the tRNAs during translation by providing peptidyl transferase activity. The tRNA then brings the necessary amino acids corresponding to the appropriate mRNA codon (Figure 2.9).

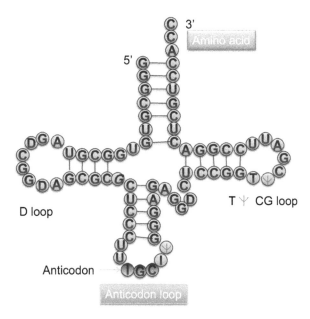

FIGURE 2.8   Transfer RNA (tRNA) showing the CG loop and D loop and the amino acid site and anticodon site.

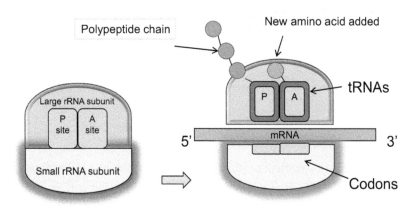

FIGURE 2.9   Ribosomal RNA (rRNA) showing the formation of a polypeptide chain with the help of tRNAs.

2.5.2.2.4 snRNA

snRNA is a class of small RNA molecules that is found within the nucleus of eukaryotic cells. They are transcribed by RNA polymerase II or RNA polymerase III and are involved in a variety of important processes such as *RNA splicing* (removal of introns from heterogeneous nuclear RNA [hnRNA]), regulation of transcription factors (7SK RNA) or RNA polymerase II (B2 RNA), and maintaining the telomeres. They are always associated with specific proteins, and the complexes are referred to as *small nuclear ribonucleoproteins* (snRNP) or sometimes as *snurps*. These elements are rich in uridine content. A large group of snRNAs is known as *small nucleolar RNAs* (snoRNAs). These are small RNA molecules that play an essential role in RNA biogenesis and guide chemical modifications of rRNAs and other RNA genes (tRNA and snRNAs). They are in the nucleolus and the cajal bodies of eukaryotic cells (the major sites of RNA synthesis).

2.5.2.3 Nucleolus

The *nucleolus* is a small, typically round granular body composed of protein and RNA in the nucleus of a cell. It is usually associated with a specific chromosomal site and involved in ribosomal RNA synthesis and the formation of ribosomes. The ribosomes are eventually transferred out of the nucleus into the cytoplasm via pores in the nuclear envelope.

2.5.2.4 Chromatin

Chromatin is the complex combination of DNA, RNA, and protein that makes up chromosomes. It is found inside the nuclei in eukaryotic cells, and within the nucleoid in prokaryotic cells. It is divided between heterochromatin (condensed) and euchromatin (extended) forms. The major components of chromatin are DNA and histone proteins, although many other chromosomal proteins have prominent roles too. The functions of chromatin are to package DNA into a smaller volume to fit in the cell, to strengthen the DNA to allow mitosis and meiosis, and to serve as a mechanism to control expression and DNA replication. Chromatin contains genetic material instructions to direct cell functions. Changes in chromatin structure are affected by chemical modifications of histone proteins such as methylation (DNA and proteins) and acetylation (proteins), and by non-histone DNA-binding proteins.

2.6   EPIGENETICS

Epigenetics is the science of cellular and physiological traits that are heritable by daughter cells and not caused by changes in the DNA sequence. In addition, epigenetics defines the study of stability and long-term alterations in the transcriptional potential of a cell. These modifications may or may not be genetic in nature, although the use of the term epigenetic to designate processes that are not heritable is argumentative. The term epigenetics also denotes the variations in the genome that do not involve a change in the nucleotide sequence. The mechanisms that produce such changes are DNA methylation and histone modification. Gene expression can be controlled through the action of repressor proteins that attach to silencer regions of the DNA. These epigenetic changes may last through cell divisions for the duration of the cell's life and may also last for multiple generations. One example of an epigenetic change in eukaryotic organisms is the process of cellular differentiation.

2.6.1   HISTORICAL VIEWPOINT

It all started during the mid-twentieth century with scientists Conrad H. Waddington and Ernst Hadorn, who did extensive research focused on merging genetics and developmental biology, and has evolved into a new field, which we currently call epigenetics. The word epigenetics was coined by Waddington in 1942 and originally described the influence of genetic processes on development. During the 1990s, there was a renewed interest in genetic assimilation that led to the elucidation of the molecular basis of Waddington's observations in which environmental stress caused genetic assimilation in fruit flies. Subsequently, research efforts focused on unraveling the epigenetic mechanisms related to these types of changes. Presently, DNA methylation is one of the most largely studied epigenetic modifications. The renewed interest in epigenetics has led to new findings about the relationship between epigenetic variations and many diseases such as cancers, immune disorders, mental retardation-associated disorders, and pediatric and psychiatric disorders.

2.6.2 CLINICAL IMPLICATIONS

Epigenetics changes may be reflected at various stages throughout a person's life; for example, several studies have provided evidence that prenatal and early postnatal environmental factors influence human embryos and may cause the development of various chronic diseases and behavioral disorders in adulthood. For example, studies have shown that children born during the period of the famine from 1944 to 1945 in Holland have shown increased rates of coronary heart disease and obesity after maternal exposure to famine during pregnancy. Similarly, adults that were prenatally exposed to famine conditions have also been reported to have a higher frequency of schizophrenia. Cancer was the first human disease to be linked to epigenetics, for example, research performed by Feinberg and Vogelstein in 1983, found that genes of colorectal cancer cells were significantly hypo-methylated compared with normal tissues. It has been reported that DNA hypo-methylation can activate oncogenes and initiate chromosome instability; however, DNA hyper-methylation initiates silencing of tumor suppressor genes. The buildup of genetic and epigenetic faults can convert a normal cell into a metastatic tumor cell. Furthermore, DNA methylation patterns may cause abnormal expression of cancer-associated genes. The modification of global histone patterns is linked with prostate, breast, and pancreatic cancers. Therefore, epigenetic changes or modifications can be used as biomarkers for the diagnosis of early cancer.

Several disorders are associated with epigenetics, such as Beckwith–Weidman (BWS), Prader–Willi, ATR-X, Fragile X, Rett, and Angelman syndromes. Interestingly, disorders like Prader–Willi syndrome and Angelman syndrome show an abnormal phenotype because of the absence of the paternal or the maternal copy of a gene. In these imprint disorders, there is a genetic deletion in chromosome 15 in most patients. The same gene on the corresponding chromosome cannot compensate for the deletion because it has been turned off by the methylation process. Furthermore, genetic deletions inherited from the father cause Prader–Willi syndrome, whereas those inherited from the mother cause Angelman syndrome.

2.7 GENES AND GENETICS

Recall that DNA contains the genetic information that allows all living organisms to function and reproduce. However, it is unclear how long in the 4-billion-year history of life DNA has performed this function, as it has been proposed that the earliest forms of life may have used RNA as their genetic material. RNA may have acted as the central part of early cell metabolism as it can both transmit genetic information and carry out catalysis as part of ribozymes. This ancient RNA world where nucleic acid would have been used for both catalysis and genetics may have influenced the evolution of the current genetic code based on four nucleotide bases. This would occur because the number of unique bases in such an organism is a trade-off between a small number of bases increasing replication accuracy and many bases increasing the catalytic efficiency of ribozymes. Unfortunately, there is no direct evidence of ancient genetic systems, as recovery of DNA from most fossils is impossible. This is because DNA can survive in the environment for less than 1 million years and slowly degrades into short fragments in solution. Claims for older DNA have been made, most notably a report of the isolation of a viable bacterium from a salt crystal 250 million years old, but these claims are controversial.

2.7.1 MENDELIAN GENETICS

For thousands of years, farmers and herders have been selectively breeding their plants and animals to produce more useful hybrids. It was somewhat of a hit or miss process because the actual mechanisms governing inheritance were unknown. Knowledge of these genetic mechanisms finally emerged because of careful laboratory breeding experiments carried out over the last century and a half. By the 1890s, the invention of better microscopes allowed biologists to discover the basic facts of cell division and sexual reproduction. The focus of genetic research then shifted to understanding what really happens in the transmission of hereditary traits from parents to children. Several hypotheses were suggested to explain heredity, but Gregor Mendel, a little known Central European monk, was the only one who got it more or less right. His ideas were published in 1866 but went largely unrecognized until 1900, which was long after his death. Although Mendel's research was with plants, the basic underlying principles of heredity that he discovered also apply to humans and other animals because the mechanisms of heredity are essentially the same for all complex life forms. Through the selective crossbreeding of common pea plants (*Pisum sativum*) over many generations, Mendel discovered that certain traits show up in offspring without any blending of parental characteristics. For instance, the pea flowers are either purple or white; intermediate colors do not appear in the offspring of cross-pollinated pea plants.

The observation that these traits do not show up in offspring plants with intermediate forms was critically important because the leading theory in biology at the time was that inherited traits blend from generation to generation. Most of the leading scientists in the nineteenth century accepted this "blending theory." Charles Darwin proposed another equally wrong theory known as "pangenesis." This held that hereditary "particles" in our bodies are affected by the things we do during our lifetime. These modified particles were thought to migrate via blood to the reproductive cells and subsequently could be inherited by the next generation. This was essentially a variation of Lamarck's incorrect idea of the "inheritance of acquired characteristics." Mendel picked common garden pea plants for the focus of his research because they can be grown easily in large numbers and their reproduction can be manipulated. Pea plants have both male and female reproductive organs. As a result, they can either self-pollinate themselves or cross-pollinate with another plant. In his experiments, Mendel could selectively cross-pollinate purebred plants with traits and observe the outcome over many generations. This was the basis for his conclusions about the nature of genetic inheritance. The following lists Mendel's conclusions from the result of his experiments:

- The inheritance of each trait is determined by genes that are passed on to descendants unchanged.
- An individual inherits one such gene from each parent for each trait.
- A trait may not show up in an individual but can still be passed on to the next generation.

It is important to realize that, in this experiment, the starting parent plants were homozygous for pea seed color. And they each had two identical forms (or alleles) of the gene for this trait—2 yellows (Y) or 2 greens (G). The plants in the first offspring (F1) generation were all heterozygous. In other words, they each had inherited two different alleles—one from each parent plant. It becomes clearer when we look at the actual genetic makeup, or genotype, of the pea plants instead of only the phenotype, or observable physical characteristics. Note that each of the F1 generation plants inherited a Y allele from one parent and a G allele from the other. When the F1 plants breed, each has an equal chance of passing on either Y or G alleles to each offspring. With all the seven pea plant traits that Mendel examined, one form appeared dominant over the other, which means it masked the presence of the other allele. For example, when the genotype for pea seed color is YG (heterozygous), the phenotype is yellow. However, the dominant yellow allele does not alter the recessive green one in any way. Both alleles can be passed on to the next generation unchanged.

Mendel's observations from his experiments can be summarized in two principles: the principle of segregation and the principle of independent assortment. According to the principle of segregation, for any specific trait, the pair of alleles of each parent separate and only one allele passes on from each parent to an offspring. Which allele in a parent's pair of alleles is inherited is a matter of chance. This segregation of alleles occurs during meiosis. According to the principle of independent assortment, different pairs of alleles are passed on to offspring independent of each other. The result is that new combinations of genes present in neither parent are possible. For example, a pea plant's inheritance of the ability to produce purple flowers instead of white ones does not make it more likely that it will also inherit the ability to produce yellow pea seeds in contrast to green ones. Likewise, the principle of independent assortment explains why the human inheritance of a particular eye color does not increase or decrease the likelihood of having six fingers on each hand. Today, we know this is because the genes for independently assorted traits are present in different chromosomes. These two principles of inheritance, along with the understanding of unit inheritance and dominance, were the beginnings of our modern science of genetics.

2.7.2 Modern Genetics

It was not until the late 1940s and early 1950s that most biologists accepted the evidence showing that DNA must be the chromosomal component that carries hereditary information. One of the most convincing experiments was that of Alfred Hershey and Martha Chase, who used radioactive labeling to reach this conclusion in 1952. This team of biologists grew a specific type of phage, known as T2, in the presence of two different radioactive labels so that the phage DNA incorporated radioactive phosphorus (32P), while the protein incorporated radioactive sulfur (35S). They then allowed the labeled phage particles to infect non-radioactive bacteria trying to find the label associated with the infected cell. Their analysis showed that most of the 32P-label was found inside of the cell, while most of the 35S was found outside. This suggested that the proteins of the T2 phage remained outside of the newly infected bacterium while the phage-derived DNA was injected into the cell. They then showed that the phage-derived DNA caused the infected cells to produce new phage particles. This elegant work showed, conclusively, that DNA is the molecule that holds genetic information. Meanwhile, much of the scientific world was asking questions about the physical structure of the DNA molecule, and the relationship of that structure to its complex functioning. In 1951, the then 23-year-old biologist James Watson traveled from the United States to work with Francis Crick, an English physicist at the University of Cambridge. Crick was already using the process of x-ray crystallography to study the structure of protein molecules. Together, Watson and Crick used x-ray crystallography data, produced by Rosalind Franklin and Maurice Wilkins at King's College in London, to decipher DNA's structure. The discovery of the double helix structure of DNA marks the beginning of modern biotechnology.

2.8   CELL DIVISION

In the previous section, we learned about DNA and its structural and functional attributes, especially during meiosis and mitosis when a cell undergoes division. When cells divide to yield two daughter cells, the genetic material must be divided equally so that each daughter cell contains identical DNA copies. In the next section, we will learn how DNA helps the cells to divide by meiotic and mitotic pathways.

2.8.1 Meiosis

Meiosis was discovered and described for the first time in sea urchin eggs in 1876, by noted German biologist Oscar Hertwig (1849–1922). It was described again in 1883, at the level of chromosomes in *Ascaris* worm's eggs, by Belgian zoologist Edouard Van Beneden (1846–1910). The significance of meiosis for reproduction and inheritance, however, was only described in 1890 by German biologist August Weismann (1834–1914), who noted that two cell divisions were necessary to transform one diploid cell into four haploid cells if the number of chromosomes had to be maintained. In 1911, the American geneticist Thomas Hunt Morgan (1866–1945) observed crossover in *Drosophila melano-gaster* meiosis and provided the first genetic evidence that genes are transmitted on chromosomes. Meiosis is a process of reduction division in which the number of chromosomes per cell is reduced to half. In animals, meiosis always results in the formation of gametes, whereas in other organisms, it can give rise to spores. As with mitosis, before meiosis begins, the DNA in the original cell is replicated during the S-phase of the cell cycle. Two cell divisions separate the replicated chromosomes into four haploid gametes or spores.

Meiosis is essential for sexual reproduction and therefore occurs in all eukaryotes (including single-celled organisms) that reproduce sexually. A few eukaryotes, notably the Bdelloid rotifers, have lost the ability to carry out meiosis and have acquired the ability to reproduce by parthenogenesis. Meiosis does not occur in archaea or bacteria, which reproduce via asexual processes such as binary fission. During meiosis the genome of a diploid germ cell, which is composed of long segments of DNA packaged into chromosomes, undergoes DNA replication followed by two rounds of division, resulting in four haploid cells. Each of these cells contains one complete set of chromosomes, or half of the genetic content of the original cell. If meiosis produces gametes, these cells must fuse during fertilization to create a new diploid cell, or zygote, before any new growth can occur. Thus, the division mechanism of meiosis is a reciprocal process to the joining of two genomes that occurs at fertilization. Because the chromosomes of each parent undergo genetic recombination during meiosis, each gamete, and thus each zygote, will have a unique genetic *blueprint* encoded in its DNA. Together, meiosis and fertilization constitute sexuality in the eukaryotes, and generate genetically distinct individuals in populations. In all plants, and in many protists, meiosis results in the

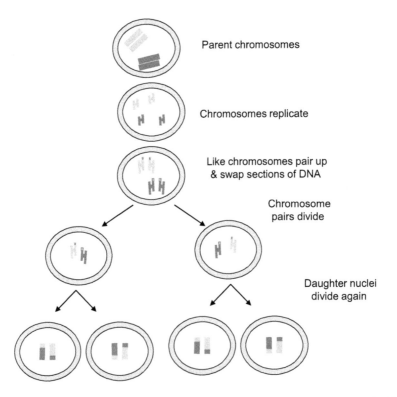

FIGURE 2.10 Stages of meiosis showing the chromosomes replicating, pairing, and swapping DNA and the formation of daughter nuclei.

formation of haploid cells that can divide vegetatively without undergoing fertilization, referred to as spores. In these groups, gametes are produced by mitosis. Meiosis uses many of the same biochemical mechanisms employed during mitosis to accomplish the redistribution of chromosomes. There are several features unique to meiosis, most importantly the pairing and genetic recombination between homologous chromosomes (Figure 2.10).

2.8.2 MITOSIS

Mitosis is the process in which a eukaryotic cell separates the chromosomes in its cell nucleus into two identical sets in two daughter nuclei. It is generally followed immediately by *cytokinesis*, which divides the nuclei, cytoplasm, organelles, and cell membrane into two daughter cells containing roughly equal shares of these cellular components. Mitosis and cytokinesis together define the *mitotic (M) phase* of the cell cycle such as the division of the mother cell into two daughter cells, genetically identical to each other and to their parent cell. Mitosis divides the chromosomes in a cell nucleus. Mitosis occurs exclusively in eukaryotic cells but occurs in different ways in different species. For example, animals undergo an "open" mitosis in which the nuclear envelope breaks down before the chromosomes separate, whereas fungi such as *Aspergillus nidulans* and *Saccharomyces cerevisiae* (yeast) undergo a "closed" mitosis in which chromosomes divide within an intact cell nucleus. Prokaryotic cells, which lack a nucleus, divide using binary fission. The process of mitosis is complex and highly regulated. The sequence of events is divided into phases that correspond

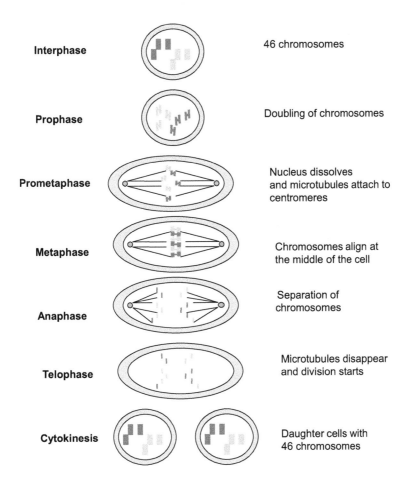

FIGURE 2.11   Stages of mitosis.

to the completion of one set of activities and the start of the next. These stages are the prophase, prometaphase, metaphase, anaphase, and telophase. During the process of mitosis, the pairs of chromosomes condense and attach to fibers that pull the sister chromatids to opposite sides of the cell. The cell then divides in cytokinesis to produce two identical daughter cells. Because cytokinesis usually occurs in conjunction with mitosis, "mitosis" is often used interchangeably with "mitotic phase." However, there are many cells in which mitosis and cytokinesis occur separately, forming single cells with multiple nuclei. This mostly occurs among the fungi and slime molds but is also found in various other groups. Even in animals, cytokinesis and mitosis may occur independently such as during certain stages of fruit fly embryonic development. Errors in mitosis can either kill a cell through apoptosis or cause mutations that may lead to cancer (Figure 2.11).

2.9 DNA REPLICATION

DNA *replication*, the basis for biological inheritance, is a fundamental process occurring in all living organisms to copy their DNA. This process is "semi-conservative" in that each strand of the original double-stranded DNA molecule serves as a template for the reproduction of the complementary strand. Hence, following DNA replication, two identical DNA molecules have been produced from a single double-stranded DNA molecule. Cellular proofreading and error-checking mechanisms ensure near perfect fidelity for DNA replication. In a cell, DNA replication begins at specific locations in the genome, called "origins." Unwinding of DNA at the origin, and synthesis of new strands, forms a replication fork. In addition to DNA *polymerase*, the enzyme that synthesizes the new DNA by adding nucleotides matched to the template strand, several other proteins are associated with the fork and assist in the initiation and continuation of DNA synthesis. DNA replication can also be performed *in vitro* (outside a cell). DNA polymerases, isolated from cells, and artificial DNA primers are used to initiate DNA synthesis at known sequences in a template molecule. The *polymerase chain reaction* (PCR), a common laboratory technique, employs such artificial synthesis in a cyclic manner to amplify a specific target DNA fragment from a pool of DNA (Figure 2.12).

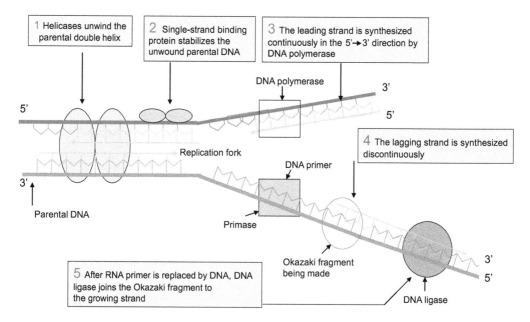

FIGURE 2.12   Process of DNA replication; five steps are shown.

2.9.1 ROLE OF DNA POLYMERASE IN REPLICATION

DNA polymerases are a family of enzymes that carry out all forms of DNA replication. A DNA polymerase can only extend an existing DNA strand paired with a template strand; it cannot begin the synthesis of a new strand. To begin synthesis of a new strand, a short fragment of DNA or RNA, called a *primer*, must be created and paired with the DNA template. Once a primer pairs with the DNA to be replicated, DNA polymerase synthesizes a new strand of DNA by extending the 3′ end of an existing nucleotide chain, adding new nucleotides matched to the template strand one at a time via the creation of phosphodiester bonds. The energy for this process of DNA polymerization comes from two of the three total phosphates attached to each unincorporated base. Free bases with their attached phosphate groups are called *nucleoside triphosphates*. When a nucleotide is being added to a growing DNA strand, two of the phosphates are removed and the energy produced creates a phosphodiester (chemical) bond that attaches the remaining phosphate to the growing chain. The energetics of this process also help explain the direction of synthesis; if DNA were synthesized in the 3′—5′ direction, the energy for the process would come from the 5′ end of the growing strand rather than from free nucleotides. DNA polymerases are generally extremely accurate, making less than one error for every 10^7 nucleotides added. Even so, some DNA polymerases also have proofreading ability; they can remove nucleotides from the end of a strand to correct mismatched bases. If the 5′ nucleotide needs to be removed during proofreading, the triphosphate end is lost. Hence, the energy source that usually provides energy to add a new nucleotide is also lost (Figure 2.13)

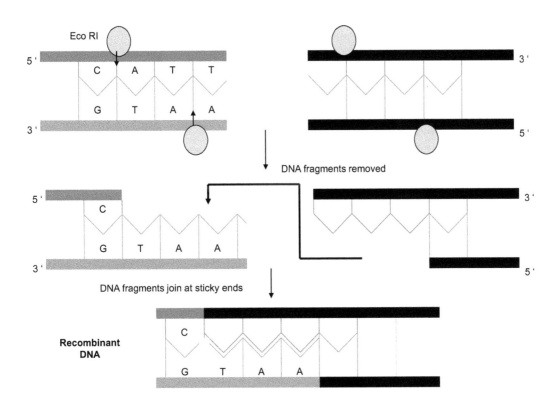

FIGURE 2.13   Role of restriction enzyme in making recombinant DNA.

2.9.2 DNA Replication within the Cell

For a cell to divide, it must first replicate its DNA. This process is initiated within the DNA, known as "origins," which are targeted by proteins that separate the two strands and initiate DNA synthesis. Origins contain DNA sequences recognized by replication initiator proteins. "These initiator proteins recruit other proteins to separate the two DNA strands at the origin, forming a bubble, and initiate replication forks." Initiator proteins recruit other proteins to separate the DNA strands at the origin, forming a bubble. Origins tend to be "AT-rich" (rich in adenine and thymine bases) to assist this process, because A-T base pairs have two hydrogen bonds (rather than the three formed in a C-G pair). Strands rich in these nucleotides are generally easier to separate because of the few flexibility/ many durability relationships found in hydrogen bonding. Once strands are separated, RNA primers are created on the template strands. More specifically, the leading strand receives one RNA primer per active origin of replication whereas the lagging strand receives several. These several fragments of RNA primers found on the lagging strand of DNA are called *Okazaki fragments*, named after their discoverer. DNA polymerase extends the leading strand in one continuous motion and the lagging strand in a discontinuous motion (because of the Okazaki fragments). RNase removes the RNA fragments used to initiate replication by the DNA polymerase, and another DNA polymerase enters to fill the gaps. When this is complete, a single nick on the leading strand and several nicks on the lagging strand can be found. Ligase works to fill in these nicks, thus completing the newly replicated DNA molecule. As DNA synthesis continues, the original DNA strands continue to unwind on each side of the bubble, forming replication forks. In bacteria, which have a single origin of replication on their circular chromosome, this process eventually creates a "theta structure." In contrast, eukaryotes have longer linear chromosomes and initiate replication at multiple origins within these.

2.9.2.1 Replication Fork

When replicating, the original DNA splits in two, forming two "prongs" that resemble a fork, hence the name *replication fork*. DNA has a ladder-like structure. Now, imagine a ladder broken in half vertically, along the steps. Each half of the ladder now requires a new half to match it. Because DNA polymerase can only synthesize a new DNA strand in a 5′—3′ manner, the process of replication goes differently for the two strands comprising the DNA double helix.

2.9.2.2 Leading Strand

The *leading strand* is that strand of the DNA double helix that is orientated in a 5′—3′ manner. On the leading strand, a polymerase "reads" the DNA and continuously adds nucleotides to it. This polymerase is DNA polymerase III (DNA Pol III) in prokaryotes and presumably Pol ε in eukaryotes.

2.9.2.3 Lagging Strand

The *lagging strand* is that strand of the DNA that is orientated in a 3′—5′ manner. Since its orientation is opposite to the working orientation of DNA polymerase III, which is in a 5′—3′ manner, replication of the lagging strand is more complicated than that of the leading strand. On the lagging strand, primase "reads" the DNA and adds RNA to it in short, separated segments. In eukaryotes, primase is intrinsic to Pol αDNA. DNA polymerase III or Pol δ lengthens the primed segments, forming Okazaki fragments. Primer removal in eukaryotes is also performed by Pol δ. In prokaryotes, DNA polymerase I "reads" the fragments, removes the RNA using its flap endonuclease domain, and replaces the RNA nucleotides with DNA nucleotides. This is necessary because RNA and DNA use slightly different kinds of nucleotides. DNA ligase joins the fragments together.

2.9.3 Regulation of DNA Replication

In eukaryotes, DNA replication is controlled within the context of the cell cycle. As the cell grows and divides, it progresses through stages in the cell cycle. DNA replication occurs during the S phase (Synthesis phase). The progress of the eukaryotic cell through the cycle is controlled by cell-cycle checkpoints. Progression through checkpoints is controlled through complex interactions between various proteins, including cyclins and cyclin-dependent kinases. The G1/S checkpoint (or restriction checkpoint) regulates whether eukaryotic cells enter the process of DNA replication and subsequent division. Cells that do not proceed through this checkpoint are quiescent in the "G0" stage and do not replicate their DNA. Replication of chloroplast and mitochondrial genomes occur independent of the cell cycle, through the process of D-loop replication.

Most bacteria do not go through a well-defined cell cycle and instead continuously copy their DNA. During rapid growth, this can result in multiple rounds of replication occurring concurrently. Within the well-characterized bacteria *E. coli*, regulation of DNA replication can be achieved through several mechanisms, including the hemimethylation and sequestering of the origin sequence, the ratio of ATP to ADP, and the levels of protein DnaA. All these control the process of initiator proteins binding to the origin sequences. Because *E. coli* methylates GATC DNA sequences, DNA synthesis results in hemimethylated sequences. This hemimethylated DNA is recognized by a protein (SeqA) which binds and sequesters the origin sequence. In addition, DnaA (required for initiation of replication) binds less well to hemimethylated DNA. As a result, newly replicated origins are prevented from immediately initiating another round of DNA replication. ATP builds up when the cell is in a rich medium, triggering DNA replication once the cell has reached a specific size. ATP competes with ADP to bind with DnaA, and the DnaA–ATP complex is able to initiate replication. A certain number of DnaA proteins are also required for DNA replication. Each time the origin is copied, the number of binding sites for DnaA doubles, requiring the synthesis of more DnaA to enable another initiation of replication.

2.9.4 TERMINATION OF REPLICATION

The chromosomes in bacteria are circular, so termination of replication occurs when the two replication forks meet each other on the opposite end of the parental chromosome. *E. coli* regulates this process using termination sequences which, when bound by the Tus protein, enable only one direction of replication fork to pass through. As a result, the replication forks are constrained to always meet within the termination region of the chromosome. Eukaryotes initiate DNA replication at multiple points in the chromosome, so replication forks meet and terminate at many points in the chromosome. These are not known to be regulated in any manner. Because eukaryotes have linear chromosomes, DNA replication often fails to synthesize to the very end of the chromosomes (telomeres), resulting in telomere shortening. This is a normal process in somatic cells. Here, cells are only able to divide a certain number of times before the DNA loss prevents further division. This is known as the *Hayflick limit*. Within the germ cell line, which passes DNA to the next generation, the enzyme telomerase extends the repetitive sequences of the telomere region to prevent degradation. Telomerase can become mistakenly active in somatic cells, sometimes leading to cancer formation.

2.10 DNA INTERACTIONS WITH PROTEINS

Besides DNA's role in protein synthesis, the function of DNA depends on interactions with proteins, and these protein interactions can be nonspecific or the protein can bind specifically to a single DNA sequence. Enzymes can also bind to DNA and of these, the polymerases that copy the DNA base sequence in transcription and DNA replication are particularly important. Structural proteins that bind DNA are well-understood examples of nonspecific DNA—protein interactions. Within chromosomes, DNA is held in complexes with structural proteins. These proteins organize the DNA into chromatin. In eukaryotes, this structure involves DNA binding to histones, whereas in prokaryotes, multiple types of proteins are involved. The histones form a disk-shaped complex called a *nucleosome*, which contains two complete turns of double-stranded DNA wrapped around its surface. These nonspecific interactions are formed through basic residues in the histones making ionic bonds to the acidic sugar-phosphate backbone of the DNA and are therefore largely independent of the base sequence. Chemical modifications of these basic amino acid residues include methylation, phosphorylation, and acetylation. These chemical changes alter the strength of the interaction between the DNA and the histones, making the DNA more or less accessible to transcription factors (TFs) and changing the rate of transcription. Other nonspecific DNA-binding proteins in chromatin include the high-mobility group proteins, which bind to bent or distorted DNA. These proteins are important in bending arrays of nucleosomes and arranging them into the larger structures that make up chromosomes.

A distinct group of DNA-binding proteins is one that specifically binds single-stranded DNA. In humans, replication protein A is the best-understood member of this group and is used in processes where the double helix is separated, including DNA replication, recombination, and DNA repair. These binding proteins seem to stabilize single-stranded DNA and protect it from forming stem-loops or being degraded by nucleases. In contrast, other proteins have evolved to bind to particular DNA sequences. The most intensively studied of these are the various TFs, which are proteins that regulate transcription. Each TF binds to one particular set of DNA sequences and activates or inhibits the transcription of genes that have these sequences close to their promoters. The TFs can do this in two ways. They can bind the RNA polymerase responsible for transcription directly or through other mediator proteins. This locates the polymerase at the promoter and allows it to begin transcription. Alternatively, TFs can bind the enzymes that modify the histones at the promoter. This will change the accessibility of the DNA template to the polymerase. As these DNA targets can occur throughout an organism's genome, changes in the activity of one type of TF can affect thousands of genes. Consequently, these proteins are often the targets of the signal transduction processes that control responses to environmental changes or cellular differentiation and development. The specificity of these TF interactions with DNA come from the proteins making multiple contacts to the edges of the DNA bases, allowing them to read the DNA sequence. Most of these base interactions are made in the major groove, where the bases are most accessible.

2.11   DNA-MODIFYING ENZYMES

Nucleases are enzymes that cut DNA strands by catalyzing the hydrolysis of the phosphodiester bonds. Nucleases that hydrolyze nucleotides from the ends of DNA strands are called *exonucleases*, whereas *endonucleases* cut within strands. The most frequently used nucleases in molecular biology are the *restriction endonucleases*, which cut DNA at specific sequences. For instance, the EcoRV enzyme recognizes the 6-base sequence 5′-GAT|ATC-3′ and makes a cut at the vertical line. In nature, these enzymes protect bacteria against phage infection by digesting the phage DNA when it enters the bacterial cell, acting as part of the restriction modification system. In technology, these sequence-specific nucleases are used in molecular cloning and DNA fingerprinting. Enzymes called DNA ligases can rejoin cut or broken DNA strands. Ligases are particularly important in lagging strand DNA replication, as they join together the short segments of DNA produced at the replication fork into a complete copy of the DNA template. They are also used in DNA repair and genetic recombination.

2.11.1   TOPOISOMERASES AND HELICASES

Topoisomerases are enzymes with both nuclease and ligase activity. These proteins change the amount of supercoiling in DNA and some of these enzymes work by cutting the DNA helix and allowing one section to rotate, thereby reducing its level of supercoiling; the enzymes then seal the DNA break. Other types of enzymes can cut one DNA helix and then pass a second strand of DNA through this break before rejoining the helix. Topoisomerases are required for many processes involving DNA, such as DNA replication and transcription. *Helicases* are proteins that are a type of molecular motor. They use the chemical energy in nucleoside triphosphates, predominantly ATP, to break hydrogen bonds between bases and unwind the DNA double helix into single strands. These enzymes are essential for most processes where enzymes need to access the DNA bases.

2.12 DNA METHYLATION

DNA methylation is one such post-synthesis modification. DNA methylation has been proven by research to be manifested in several biological processes such as regulation of imprinted genes, X chromosome inactivation, and tumor suppressor gene silencing in cancerous cells (Figure 2.14). DNA methylation also acts as a protection mechanism for bacterial pathogen DNA against the endonuclease activity that destroys any foreign DNA. The expression of genes is influenced by how the DNA is packaged in chromosomes through chromatin. Base modifications can be involved in packaging, with regions that have low or no gene expression usually containing high levels of methylation of cytosine bases. For example, cytosine methylation produces 5-methylcytosine, which is important for X-chromosome inactivation. The average level of methylation varies among organisms, for example, the worm *Caenorhabditis elegans* lacks cytosine methylation, whereas vertebrates have higher levels, with up to 1% of their DNA containing 5-methylcytosine. Despite the importance of 5-methylcytosine, it can deaminate to leave a thymine base; methylated cytosines are therefore particularly prone to mutations. Other base modifications include adenine methylation in bacteria and the glycosylation of uracil to produce the "J-base" in kinetoplastids.

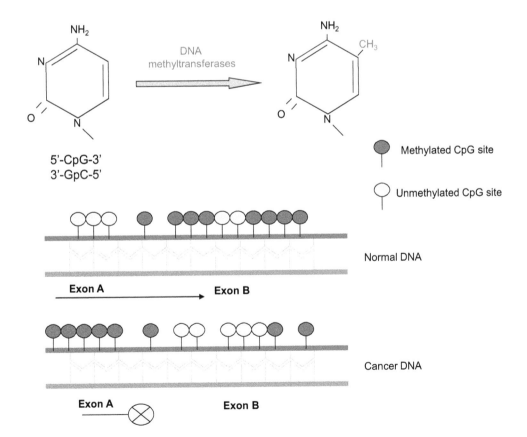

FIGURE 2.14 DNA methylation shown in both normal and cancer cells.

2.13 DNA MUTATION

DNA mutation is a change in the sequence of DNA. It can be caused by copying errors in the genetic material during cell division, by exposure to ultraviolet/ionizing radiation, chemical mutagens, viruses, or by cellular processes such as hyper-mutation (Figure 2.15). It can also be induced by the organism itself. In multicellular organisms with dedicated reproductive cells, mutations can be subdivided into *germ line mutations*, which can be passed on to descendants through the reproductive cells, and *somatic mutations*, which involve cells outside the dedicated reproductive group and which are not usually transmitted to descendants. If an organism can reproduce asexually through mechanisms such as budding, the distinction can become blurred. For example, plants can sometimes transmit somatic mutations to their descendants asexually or sexually where flower buds develop in somatically mutated parts of plants. A new mutation that was not inherited from either parent is called a *de novo* mutation. The source of the mutation is unrelated to the consequence, although the consequences are related to which cells were mutated. DNA can be damaged by many different sorts of mutagens, which change the DNA sequence. Mutagens include oxidizing agents, alkylating agents, and also high-energy electromagnetic radiation, such as ultraviolet (UV) light and x-rays. The type of DNA damage produced depends on the type of mutagen. For example, UV light can damage DNA by producing thymine dimers, which are cross-links between pyrimidine bases. On the other hand, oxidants such as free radicals or hydrogen peroxide produce multiple forms of damage including base modifications, particularly of guanosine and double-strand breaks.

A typical human cell contains about 150,000 bases that have suffered oxidative damage. Of these oxidative lesions, the most dangerous are double-strand breaks as these are difficult to repair and can produce point mutations, insertions/deletions from the DNA sequence, and chromosomal translocations. Many mutagens fit into the space between two adjacent base pairs in a process called *intercalation*. Most intercalators are aromatic and planar molecules such as ethidium bromide, daunomycin, and doxorubicin. For an intercalator to fit between base pairs the bases must separate, distorting the DNA strands by unwinding of the double helix. This inhibits both transcription and

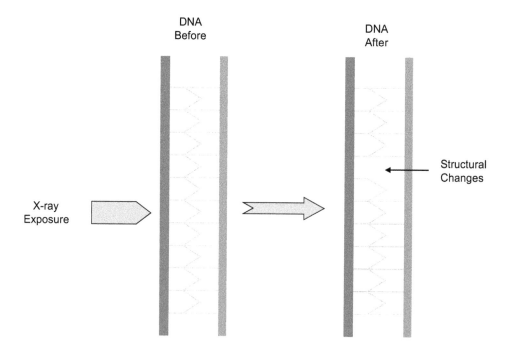

FIGURE 2.15 X-ray exposure causes structural changes in DNA

DNA replication, causing toxicity and mutations. As a result, DNA intercalators are often carcinogens. Some well-known examples of DNA intercalators are benzo[*a*]pyrene diol epoxide, acridines, aflatoxin, and ethidium bromide. Nevertheless, because of their ability to inhibit DNA transcription and replication, other similar toxins are also used in chemotherapy to inhibit rapidly growing cancer cells.

2.14 POLYMERASE CHAIN REACTION

PCR is a technique to amplify a single or a few copies of a piece of DNA across several orders of magnitude, generating thousands of millions of copies of a particular DNA sequence. PCR was developed in 1983 by Kary Mullis. Now, it is a common and indispensable technique used in medical and biological research laboratories for a variety of applications such as DNA cloning for sequencing, DNA-based phylogeny, functional analysis of genes, diagnosis of hereditary diseases, identification of genetic fingerprints, and detection and diagnosis of infectious diseases. The method relies on thermal cycling, consisting of cycles of repeated heating and cooling of the reaction for DNA melting and enzymatic replication of the DNA. Primers that are basically short DNA fragments containing sequences complementary to the target region along with a DNA polymerase are key components to enable selective and repeated amplification. As the PCR progresses, the DNA generated is itself used as a template for replication, setting in motion a chain reaction in which the DNA template is exponentially amplified. PCR can also be extensively modified to perform a wide array of genetic manipulations (Figure 2.16).

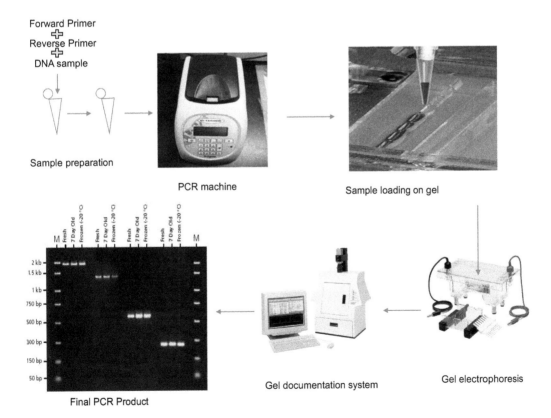

FIGURE 2.16   PCR is a multi-step process where forward and reverse primers are mixed with the DNA sample and amplified in the PCR machine and then run on the gel electrophoresis and read on the gel documentation system.

Almost all PCR applications employ a heat-stable DNA polymerase, such as *Taq polymerase*, an enzyme originally isolated from the bacterium *Thermus aquaticus*. This DNA polymerase enzymatically assembles a new DNA strand from nucleotides by using single-stranded DNA as a template and DNA oligonucleotides (also called DNA primers), which are required for initiation of DNA synthesis. The vast majority of PCR methods use thermal cycling, that is, alternately heating and cooling the PCR sample in a defined series of temperature steps. These thermal cycling steps are necessary to physically separate the two strands in a DNA double helix at a high temperature in a process called *DNA melting*. At a lower temperature, each strand is then used as the template in DNA synthesis by the DNA polymerase to selectively amplify the target DNA. The selectivity of the PCR results from the use of primers that are complementary to the DNA region targeted for amplification under specific thermal cycling conditions (Figure 2.17).

Most PCR methods typically amplify DNA fragments of up to ~10 kb pairs, although some techniques allow for amplification of fragments up to 40 kb in size.

The PCR is commonly carried out in a reaction volume of 10–200 μL in small reaction tubes (0.2–0.5 mL volume) in a thermal cycler. The thermal cycler heats and cools the reaction tubes to achieve the temperatures required at each step of the reaction. Many modern thermal cyclers make use of the *Peltier effect,* which permits both heating and cooling of the block holding the PCR tubes simply by reversing the electric current. Thin-walled reaction tubes permit favorable thermal conductivity to allow for rapid thermal equilibration. Most thermal cyclers have heated lids to prevent condensation at the top of the reaction tube. Older thermocyclers lacking a heated lid require a layer of oil on top of the reaction mixture or a ball of wax inside the tube.

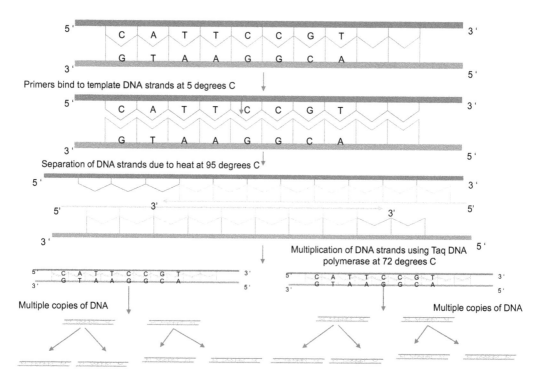

FIGURE 2.17 Amplification of DNA fragment: (1) primers binds to template DNA strands, (2) separation of the DNA strands, (3) multiplication of the DNA strands using Taq DNA polymerase, and (4) making multiple copies of the DNA.

2.14.1 APPLICATIONS OF PCR

2.14.1.1 Diagnostic Assay

For the past few years, there has been a tremendous demand for genetic-based diagnostic tests for almost all diseases. The reason for this, obviously, is the rapidness and the accuracy of the data generated by the PCR technique. PCR assays can be performed directly on genomic DNA samples to detect translocation-specific malignant cells at a sensitivity that is at least 10,000-fold higher than other methods. PCR also permits identification of non-cultivatable or slow-growing microorganisms such as mycobacteria, anaerobic bacteria, or viruses from tissue culture assays and animal models. The basis for PCR diagnostic applications in microbiology is the detection of infectious agents and the discrimination of nonpathogenic from pathogenic strains by specific genes. Viral DNA can likewise be detected by PCR. The primers used need to be specific to the targeted sequences in the DNA of the virus, and the PCR can be used for diagnostic analyses or DNA sequencing of the viral genome. The high sensitivity of the PCR permits virus detection soon after infection and even before the onset of disease. Such early detection may give physicians a significant lead in treatment. The amount of virus (or viral load) in a patient can also be quantified by PCR-based DNA quantitation techniques.

2.14.1.2 Genetic Engineering

Methods have been developed to purify DNA from organisms, such as phenol-chloroform extraction, and to manipulate it in the laboratory, such as restriction digests and PCR. Modern biology and biochemistry make intensive use of these techniques in recombinant DNA technology. *Recombinant DNA* is a man-made DNA sequence that has been assembled from other DNA sequences. It can be transferred into organisms in the form of plasmids or in the appropriate format by using a viral vector. The genetically modified organisms produced can be used to cultivate in agriculture or to produce products, such as recombinant proteins, that are used in medical research.

2.14.1.3 Forensic DNA Profiling

Forensic scientists can use the DNA in blood, semen, skin, saliva, or hair found at a crime scene to identify a matching DNA of an individual, such as a perpetrator. This process is called *genetic fingerprinting*, or more accurately, *DNA profiling*. In DNA profiling, the lengths of variable sections of repetitive DNA, such as short tandem repeats and minisatellites, are compared between people. This method is usually an extremely reliable technique for identifying a matching DNA. However, identification can be complicated if the scene is contaminated with DNA from several people. DNA profiling was developed in 1984 by British geneticist Sir Alec Jeffrey and was first used in forensic science to convict Colin Pitchfork in the 1988 Enderby murder case. People convicted of certain types of crimes may be required to provide a sample of DNA for a database. This has helped investigators solve old cases in which only a DNA sample was obtained from the scene. DNA profiling can also be used to identify victims of mass casualty incidents. On the other hand, many convicted people have been released from prison on the basis of DNA techniques, which were not available when a crime had originally been committed.

PROBLEMS

Section A: Descriptive Type

Q1. What is the cell theory?
Q2. Describe the characteristics of a prokaryotic cell.
Q3. Explain Mendelian genetics.
Q4. Explain supercoiling in a DNA molecule.

Q5. Describe the role of DNA polymerase in replication.
Q6. What are topoisomerases and helicases?
Q7. How does DNA methylation occur?
Q8. What is a PCR?
Q9. How is forensic DNA profiling done with a PCR tool?

Section B: Multiple Choices

Q1. What is the estimated number of cells in humans?
  a. 100 billion
  b. 1000 billion
  c. 100 trillion
  d. 5 trillion
Q2. Who was the first to study the internal structure of a cell?
  a. Robert Hooke
  b. Antonie Leeuwenhoek
  c. Henri Dutrochet
  d. Charles Darwin
Q3. Do viruses fit in the cell theory concept?
  a. Yes
  b. No
Q4. Prokaryotes carry extra-chromosomal DNA molecules which are called . . .
  a. Nuclei
  b. Plasmids
  c. Mitochondria
  d. Ribosomes
Q5 In eukaryotes, non-nuclear DNA is located in the . . .
  a. Endoplasmic reticulum
  b. Mitochondria
  c. Golgi bodies
  d. Chromatin
Q6. Cell surface membranes contain receptor proteins that allow cells to detect external sig-naling molecules such as hormones. True/False
Q7. What is common to mitochondria and chloroplast?
  a. Both do not contain their own genome.
  b. Both contain their own genome.
  c. Both are present in prokaryotes.
Q8. Except for these, all living organisms have genetic information stored in their DNA.
  a. Retroviruses
  b. Bacteria
  c. Fungi
Q9. Nuclear DNA is linear whereas mitochondria DNA is circular. True/False
Q10. Glycogen is a polysaccharide used by animals to store energy. True/False
Q11. Mendel observed that organisms inherit traits called . . .
  a. DNA
  b. Genes
  c. Proteins
Q12. One of the major differences between DNA and RNA is . . .
  a. Protein
  b. Hormones
  c. Sugar
  d. Phosphate

Q13. When DNA is twisted in the direction of a helix, this is called what type of supercoiling?
   a. Positive
   b. Negative
   c. Linear
Q14. Messenger RNA encodes for . . .
   a. Gene expression
   b. Protein synthesis
   c. Both protein synthesis and gene expression
Q15. What is the protein manufacturing machine of all living cells?
   a. Ribosome
   b. Golgi bodies
   c. Endoplasmic reticulum
Q16. Meiosis is a process of reduction division in which the number of chromosomes increases to double. True/False
Q17. Replication of DNA is done by the enzyme . . .
   a. Polymerase
   b. Endonuclease
   c. Exonuclease
   d. Telomerase
Q18. The structural change in the DNA sequence is called . . .
   a. DNA methylation
   b. DNA mutation
   c. DNA replication
Q19. Who invented the PCR?
   a. James Watson
   b. Kary Mullis
   c. Ian Wilmut
Q20. Nested PCR is used to increase the specificity of amplification of . . .
   a. RNA
   b. DNA
   c. Both DNA and RNA
   d. None of the above

Section C: Critical Thinking

Q1. To identify the real culprit among a group of crime suspects, what technique can be used to establish the identity of the culprit? Explain with suitable examples.
Q2. Is it possible to study the genetic information of an individual by working with mRNA only? Explain.
Q3. What would be the status of gene expression if mRNA is not available?
Q4. What will happen if nuclear DNA is circular in shape and mitochondrial DNA is linear in shape?

ASSIGNMENT

With the help of your course instructor, organize special lectures on the origin of DNA and submit your analysis in the form of a written assignment.

REFERENCES AND FURTHER READING

Alberts, B., Johnson, A., Lewis, J. et al. *Molecular Biology of the Cell*, 4th edn. Garland Science, New York, 2002.

Aldaye, F.A., Palmer, A.L., and Sleiman, H.F. Assembling materials with DNA as the guide. *Science* 321: 1795–1799, 2008.

Ananthakrishnan, R., and Ehrlicher, A. The forces behind cell movement. *Int. J. Biol. Sci.* 3: 303–317, 2007.

Baianu, I.C. X-ray scattering by partially disordered membrane systems. *Acta Cryst.* A34(5): 751–753, 1978.

Baianu, I.C. Structural order and partial disorder in biological systems. *Bull. Math. Biol.* 42(4): 137–141, 1980.

Bickle, T., and Kruger, D. Biology of DNA restriction. *Microbiol. Rev.* 57: 434–450, 1993.

Bird, A. DNA methylation patterns and epigenetic memory. *Genes Dev.* 16: 6–21, 2002.

Burt, D.W. Origin and evolution of avian microchromosomes. *Cytogenet. Genome. Res.* 96: 97–112, 2002.

Campbell, N.A., Williamson, B., and Heyden, R.J. *Biology: Exploring Life.* Pearson Prentice Hall, Boston, MA, 2006.

The ENCODE Project Consortium. Identification and analysis of functional elements in 1% of the human genome by the ENCODE pilot project. *Nature* 447(7146): 799–816, 2007.

Leslie, A.G., Arnott, S., Chandrasekaran, R., and Ratliff, R.L. Polymorphism of DNA double helices. *J. Mol. Biol.* 143: 49–72, 1980.

Li, Z., Van Calcar, S., Qu, C., Cavenee, W., Zhang, M., and Ren, B. A global transcriptional regulatory role for c-Myc in Burkitt's lymphoma cells. *Proc. Natl. Acad. Sci. USA* 100: 8164–8169, 2003.

Lindahl, T. Instability and decay of the primary structure of DNA. *Nature* 362(6422): 709–715, 1993.

Luger, K., Mäder, A., Richmond, R., Sargent, D., and Richmond, T. Crystal structure of the nucleosome core particle at 2.8 A resolution. *Nature* 389: 251–260, 1997.

Maddox, B. The double helix and the wronged heroine [PDF]. *Nature* 421: 407–408, 2003.

Makalowska, I., Lin, C., and Makalowski, W. Overlapping genes in vertebrate genomes. *Comput. Biol. Chem.* 29: 1–12, 2005.

Mandelkern, M., Elias, J., Eden, D., and Crothers, D. The dimensions of DNA in solution. *J. Mol. Biol.* 152: 153–161, 1981.

Martinez, E. Multi-protein complexes in eukaryotic gene transcription. *Plant. Mol. Biol.* 50: 925–947, 2002.

Mendell, J.E., Clements, K.D., Choat, J.H., and Angert, E.R. Extreme polyploidy in a large bacterium. *Proc. Natl. Acad. Sci. USA* 105: 6730–6734, 2008.

Ménétret, J.F., Schaletzky, J., Clemons, W.M. et al. Ribosome binding of a single copy of the SecY complex: Implications for protein translocation. *Mol. Cell.* 28: 1083–1092, 2007.

Michie, K., and Löwe, J. Dynamic filaments of the bacterial cytoskeleton. *Annu. Rev. Biochem.* 75: 467–492, 2006.

Nakabachi, A., Yamashita, A., Toh, H., Ishikawa, H., Dunbar, H., Moran, N., and Hattori, M. The 160-kb genome of the bacterial endosymbiont carsonella. *Science* 314: 267, 2006.

Neale, M.J., and Keeney, S. Clarifying the mechanics of DNA strand exchange in meiotic recombination. *Nature* 442: 153–158, 2006.

Nickle, D., Learn, G., Rain, M., Mullins, J., and Mittler, J. Curiously modern DNA for a 250 million-year-old bacterium. *J. Mol. Evol.* 54: 134–137, 2002.

Painter, T.S. The spermatogenesis of man. *Anat. Res.* 23: 129, 1922.

Thanbichler, M., and Shapiro, L. Chromosome organization and segregation in bacteria. *J. Struct. Biol.* 156(2): 292–303, 2006.

Thanbichler, M., Wang, S.C., and Shapiro, L. 2005. The bacterial nucleoid: A highly organized and dynamic structure. *J. Cell. Biochem.* 96(3): 506–521, 2005.

Thomas, J. HMG1 and 2: Architectural DNA-binding proteins. *Biochem. Soc. Trans.* 29(Pt 4): 395–401, 2001.

Tjio, J.H., and Levan, A. The chromosome number of man. *Hereditas* 42: 1–6, 1956.

Tuteja, N., and Tuteja, R. Unraveling DNA helicases. Motif, structure, mechanism and function. *Eur. J. Biochem.* 271(10): 1849–1863, 2004.

Valerie, K., and Povirk, L. Regulation and mechanisms of mammalian double-strand break repair. *Oncogene* 22(37): 5792–5812, 2003.

Watson, J.D., and Crick, F.H.C. A structure for deoxyribose nucleic acid (PDF). *Nature* 171: 737–738, 1953.

3 Proteins and Proteomics

LEARNING OBJECTIVES

- Define protein and explain its significance
- Discuss the role of protein in cell signaling
- Discuss the structural and functional attributes of protein
- Explain the process of protein biosynthesis
- Explain different methods of protein prediction models
- Explain the significance of protein folding and protein modification
- Discuss protein transport and degradation
- Discuss genetic regulation of protein synthesis
- Discuss tools and techniques used to analyze proteins

3.1 INTRODUCTION

Proteomics is the science of proteins, and it is much more complicated than genomics mostly because although an organism's genome is constant, the proteome differs from cell to cell and from time to time. The *proteome* is the entire set of proteins expressed by a genome, cell, tissue, or organism at a certain time. This is because distinct genes are expressed in distinct cell types. This means that even the basic set of proteins that are produced in a cell needs to be determined. Nowadays, we keep hearing of high-protein or low-protein diets. But why do we need to keep our protein level under control? This is because proteins are very important for body functions, and any deficiency or malfunction in protein may lead to serious ailments. The proteome is larger than the genome, especially in eukaryotes, in the sense that there are more proteins than genes. This is because of alternative splicing of genes and posttranslational modifications like glycosylation or phosphorylation. Moreover, the proteome has at least two levels of complexity lacking in the genome. although the genome is defined by the sequence of nucleotides, the proteome cannot be limited to the sum of the sequences of the proteins present. Knowledge of the proteome requires knowledge of (1) the structure of the proteins in the proteome and (2) the functional interaction between the proteins.

Proteins are the primary components of numerous body tissues such as muscle tissues. They help to increase strength, improve athletic performance, and develop muscles. Proteins also make up the outer layers of hair, nails, and skin. The most important function of proteins is to build up, maintain, and replace the tissues in the body. Muscles, organs, and some hormones (insulin) are mostly made up of proteins. Proteins also make up antibodies and hemoglobin (responsible for delivering oxygen to the blood cells). They make up half the dry weight of an *Escherichia coli* cell. On the other hand, they make up 3% of a DNA molecule and 20% of an RNA molecule.

The main characteristic of proteins that allows their diverse set of functions is their ability to specifically and tightly bind with other molecules. The region of the protein responsible for binding with another molecule is called the *binding site*. It is often a depression or "pocket" on the molecular surface of a protein. This binding ability is mediated by the tertiary structure of the protein, which defines the binding site pocket, and by the chemical properties of the surrounding amino acids' side chains. Protein binding can be extraordinarily tight and specific. For example, the ribonuclease inhibitor protein binds to human angiogenin with a sub-femtomolar dissociation constant ($<10^{-15}$ M) but does not bind at all to its amphibian homolog onconase (>1 M). Extremely minor chemical changes such as the addition of a single methyl group to a binding partner can sometimes suffice to nearly eliminate binding. For example, the aminoacyl tRNA synthetase specific to the amino acid valine discriminates against the very similar side chain of the amino acid isoleucine. Proteins can bind to other proteins as well as to small-molecule substrates. When proteins bind specifically to other copies of the same molecule, they can oligomerize to form fibrils. This process occurs often in structural proteins that consist of globular monomers that self-associate to form rigid fibers. Protein-to-protein interactions also regulate enzymatic activity, control progression through the cell cycle, and allow the assembly of large protein complexes that carry out many closely related reactions with a common biological function. The ability of binding partners to induce conformational changes in proteins allows the construction of enormously complex signaling networks. Importantly, as interactions between proteins are reversible and depend heavily on the availability of different groups of partner proteins to form aggregates that can carry out discrete sets of functions, the study of the interactions between specific proteins is a key to understanding important aspects of cellular functions and ultimately the properties that distinguish specific cell types. Before we learn more about proteins and proteomics, let's quickly learn about the other macromolecules that are equally important for our body functions—carbohydrates, nucleic acids, and lipids.

3.2   SIGNIFICANCE OF PROTEINS

Proteins play a key role in food intake regulation through satiety related to diet-induced thermogenesis. Proteins also play a key role in body weight regulation through their effect on thermogenesis and body composition. In this section, we have analyzed various applications of proteins in detail.

3.2.1   PROTEINS FOR BODY FUNCTIONS

There are 20 different identified types of amino acids that are necessary for normal body functions. Of these amino acids, 14 are produced by the body and 6 are ingested through the foods we eat. Those amino acids that the body can produce are called nonessential amino acids and those that the body cannot produce are called essential amino acids.

3.2.1.1   Nonessential Amino Acids

The following lists the nonessential amino acids with some of their functions, benefits, and side effects:

- **Alanine**: It removes toxic substances released from the breakdown of muscle protein during intense exercise. An excessive alanine level in the body is associated with chronic fatigue.
- **Cysteine**: It is a component of the protein type abundant in nails, skin, and hair. It also acts as an antioxidant (free radical scavenger) and has synergetic effect when taken along with other antioxidants such as vitamin E and selenium.
- **Cystine**: The same as cysteine; it aids in the removal of toxins and the formation of skin.
- **Glutamine**: It promotes healthy brain function. It is also necessary for the synthesis of RNA and DNA molecules.
- **Glutathione**: It is an antioxidant and has an antiaging effect. It is useful in the removal of toxins.
- **Glycine**: It is a component of skin and is beneficial for wound healing. It also acts as a neurotransmitter. High levels of glycine in the body may cause fatigue.
- **Histidine**: It is important in the synthesis of red and white blood cells and as a precursor for histamine, which is good for sexual arousal. It also improves blood flow. High levels of histidine may cause stress and anxiety.
- **Serine**: It is a constituent of brain proteins. It aids in the synthesis of immune system proteins and helps improve muscle development.

- **Taurine**: It is necessary for proper brain functioning and synthesis of amino acids. It is also important in the assimilation of mineral nutrients such as magnesium, calcium, and potassium.
- **Threonine**: It balances the protein level in the body and promotes the immune system. It is also beneficial for the synthesis of tooth enamel and collagen.
- **Asparagine**: It helps promote equilibrium in the central nervous system, thus balancing the emotional state.
- **Aspartic acid**: It enhances stamina, aids in removal of toxins and ammonia from the body, and is beneficial in the synthesis of proteins involved in the immune system.
- **Proline**: It plays a role in intracellular signaling.
- **L-arginine**: It plays a role in blood vessel relaxation and removal of excess ammonia from the body.

3.2.1.2 Essential Amino Acids

The following lists the eight amino acids that are generally essential in humans.

- **Phenylalanine**: Phenylalanine (abbreviated as Phe or F) is an α-amino acid with the formula $HO_2CCH(NH_2)CH_2C_6H_5$. This essential amino acid is classified as nonpolar because of the hydrophobic nature of the benzyl side chain. The codons for L-phenylalanine are UUU and UUC. L-Phenylalanine (LPA) is an electrically neutral amino acid, one of the 20 common amino acids used to biochemically form proteins, coded for by DNA. Phenylalanine is structurally closely related to dopamine, epinephrine (adrenaline), and tyrosine. Phenylalanine is found naturally in the breast milk of mammals. It is manufactured for food and drink products and is also sold as a nutritional supplement for its reputed analgesic and antidepressant effects. It is a direct precursor to the neuromodulator phenylethylamine, a commonly used dietary supplement.
- **Valine**: Valine (abbreviated as Val or V) is an α-amino acid with the chemical formula $HO_2CCH(NH_2)CH(CH_3)_2$. L-Valine is one of 20 proteinogenic amino acids. Its codons are GUU, GUC, GUA, and GUG. This essential amino acid is classified as nonpolar. Human dietary sources include cottage cheese, fish, poultry, peanuts, sesame seeds, and lentils. Along with leucine and isoleucine, valine is a branched-chain amino acid. It is named after the plant valerian. In sickle-cell disease, valine substitutes for the hydrophilic amino acid glutamic acid in hemoglobin. Because valine is hydrophobic, the hemoglobin does not fold correctly.
- **Threonine**: Threonine is an α-amino acid with the chemical formula $HO_2CCH(NH_2)$ $CH(OH)CH_3$. Its codons are ACU, ACA, ACC, and ACG. This essential amino acid is classified as polar. Together with serine and tyrosine, threonine is one of three proteinogenic amino acids bearing an alcohol group. The threonine residue is susceptible to numerous posttranslational modifications (PTMs). Basically, PTM is a step in protein biosynthesis. Proteins are created by ribosomes translating mRNA into polypeptide chains. The polypeptide chains undergo PTMs such as folding, cutting, and other processes before becoming the mature protein product, whereas the hydroxyl side chain undergo O-linked glycosylation. In addition, threonine residues undergo phosphorylation through the action of a threonine kinase. In its phosphorylated form, it is referred to as phosphothreonine.
- **Tryptophan**: Tryptophan (abbreviated as Trp and sold as Tryptan) is one of the 20 standard amino acids as well as an essential amino acid in the human diet. It is encoded in the standard genetic code as the codon UGG. Only the L-stereoisomer of tryptophan is used in structural or enzyme proteins, but the D-stereoisomer is occasionally found in naturally produced peptides (such as the marine venom peptide contryphan). The distinguishing structural characteristic of tryptophan is that it contains an indole functional group.

- **Isoleucine**: Isoleucine (abbreviated as Ile or I) is an α-amino acid with the chemical formula $HO_2CCH(NH_2)CH(CH_3)CH_2CH_3$. The codons of these essential amino acids are AUU, AUC, and AUA. With a hydrocarbon side chain, isoleucine is classified as a hydrophobic amino acid. Together with threonine, isoleucine is one of two common amino acids that have a chiral side chain. Four stereoisomers of isoleucine are possible, including two possible diastereomers of L-isoleucine. However, isoleucine present in nature exists in an enantiomeric form, (2S, 3S)-2-amino-3-methylpentanoic acid.
- **Methionine**: Methionine is one of only two amino acids encoded by a single codon (AUG) in the standard genetic code (tryptophan, encoded by UGG, is the other). The codon AUG is also the "Start" message for a ribosome that signals the initiation of protein translation from mRNA. Consequently, methionine is incorporated into the N-terminal position of all proteins in eukaryotes and archaea during translation, although it is usually removed by PTM.
- **Leucine**: Leucine (abbreviated as Leu or L) is an α-amino acid with the chemical formula $HO_2CCH(NH_2)CH_2CH(CH_3)_2$. The codons of this essential amino acid are UUA, UUG, CUU, CUC, CUA, and CUG. With a hydrocarbon side chain, leucine is classified as a hydrophobic amino acid. It has an isobutyl R group. Leucine is a major component of the subunits in ferritin, astacin, and other "buffer" proteins.
- **Lysine**: Lysine is an α-amino acid with the chemical formula $HO_2CCH(NH_2)(CH_2)4NH_2$. The codons of this essential amino acid are AAA and AAG. Lysine is a base, as are arginine and histidine. The ε-amino group often participates in hydrogen bonding and as a general base in catalysis. Common PTMs include methylation of the ε-amino group, giving methyl-, dimethyl-, and trimethyl lysine. The latter occurs in calmodulin. Other PTMs at lysine residues include acetylation and ubiquitination. Collagen contains hydroxylysine which is derived from lysine by lysyl hydroxylase. O-Glycosylation of lysine residues in the endoplasmic reticulum or Golgi apparatus is used to mark certain proteins for secretion from the cell.

3.2.1.3   Other Amino Acids

Essential amino acids are called essential not because they are more important than the others, but because the body does not synthesize them, making it essential to include them in one's diet to obtain them. On the other hand, the amino acids arginine, cysteine, glycine, glutamine, histidine, proline, serine, and tyrosine are considered conditionally essential, meaning they are not normally required in the diet but must be supplied exogenously to specific populations who do not synthesize them in adequate amounts. For example, individuals living with phenylketonuria (PKU) disease must keep their intake of phenylalanine extremely low to prevent mental retardation and other metabolic complications. However, phenylalanine is the precursor for tyrosine synthesis. Without phenylalanine, tyrosine cannot be made and so tyrosine becomes essential in the diet of PKU patients.

3.2.2   Proteins as Enzymes

The best-known role of proteins in the cell is as enzymes, which catalyze chemical reactions. Enzymes are usually highly specific and accelerate only one or a few chemical reactions. Enzymes carry out most of the reactions involved in metabolism as well as DNA manipulation in processes such as DNA replication, DNA repair, and transcription. Some enzymes act on other proteins to add or remove chemical groups in PTMs (such as folding, cutting, and other processes) to produce the mature protein product. About 4000 reactions are known to be catalyzed by enzymes. The rate of acceleration conferred by enzymatic catalysis is often enormous, as much as a 10^{17}-fold increase in rate over the uncatalyzed reaction rate in the case of orotate decarboxylase (78 million years without the enzyme, 18 ms with the enzyme). The molecules bound and acted upon by enzymes are called *substrates.* Although enzymes can consist of hundreds of amino acids, it is usually only a small fraction of the residues that meet the substrate, and an even smaller fraction, 3–4 residues on the average, are directly involved in catalysis. The region of the enzyme that binds the substrate and contains the catalytic residues is known as the *active site.*

3.2.3 PROTEINS IN CELL SIGNALING AND LIGAND BINDING

Many proteins are involved in the process of cell signaling and signal transduction. Some proteins such as insulin are extracellular proteins that transmit a signal from the cell in which they were synthesized to other cells in distant tissues (Figure 3.1). Others are membrane proteins that act as receptors whose main function is to bind a signaling molecule and induce a biochemical response in the cell. Many receptors have a binding site exposed on the cell surface and an effector domain within the cell, which may have enzymatic activity or may undergo a conformational change detected by other proteins within the cell. *Antibodies* are protein components of the adaptive immune system whose main function is to bind antigens, or foreign substances in the body, and target them for destruction. Antibodies can be secreted into the extracellular environment or anchored in the membranes of specialized B cells known as *plasma cells.* Whereas enzymes are limited in their binding affinity for their substrates by the necessity of conducting their reaction, antibodies have no such constraint. An antibody's binding affinity to its target is extraordinarily high.

Many ligand transport proteins bind small biomolecules and transport them to other locations in the body of a multicellular organism. These proteins must have a high binding affinity when their ligand is present in high concentrations but must also release the ligand when it is present at low concentrations in the target tissues. The canonical example of a ligand-binding protein is hemoglobin, which transports oxygen from the lungs to other organs and tissues in all vertebrates and has close homologs in every biological kingdom. *Lectins* are sugar-binding proteins that are highly specific for their sugar moieties. Lectins typically play a role in biological recognition phenomena involving cells and proteins. Receptors and hormones are highly specific binding proteins. Transmembrane

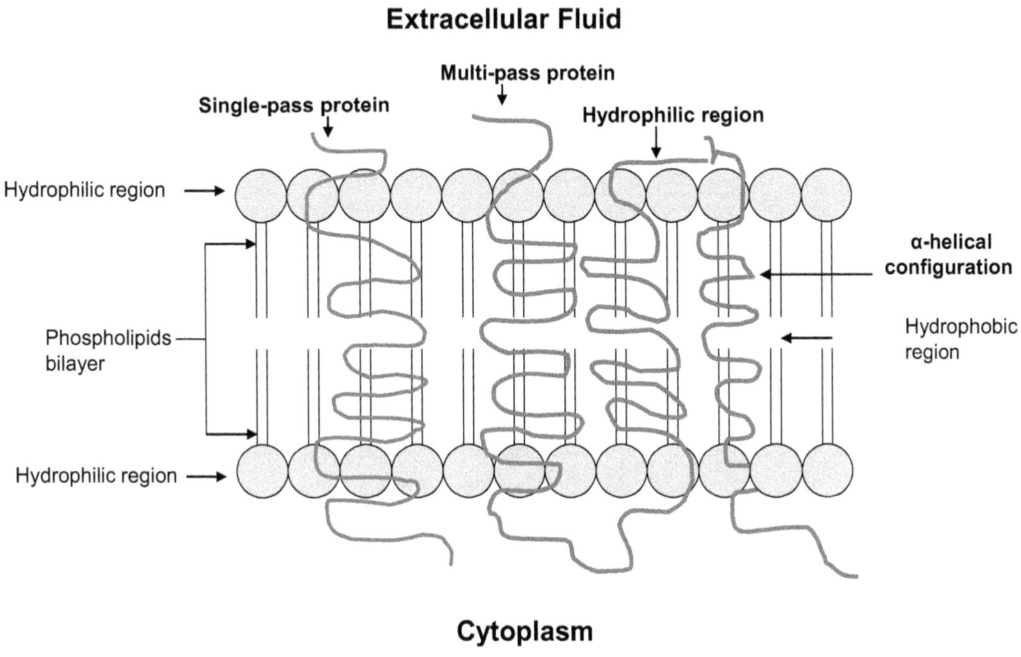

FIGURE 3.1 Cell signaling through proteins shows single-pass protein and multi-pass protein.

proteins can also serve as ligand transport proteins that alter the permeability of the cell membrane to small molecules and ions. The membrane alone has a hydrophobic core through which polar or charged molecules cannot diffuse. Membrane proteins contain internal channels that allow such molecules to enter and exit the cell. Many ion channel proteins are specialized to select for only a specific ion. For example, potassium and sodium channels often discriminate for only one of the two ions.

3.2.4 STRUCTURAL PROTEINS

Structural proteins confer stiffness and rigidity to otherwise fluid biological components. Most structural proteins are fibrous proteins. For example, actin and tubulin are globular and soluble as monomers, but polymerize to form long, stiff fibers that comprise the cytoskeleton, which allows the cell to maintain its shape and size. Collagen and elastin are critical components of connective tissue such as cartilage, whereas keratin is found in hard or filamentous structures such as hair, nails, feathers, hooves, and some animal shells. Other proteins that serve structural functions are motor proteins such as myosin, kinesin, and dynein, which can generate mechanical forces. These proteins are crucial for cellular motility of single-celled organisms and the sperm of many multicellular organisms that reproduce sexually. They also generate the forces exerted by contracting muscles.

3.3 PROTEIN BIOSYNTHESIS

After learning various attributes of proteins, the next question that arises is how our body makes these proteins. In this section, we will learn how our body makes proteins in a step-by-step manner. Recall that a single gene expression results in the formation of protein, and protein synthesis is a multistep process that involves not only DNA but also mRNA, tRNA, and rRNA in a well-coordinated manner (Figure 3.2). Protein synthesis starts with gene expression (or *transcription*

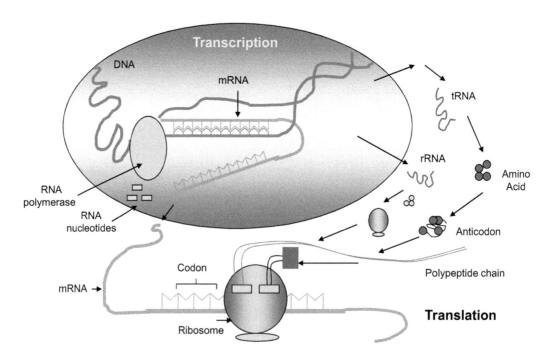

FIGURE 3.2 Process of protein biosynthesis shows gene expression, transcription, and translation and protein synthesis.

phase), leading to translation and to the final protein formation (Figure 3.2). Also, recall that gene expression is used by all known life—eukaryotes (including multicellular organisms), prokaryotes (bacteria and archaea), and viruses—to generate the macromolecular machinery for life. Several steps in the gene expression process may be modulated, including the transcription, RNA splicing, translation, and PTM of a protein.

3.3.1 Transcription Stage

Transcription, also called RNA synthesis, is the process of creating an equivalent RNA copy of a sequence of DNA. Both RNA and DNA are nucleic acids, which use base pairs of nucleotides as a complementary language that can be converted back and forth from DNA to RNA in the presence of the correct enzymes. During transcription, a DNA sequence is read by an RNA polymerase, which produces a complementary and antiparallel RNA strand. As opposed to DNA replication, transcription results in an RNA compliment that includes uracil (U) in all instances where thymine (T) would have occurred in a DNA compliment. Transcription is the first step leading to gene expression. The stretch of DNA transcribed into an RNA molecule is called a *transcription unit*, which encodes at least one gene. If the gene transcribed encodes for a protein, the result of the transcription is a messenger RNA (mRNA), which will then be used to create that protein via the process of translation.

A DNA transcription unit encoding for a protein contains not only the sequence that will eventually be directly translated into the protein but also regulatory sequences that direct and regulate the synthesis of that protein. The regulatory sequence before the coding sequence is called the *five-prime untranslated region* (5′UTR) and is also known as the *upstream process*. The sequence following the coding sequence is called the *three prime untranslated region* (3′UTR) and is also known as the *downstream process*. Transcription has some proofreading mechanisms, but they are less effective than the controls for copying DNA. Therefore, transcription has a lower copying fidelity than DNA replication. As in DNA replication, DNA is read from 3′—5′ during transcription. Meanwhile, the complementary RNA is created from the 5′—3′ direction. Although DNA is arranged as two antiparallel strands in a double helix, only one of the two DNA strands, called the *template strand*, is used for transcription. This is because RNA is only single-stranded, as opposed to double-stranded DNA. The other DNA strand is called the *coding strand* because its sequence is the same as the newly created RNA transcript except for the substitution of uracil for thymine. The use of only the 3′—5′ strand eliminates the need for the Okazaki fragments seen in DNA replication (Figure 3.2).

3.3.2 Transcription in Prokaryotes and Eukaryotes

One major difference between prokaryotes and eukaryotes is the existence of membrane-bound structures within eukaryotes, including a cell nucleus with a nuclear membrane that encapsulates cellular DNA. In prokaryotes, mRNA usually remains unmodified whereas eukaryotic mRNA is heavily processed through RNA splicing, 5′ end capping (5′ cap), and the addition of a polyA tail (Figure 3.3).

3.3.3 Stages of Transcription

Transcription is divided into five stages: pre-initiation, initiation, promoter clearance, elongation, and termination. We will describe each step to understand the beginning of protein synthesis (Figure 3.4).

3.3.3.1 Pre-Initiation

In eukaryotes, RNA polymerase, and therefore the initiation of transcription, requires the presence of a core promoter sequence in the DNA. Promoters are regions of DNA that promote transcription

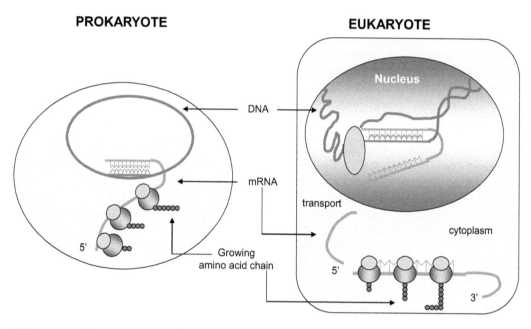

FIGURE 3.3   Transcription in prokaryote and eukaryote cells.

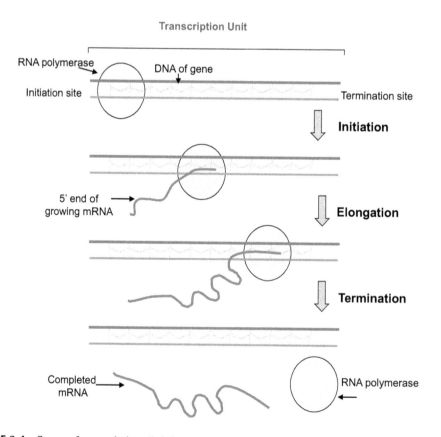

FIGURE 3.4   Stages of transcription: (1) initiation, (2) elongation, and (3) termination.

and are found around 10–35 base pairs upstream from the start site of transcription. Core promoters are sequences within the promoter that are essential for transcription initiation. RNA polymerase can bind to core promoters in the presence of various specific TFs. The most common type of core promoter in eukaryotes is a short DNA sequence known as a *TATA box*. The TATA box, as a core promoter, is the binding site for a TF known as *TATA binding protein* (TBP), which is itself a sub-unit of another TF called *transcription factor II D* (TFIID). After TFIID binds to the TATA box via the TBP, five more TFs and RNA polymerase combine around the TATA box in a series of stages to form a pre-initiation complex. One TF, DNA helicase, has helicase activity and so is involved in the separating of opposing strands of double-stranded DNA to provide access to a single-stranded DNA template. However, only a low or basal rate of transcription is driven by the pre-initiation complex alone. Other proteins known as activators and repressors, along with any associated coactivators or corepressors, are responsible for modulating the transcription rate. The transcription pre-initiation in archaea, formerly a domain of prokaryotes, is essentially homologous to that of eukaryotes but is much less complex. The archaeal pre-initiation complex assembles at a TATA-box binding site. However, in archaea, this complex is composed of only RNA polymerase II, TBP, and TFB (the archaeal homologue of eukaryotic transforming factor II B (TFIIB).

3.3.3.2 Initiation
In bacteria, transcription begins with the binding of RNA polymerase to the promoter in DNA. RNA polymerase is a core enzyme consisting of five subunits: 2 α subunits, 1 β subunit, 1 β′ sub-unit, and 1 ω subunit. At the start of initiation, the core enzyme is associated with a sigma factor (number 70) that aids in finding the appropriate −35 and −10 base pairs downstream of promoter sequences. Transcription initiation is more complex in eukaryotes. Eukaryotic RNA polymerase does not directly recognize the core promoter sequences. Instead, a collection of proteins called TFs mediates the binding of RNA polymerase and the initiation of transcription. Only after certain TFs are attached to the promoter does the RNA polymerase bind to it. The completed assembly of TFs and RNA polymerase bind to the promoter, forming a transcription initiation complex. Transcription in the archaea domain is like transcription in eukaryotes.

3.3.3.3 Promoter Clearance
After the first bond is synthesized, the RNA polymerase must clear the promoter. During this time, there is a tendency to release the RNA transcript and produce truncated transcripts. This is called *abortive initiation* and is common for both eukaryotes and prokaryotes. Abortive initiation continues to occur until the σ factor rearranges, resulting in the transcription elongation complex (which gives a 35 base pair moving footprint). The σ factor is released before 80 nucleotides of mRNA are synthesized. Once the transcript reaches approximately 23 nucleotides, it no longer slips, and elongation can occur. This, like most of the remainder of transcription, is an energy-dependent process, consuming ATP. Promoter clearance coincides with phosphorylation of serine 5 on the carboxyl terminal domain of RNA Pol in prokaryotes, which is phosphorylated by transcription factor II H (TFIIH).

3.3.3.4 Elongation
One strand of DNA, the template strand (or noncoding strand), is used as a template for RNA synthesis. As transcription proceeds, RNA polymerase traverses the template strand and uses base pairing complementary with the DNA template to create an RNA copy. Although RNA polymerase traverses the template strand from 3′ → 5′, the coding (non-template) strand and newly formed RNA can also be used as reference points, so transcription can be described as occurring 5′ → 3′. This produces an RNA molecule from 5′ → 3′, an exact copy of the coding strand except that thymines are replaced with uracils and the nucleotides are composed of a ribose (5-carbon) sugar where DNA has deoxyribose in its sugar-phosphate backbone. Unlike DNA replication, mRNA transcription can involve multiple RNA polymerases on a single DNA template and multiple rounds of transcription (amplification of specific mRNA), so many mRNA molecules can be rapidly produced from a single

copy of a gene. Elongation also involves a proofreading mechanism that can replace incorrectly incorporated bases. In eukaryotes, this may correspond to short pauses during transcription that allow appropriate RNA editing factors to bind. These pauses may be intrinsic to the RNA polymerase or because of the chromatin structure.

3.3.3.5 Termination

Bacteria use two different strategies for transcription termination: Rho-independent and Rho-dependent transcription termination. In Rho-independent transcription termination, RNA transcription stops when the newly synthesized RNA molecule forms a G-C rich hairpin loop followed by a run of Us, which makes it detach from the DNA template. In the Rho-dependent type of termination, a protein factor called "Rho" destabilizes the interaction between the template and the mRNA, thus releasing the newly synthesized mRNA from the elongation complex. Transcription termination in eukaryotes is less understood but involves cleavage of the new transcript followed by template-independent addition of As at its new 3′ end, in a process called *polyadenylation* (Figure 3.4).

3.3.4 TRANSLATION

Translation is the second stage of protein biosynthesis and is a part of the overall process of gene expression. It is the production of proteins by decoding mRNA produced in transcription. It also occurs in the cytoplasm where the ribosomes are located. Ribosomes are made of a small and a large subunit that surround the mRNA. In translation, mRNA is decoded to produce a specific polypeptide according to the rules specified by the genetic code. This uses an mRNA sequence as a template to guide the synthesis of a chain of amino acids that form a protein. Many types of transcribed RNA, such as transfer RNA, ribosomal RNA, and small nuclear RNA are not necessarily translated into an amino acid sequence (Figure 3.5).

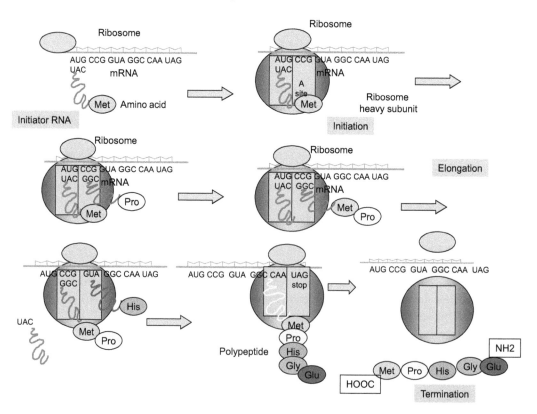

FIGURE 3.5   Different phases of translation during protein synthesis.

Translation proceeds in four phases: activation, initiation, elongation, and termination, and all describe the growth of the amino acid chain or polypeptide that is the product of translation. Amino acids are brought to ribosomes and assembled into proteins. In activation, the correct amino acid is covalently bonded to the correct tRNA. Although this is not technically a step in translation, it is required for translation to proceed. The amino acid is joined by its carboxyl group to the 3′ OH of the tRNA via an ester bond. When the tRNA has an amino acid linked to it, it is considered "charged." Initiation involves the small subunit of the ribosome binding to the 5′ end of mRNA with the help of initiation factors (IFs). Termination of the polypeptide happens when the A site of the ribosome faces a stop codon (UAA, UAG, or UGA). When this happens no tRNA can recognize it, but a releasing factor can recognize nonsense codons and causes the release of the polypeptide chain. The 5′ end of the mRNA gives rise to the protein's N-terminus, and the direction of translation can therefore be stated as N → C. A number of antibiotics act by inhibiting translation. Some of these are anisomycin, cycloheximide, chloramphenicol, tetracycline, streptomycin, erythromycin, and puromycin. Prokaryotic ribosomes have a different structure from that of eukaryotic ribosomes, and thus antibiotics can specifically target bacterial infections without any detriment to a eukaryotic host's cells. The mRNA carries genetic information encoded as a ribonucleotide sequence from the chromosomes to the ribosomes. The ribonucleotides are "read" by translational machinery in a sequence of nucleotide triplets called codons. Each of these triplets codes for a specific amino acid. The ribosome and tRNA molecules translate this code to a specific sequence of amino acids. The ribosome is a multisubunit structure containing rRNA and proteins. It is the "factory" where amino acids are assembled into proteins. tRNAs are small noncoding RNA chains (74–93 nucleotides) that transport amino acids to the ribosome. tRNAs have a site for amino acid attachment called an anticodon. The anticodon is a tRNA triplet complementary to the mRNA triplet that codes for their cargo amino acid.

Aminoacyl tRNA synthetase catalyzes the bonding between specific tRNAs and the amino acids that their anticodon sequences call for. The product of this reaction is an aminoacyl-tRNA molecule. This aminoacyl-tRNA travels inside the ribosome, where mRNA codons are matched through complementary base pairing to specific tRNA anticodons. The amino acids that the tRNAs carry are then used to assemble a protein. The rate of translation varies. It is significantly higher in prokaryotic cells, up to 17–21 amino acid residues per second, than in eukaryotic cells, up to 6–7 amino acid residues per second (Figure 3.5).

3.3.4.1   Post-Translational Modification

PTM is the chemical modification of a protein after its translation. It is one of the later steps in protein biosynthesis. After translation, the PTM of amino acids extends the range of functions of the protein by attaching to it other biochemical functional groups such as acetate, phosphate, various lipids, and carbohydrates; by changing the chemical nature of an amino acid; or by making structural changes, like the formation of disulfide bridges. In addition, enzymes may remove amino acids from the amino end of the protein or cut the peptide chain in the middle. For instance, the peptide hormone insulin is cut twice after disulfide bonds are formed, and a pro-peptide is removed from the middle of the chain. The resulting protein consists of two polypeptide chains connected by disulfide bonds. In addition, most nascent polypeptides start with the amino acid methionine because the "start" codon on mRNA also codes for this amino acid. This amino acid is usually taken off during PTM. Other modifications, like phosphorylation, are part of common mechanisms for controlling the behavior of a protein such as activating or inactivating an enzyme. PTM of proteins is detected by *mass spectrometry* (MS) or *eastern blotting*. The structure of protein is illustrated in Figure 3.6.

3.4   PROTEIN FOLDING

Protein folding is the physical process by which a polypeptide folds into its characteristic and functional three-dimensional structure from a random coil (Figure 3.7). Each protein exists as an unfolded polypeptide or random coil when translated from a sequence of mRNA to a linear chain

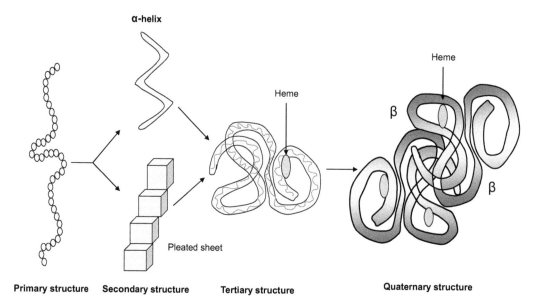

FIGURE 3.6 Different protein structures: (1) primary structure, (2) secondary structure, (3) tertiary structure, and (4) quaternary structure.

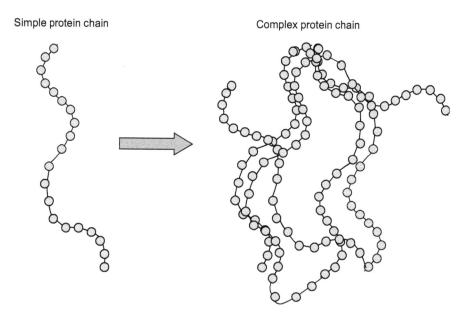

FIGURE 3.7 Process of protein folding.

of amino acids. This polypeptide lacks any developed three-dimensional structure. However, amino acids interact with each other to produce a well-defined three-dimensional structure, the folded protein, known as the *native state*. The resulting three-dimensional structure is determined by the amino acid sequence. For many proteins, the correct three-dimensional structure is essential for normal function and failure to fold into the intended shape usually produces inactive proteins with different properties, including toxic prions. Several neurodegenerative and other diseases are believed to result from the accumulation of incorrectly folded proteins. Aggregated proteins are associated

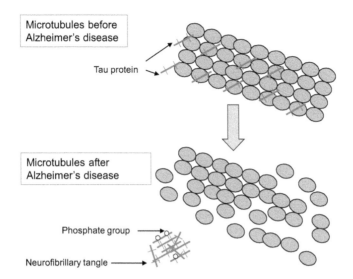

FIGURE 3.8 Protein tangling in Alzheimer's disease.

with prion-related illnesses such as Creutzfeldt–Jakob disease and bovine spongiform encephalopathy (mad cow disease) and amyloid-related illnesses such as Alzheimer's disease (Figure 3.8) and familial amyloid cardiomyopathy or polyneuropathy, as well as intracytoplasmic aggregation diseases such as Huntington's and Parkinson's diseases. These age-related degenerative diseases are associated with the multimerization of misfolded proteins into insoluble, extracellular aggregates and/or intracellular inclusions including cross-β-sheet amyloid fibrils. It is not clear whether the aggregates are the cause or merely a reflection of the loss of protein homeostasis or the balance between synthesis, folding, aggregation, and protein turnover. The excessive misfolding and degradation of protein leads to several proteopathy diseases such as antitrypsin-associated emphysema, cystic fibrosis, and the lysosomal storage diseases, where loss of function is the origin of the disorder. You might be wondering how to treat the problems associated with protein dysfunction, when protein replacement therapy has historically been used to correct the defect. An emerging approach now is to use pharmaceutical chaperones to fold mutated proteins to render them functional. One of the basic problems in protein folding is how to study their structures. With the help of current advances in scientific techniques, they can now be studied by using several methods as described in the following section.

3.5 PROTEIN MODIFICATION

When protein synthesis is completed, proteins undergo structural alteration that affects various physiological functions of the body. These structural alterations, also called *modifications*, are caused by phosphorylation, ubiquitination, and other modifications that we will describe briefly in the following subsections.

3.5.1 PROTEIN MODIFICATION BY PHOSPHORYLATION

During cell signaling, many enzymes and structural proteins undergo phosphorylation. The addition of a phosphate to amino acids, most commonly serine and threonine mediated by serine or threonine kinases, or more rarely tyrosine mediated by tyrosine kinases, causes a protein to become a target for binding or interacting with a distinct set of other proteins that recognize the phosphorylated domain. Because protein phosphorylation is one of the most-studied protein modifications, many "proteomics" efforts are geared to determining the set of phosphorylated proteins in a specific cell or tissue-type under specific circumstances.

3.5.2 PROTEIN MODIFICATION BY UBIQUITINATION

Ubiquitin is a small protein that can be affixed to certain protein substrates by enzymes called *E3 ubiquitin ligases*. Determining which proteins are poly-ubiquitinated can clarify how protein pathways are regulated. This is therefore an additional legitimate "proteomic" study. Similarly, once it is determined what substrates are ubiquitinated by each ligase, determining the set of ligases expressed in a specific cell type will be helpful.

3.5.3 ADDITIONAL MODIFICATIONS

Listing all the protein modifications that might be studied in a "proteomics" project would require a discussion of most of biochemistry. Therefore, a short list will serve here to illustrate the complexity of the problem. In addition to phosphorylation and ubiquitination, proteins can be subjected to methylation, acetylation, glycosylation, oxidation, nitrosylation, etc. Some proteins undergo these modifications, which nicely illustrate the potential complexity one must deal with when studying protein structure and function.

3.6 PROTEIN TRANSPORT

Upon successful synthesis of proteins, the next obvious step will be to transport the proteins for the required or assigned function. For that reason, proteins must be transported to an organ or tissue.

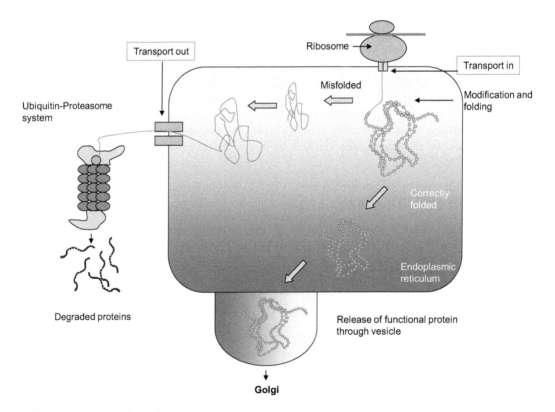

FIGURE 3.9 Mechanism of protein transport.

Many proteins are destined for other parts of the cell than the cytosol and a wide range of signaling sequences are used to direct proteins to where they are supposed to be. In prokaryotes, this is normally a simple process because of the limited compartmentalization of the cell. However, in eukaryotes, there is a great variety of different targeting processes to ensure the protein arrives at the correct organelle. Not all proteins remain within the cell and many are exported, such as digestive enzymes, hormones, and extracellular matrix proteins. In eukaryotes, the export pathway is well developed and the main mechanism for the export of these proteins is translocation to the endoplasmic reticulum, followed by transport via the Golgi apparatus (Figure 3.9).

3.7   PROTEIN DYSFUNCTION AND DEGRADATION

Protein molecules are continually synthesized and degraded in all living organisms. The concentration of individual cellular proteins is determined by a balance between the rates of synthesis and degradation, which in turn are controlled by a series of regulated biochemical mechanisms. Differences in the rates of protein synthesis and breakdown result in cellular and tissue atrophy (loss of proteins from cells) and hypertrophy (increase in protein content of cells). The degradation rates of proteins are important in determining their cellular concentrations. Protein degradation exhibits first-order kinetics unlike protein synthesis, which is zero-order. Protein degradation is energy-dependent, requiring ATP, and is limited by the concentration of the reactants, whereas protein synthesis cannot be completed in the absence of any one of the necessary reactants.

Proteins break down at rates ranging from 100% per hour to <10% per hour and their half-lives (time taken for the loss of half the protein molecules) vary between 24 and 72 h. Regulatory enzymes and proteins have much shorter half-lives on the order of 5–120 min. Protein breakdown can take place in the mitochondria, chloroplasts, the lumen of the endoplasmic reticulum, and the endosomes, but most commonly occurs in one of two major sites of intracellular proteolysis, lysosomes and the cytosol. The individual degradation rates of proteins vary within a single organelle or cell compartment and from compartment to compartment, owing to either differing sensitivity to local proteases or differing rates of transfer to the cytosol or lysosomes. The range of protein degradation rates within a single organelle is limited, suggesting that the proteins may be treated as groups or families (Figure 3.10).

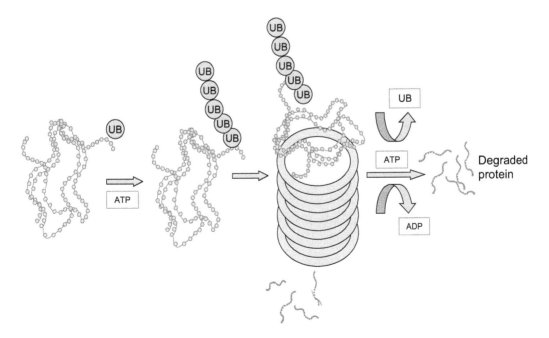

FIGURE 3.10   Protein degradation.

Most nonselective protein degradation takes place in the lysosomes, where changes in the supply of nutrients and growth factors can influence the rates of protein breakdown. Proteins enter lysosomes by *macroautophagy*, which is the enclosure of a volume of the cytoplasm by an intracellular membrane. The rates of lysosomal degradation can vary greatly with cell type and conditions, ranging from <1%/h of total cell protein to 5%–10%/h. The lysosomal degradation of some cytosolic proteins increases in cells deprived of nutrients. It is assumed that the proteins undergoing enhanced degradation are of limited importance for cell viability and can be sacrificed to support the continuing synthesis of key proteins. Short-lived regulatory proteins are degraded in the cytosol by local proteolytic mechanisms. All short-lived proteins are thought to contain recognition signals that mark them for early degradation. One commonly employed method is the selective labeling of targeted proteins by ubiquitin (UB) molecules. Ubiquitin, a protein of 76 amino acids, binds covalently to available lysine residues on target proteins, which are then recognized by proteases. Ubiquitin is a small (8.5 kDa) regulatory protein that has been found in almost all tissues (ubiquitously) of eukaryotic organisms. It was discovered in 1975 and further characterized throughout the 1970s and 1980s. There are four genes in the human genome that produce ubiquitin; UBB, UBC, UBA52, and RPS27A.

3.8   REGULATION OF PROTEIN SYNTHESIS

To control the process of protein synthesis in the body, the function of genes must be regulated with great accuracy and precision. The total amount of DNA present in a cell may contain from a few to thousands of genes. Although the different types of cells in the body of a multicellular organism differ in structure and function, their genes are identical as all the cells are ultimately derived from the zygote. The problem therefore is how do cells with identical genetic complements differ so much in structure and function? The answer is that not all genes are active at one time. As development proceeds, certain genes become active while others become inactive, that is, the genes are "switched on" and "switched off" at different times. This process is called differential gene action. When genes are active, they direct the formation of enzymes, which affect certain traits. The metabolic products formed may repress synthesis of enzymes (feedback or end product inhibition). Thus, the enzyme synthesis is induced and repressed at different times. Although a cell has the genes to produce hundreds of enzymes, only the enzymes required at a time are produced. This control mechanism ensures that the cell is not flooded with unnecessary enzymes. A hypothesis to explain induction and repression of enzyme synthesis was first put forward in 1961 by Francois Jacob and Jacques Monod of the Institute Pasteur in Paris. For this and some other major contributions in biochemistry, Jacob and Monod were awarded the Nobel Prize in Medicine in 1965. The scheme proposed by these workers is called the operon model and has been the leading biological discovery of the present century, along with the elucidation of the structure of DNA by Watson and Crick (1953).

Furthermore, gene regulation drives the processes of cellular differentiation and morphogenesis, leading to the creation of different cell types in multicellular organisms in which the different types of cells may possess different gene expression profiles though they all possess the same genome sequence.

3.8.1   STAGES OF GENE EXPRESSION

Gene expression may be regulated from the DNA–RNA transcription step to PTM, RNA transport, translation, mRNA degradation.

3.8.1.1   Operon Model for Gene Regulation

The operon consists of the following components: regulator gene, promoter gene (a relatively recent concept), operator gene, structural genes, repressor, corepressor, and inducer. For example, to synthesize β-galactosidase in *E. coli*, Jacob and Monod in 1961 proposed a model based on an inducible system which is also known as the *operon model*. An *operon* consists of an operator gene, which controls the activity of protein synthesis, and several structural genes that take part in the synthesis

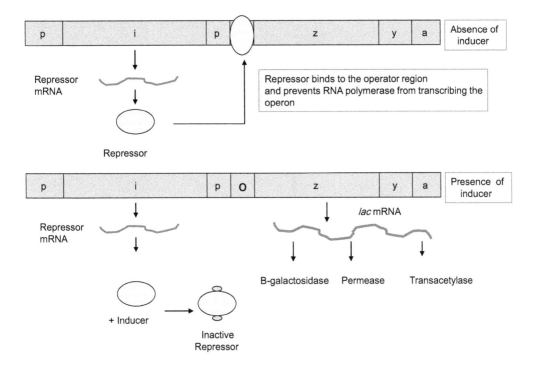

FIGURE 3.11 The *lac* Operon model 1.

of proteins. In brief, the structural genes will synthesize mRNA under the operational control of an operator gene, which in turn is under the control of a repressor molecule synthesized by a regulator gene that is not a part of the operon (Figure 3.11).

3.8.1.2 General Transcription Factors

TBP is a general TF that binds specifically to a DNA sequence called the TATA box. This DNA sequence is found about 30 base pairs upstream of the transcription start site in some eukaryotic gene promoters. TBP, along with a variety of TBP-associated factors, make up the TFIID, a general TF that in turn makes up part of the RNA polymerase II preinitiation complex. As one of the few proteins in the preinitiation complex that binds DNA in a sequence-specific manner, it helps position RNA polymerase II over the transcription start site of the gene. However, it is estimated that only 10%–20% of human promoters have TATA boxes. Therefore, TBP is probably not the only protein involved in positioning RNA polymerase II. TBP is involved in DNA melting (double strand separation) by bending the DNA by 80° (the AT-rich sequence to which it binds facilitates easy melting). The TBP is an unusual protein in that it binds the minor groove using a β sheet.

TBP is a subunit of the eukaryotic transcription factor TFIID. TFIID is the first protein to bind to DNA during the formation of the pre-initiation transcription complex of RNA polymerase II (RNA Pol II). Binding of TFIID to the TATA box in the promoter region of the gene initiates the recruitment of other factors required for RNA Pol II to begin transcription. Some of the other recruited TFs include TFIIA, TFIIB, and TFIIF. Each of these TFs is formed from the interaction of many protein subunits, indicating that transcription is a heavily regulated process. TBP is also a necessary component of RNA polymerase I and RNA polymerase II, and is, it is thought, the only common subunit required by all three of the RNA polymerases (Figure 3.12).

3.8.1.3 Enhancers

Enhancers are sites on the DNA helix that are bound to by activators to loop the DNA, bringing a specific promoter to the initiation complex.

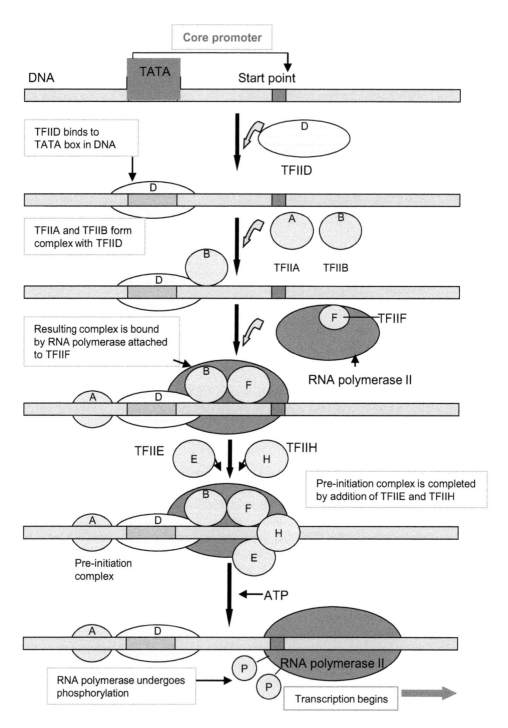

FIGURE 3.12   Protein transcription through TATA box.

3.8.1.4 Induction and Repression

Escherichia coli synthesis of β-galactosidase has been extensively studied in which lactose is converted into glucose and galactose. To understand the synthesis of β-galactosidase in *E. coli*, experiments were performed, and it was observed that if β-galactosides are not supplied to *E. coli* cells, the presence of β-galactosidase is hardly detectable but as soon as lactose is added, production of the enzyme β-galactosidase increases as much as 10,000 times. The enzyme quantity again falls as quickly as the substrate (lactose) is removed. Such enzymes whose synthesis can be induced by adding a substrate are known as inducible enzymes and the genetic systems responsible for the synthesis of such enzymes are known as inducible systems. In another situation where no amino acids are supplied from outside, *E. coli* cells can synthesize all the enzymes needed for the synthesis of different amino acids. However, if an amino acid like histidine is added, the production of histidine-synthesized enzymes declines. In such a scenario, the addition of an end product of a biosynthetic pathway will check synthesis of the enzymes needed for its biosynthesis. The enzymes whose synthesis can be repressed by adding an end product are known as repressible enzymes and their genetic systems are known as repressible systems. The substrate whose addition induced the synthesis of an enzyme (as lactose in the case of synthesis of β-galactosidase) is called an inducer. In the same way, the end product whose addition repressed the synthesis of biosynthetic enzymes is called a co-repressor. Note that in the absence of lactose, no β-galactosidase is synthesized. This would mean that in the absence of an inducer, the gene or genes responsible for the synthesis of β-galactosidase do not function.

3.9 PROTEIN PURIFICATION

To perform *in vitro* analysis, a protein must be purified away from other cellular components. This process usually begins with cell lysis, in which a cell's membrane is disrupted and its internal contents released into a solution known as a *crude lysate*. The resulting mixture can be purified using ultracentrifugation, which fractionates the various cellular components into fractions containing soluble proteins, membrane lipids and proteins, cellular organelles, and nucleic acids. Precipitation by a method known as *salting out* can concentrate the proteins from this lysate. Various types of chromatography are then used to isolate the protein or proteins of interest based on properties such as molecular weight, net charge, and binding affinity. The level of purification can be monitored using various types of gel electrophoresis (Figure 3.13) if the desired protein's molecular weight and isoelectric point are known, by spectroscopy if the protein has distinguishable spectroscopic features, or by enzyme assays if the protein has enzymatic activity. Additionally, proteins can be isolated according their charge using electrofocusing. For natural proteins, a series of purification steps may be necessary to obtain protein sufficiently pure for laboratory applications. To simplify this process, genetic engineering is often used to add chemical features to proteins that make them easier to purify without affecting their structure or activity. Here, a "tag" consisting of a specific amino acid sequence, often a series of histidine residues (a "His-tag"), is attached to one terminus of the protein. As a result, when the lysate is passed over a chromatography column containing nickel, the histidine residues ligate the nickel and attach to the column while the untagged components of the lysate pass unimpeded. Several different tags have been developed to help researchers purify specific proteins from complex mixtures.

3.10 TOOLS OF PROTEOMICS

The total complement of proteins present at a time in a cell or cell type is known as its *proteome*, and the study of such large-scale data sets defines the field of *proteomics*, named by analogy to the related field of genomics. The proteins can be measured by using western blotting techniques (Figures 3.14, 3.15) and by enzyme-linked immunoassay (ELISA) (Figure 3.16). The key experimental techniques in proteomics are described in the following text.

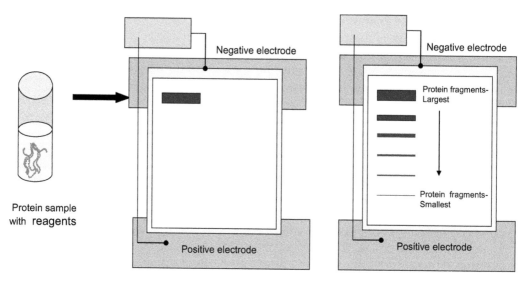

FIGURE 3.13 Gel electrophoresis of protein.

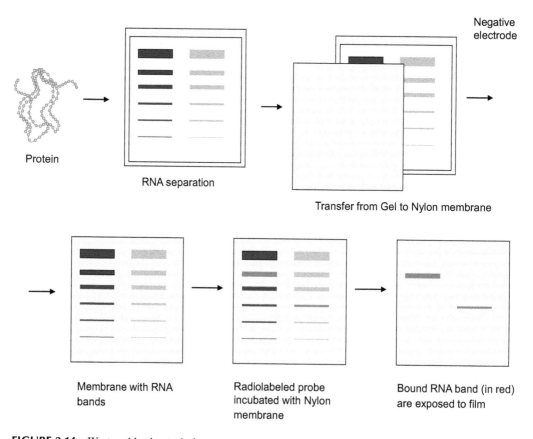

FIGURE 3.14 Western blotting technique.

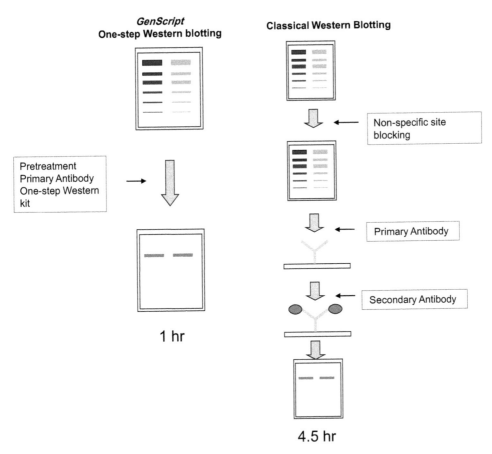

GenScript
One-step Western blotting

Classical Western Blotting

Non-specific site blocking

Pretreatment
Primary Antibody
One-step Western kit

Primary Antibody

Secondary Antibody

1 hr

4.5 hr

FIGURE 3.15 **Western blotting technique:** Comparison between classical and advanced techniques.

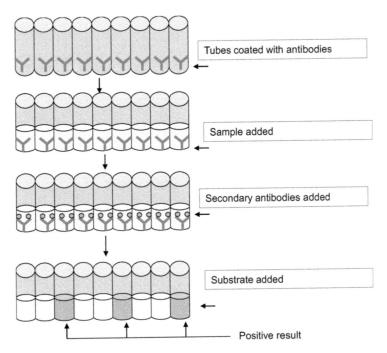

Tubes coated with antibodies

Sample added

Secondary antibodies added

Substrate added

Positive result

FIGURE 3.16 Enzyme-linked immunosorbent assay (ELISA) technique.

3.10.1 TWO-DIMENSIONAL ELECTROPHORESIS

Two-dimensional gel electrophoresis or 2-D electrophoresis is a form of gel electrophoresis commonly used to separate many proteins. Mixtures of proteins are separated by two properties in two dimensions on 2-D gels. 2-D electrophoresis begins with 1-D electrophoresis but then separates the molecules by a second property in the direction 90° from the first. In the 1-D electrophoresis, proteins are separated in one dimension, so that all the proteins will lie along a lane but separated from each other by an isoelectric point. The result is that the molecules are spread out across the 2-D gel. Because it is unlikely that two molecules will be similar in two distinct properties, molecules are more effectively separated by 2-D electrophoresis than by 1-D electrophoresis. To separate the proteins by isoelectric point is called *isoelectric focusing* (IEF). Thereby, a gradient of pH is applied to a gel and an electric potential is applied across the gel, making one end more positive than the other. At all pHs other than their isoelectric point, proteins will be charged. If they are positively charged, they will be pulled towards the more negative end of the gel and if they are negatively charged, they will be pulled to the more positive end of the gel. The proteins applied in the first dimension will move along the gel and will accumulate at their isoelectric point. That is, the point at which the overall charge of the protein is 0 (a neutral charge). The result of this is a gel with proteins spread out on its surface. These proteins can then be detected by a variety of means, but the most commonly used are silver and Coomassie staining. In this case, a silver colloid is applied to the gel. The silver binds to cysteine groups within the protein. The silver is darkened by exposure to UV light. The darkness of the silver can be related to the amount of silver and therefore the amount of protein at a given location on the gel. This measurement can only give approximate amounts but is adequate for most purposes (Figure 3.17).

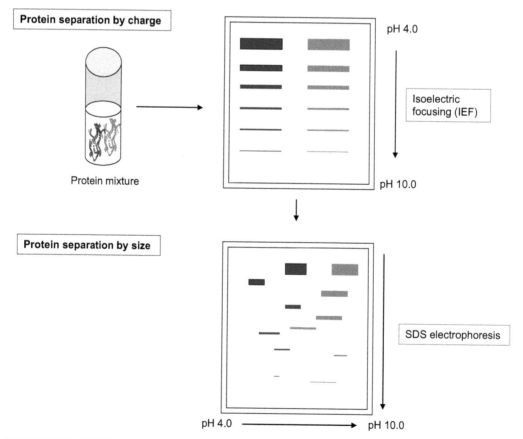

FIGURE 3.17 2D Electrophoresis.

3.10.2 MASS SPECTROMETRY

MS is an analytical technique for the determination of the elemental composition of a sample or molecule that allows rapid and high-throughput identification of proteins and sequencing of peptides most often after in-gel digestion. It is also used for elucidating the chemical structures of molecules, such as peptides and other chemical compounds. The MS principle consists of ionizing chemical compounds to generate charged molecules or molecule fragments and measurement of their mass-to-charge ratios. MS instruments consist of three modules: an *ion source*, which can convert gas phase sample molecules into ions (or, in the case of electrospray ionization (ESI), move ions that exist in solution into the gas phase); a *mass analyzer*, which sorts the ions by their masses by applying electromagnetic fields; and a *detector*, which measures the value of an indicator quantity and thus provides data for calculating the abundances of each ion present. The technique has both qualitative and quantitative uses. These include identifying unknown compounds, determining the isotopic composition of elements in a molecule, and determining the structure of a compound by observing its fragmentation. MS is now in very common use in analytical laboratories that study physical, chemical, or biological properties of a great variety of compounds. MS is an important emerging method for the characterization of proteins. The two primary methods for ionization of whole proteins are ESI and *matrix-assisted laser desorption/ionization* (MALDI). In keeping with the performance and mass range of available mass spectrometers, two approaches are used for characterizing proteins. In the first, intact proteins are ionized by either of the two techniques described earlier, and then introduced into a mass analyzer. This approach is referred to as the "top-down" strategy of protein analysis. In the second, proteins are enzymatically digested into smaller peptides using proteases such as trypsin or pepsin, either in solution or in gel after electrophoretic separation. Other proteolytic agents are also used. The collection of peptide products is then introduced into the mass analyzer. When the characteristic pattern of peptides is used for the identification of the protein, the method is called *peptide mass fingerprinting* (PMF). If the identification is performed using the sequence data determined in tandem with MS analysis, it is called *de novo sequencing*. These procedures of protein analysis are also referred to as the "bottom-up" approach.

3.10.3 PROTEIN MICROARRAY

A *protein microarray*, sometimes referred to as a *protein binding microarray*, provides a multiplex approach to identify protein–protein interactions, to identify the substrates of protein kinases, to identify TF protein-activation, or to identify the targets of biologically active small molecules. The array is a piece of glass on which different molecules of protein or specific DNA binding sequences (as capture probes for the proteins) have been affixed at separate locations in an ordered manner thus forming a microscopic array. The most common protein microarray is the *antibody microarray*, where antibodies are spotted onto the protein chip and are used as *capture molecules* to detect proteins from cell lysate solutions. Protein microarrays (also biochip, protein chip) are measurement devices used in biomedical applications to determine the presence and/or amount (referred to as relative quantitation) of proteins in biological samples, for example, blood. They have the potential to be an important tool for proteomics research. Usually different capture agents, most frequently monoclonal antibodies, are deposited on a chip surface (glass or silicon) in a miniature array. This format is often referred to as a microarray (a more general term for chip-based biological measurement devices). There are several types of protein chips, the most common being glass slide chips and nano-well arrays (Figure 3.18).

3.10.4 TWO-HYBRID SCREENING

Two-hybrid screening allows the systematic exploration of protein–protein interactions. It is also known as yeast two-hybrid system or Y2H, a molecular biology technique used to discover protein–protein interactions and protein–DNA interactions by testing for physical interactions such as

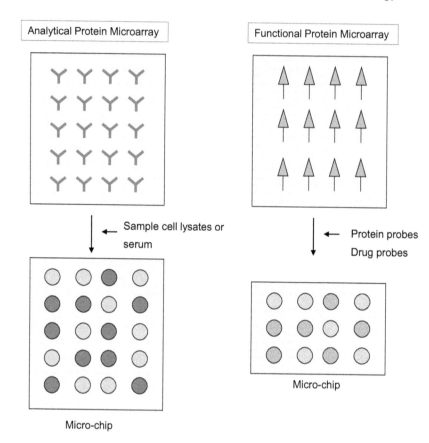

FIGURE 3.18 Protein microarray.

binding between two proteins or a single protein and a DNA molecule, respectively. The premise behind the test is the activation of downstream reporter gene(s) by the binding of a TF onto an upstream activating sequence (UAS). For two-hybrid screening, the TF is split into two separate fragments, called the *binding domain* (BD) and *activating domain* (AD). The BD is the domain responsible for binding to the UAS, whereas the AD is the domain responsible for the activation of transcription. One limitation of classic yeast two-hybrid screens is that they are limited to soluble proteins. It is therefore impossible to use them to study the protein–protein interactions between insoluble integral membrane proteins. The split-ubiquitin system provides a method for overcoming this limitation. In the split-ubiquitin system, two integral membrane proteins to be studied are fused to two different ubiquitin moieties: A C-terminal ubiquitin moiety ("Cub," residues 35–76) and an N-terminal ubiquitin moiety ("Nub," residues 1–34). These fused proteins are called the *bait* and *fish*, respectively. In addition to being fused to an integral membrane protein, the Cub moiety is also fused to a TF that can be cleaved off by ubiquitin-specific proteases. Upon bait–fish interaction, Nub- and Cub-moieties assemble, reconstituting the split ubiquitin. The reconstituted split-ubiquitin molecule is recognized by ubiquitin-specific proteases that cleave off the reporter protein, allowing it to induce the transcription of reporter genes.

3.10.5 PROTEIN STRUCTURE PREDICTION

There are various techniques available to analyze the structure of proteins because it is very important to know which protein is structurally normal or abnormal especially in the therapeutic proteins developed by using recombinant DNA technology. We will briefly describe a few techniques to analyze the protein structurally. Molecular dynamics (MD) is an important tool for studying protein folding and dynamics in silico. Because of computational cost, *ab initio* MD folding simulations with explicit water are limited to peptides and very small proteins. MD simulations of larger proteins remain restricted to dynamics of the experimental structure or its high temperature unfolding. To simulate long-time folding processes (beyond about 1 μs), like folding of small-size proteins (about 50 residues) or larger, some approximations or simplifications in protein models need to be introduced. An approach using reduced protein representation (pseudo-atoms representing groups of atoms are defined) and statistical potential is not only useful in protein structure prediction but is also capable of reproducing the folding pathways. There are distributed computing projects that use idle CPU time of personal computers to solve problems such as protein folding or prediction of protein structure. People can run these programs on their computer or PlayStation 3 to support them.

3.10.6 PEGYLATION

PEGylation is the process of covalent attachment of poly(ethylene glycol) (PEG) polymer chains to another molecule, normally a drug or therapeutic protein. PEGylation is routinely achieved by incubation of a reactive derivative of PEG with the target macromolecule. The covalent attachment of PEG to a drug or therapeutic protein can "mask" the agent of the host's immune system (reduced immunogenicity and antigenicity) and increase the hydrodynamic size (size in solution) of the agent, which prolongs its circulatory time by reducing renal clearance. PEGylation can also provide water solubility to hydrophobic drugs and proteins.

3.10.7 HIGH-PERFORMANCE LIQUID CHROMATOGRAPHY

High-performance liquid chromatography (or high-pressure liquid chromatography, HPLC) is a form of column chromatography used frequently in biochemistry and analytical chemistry to separate, identify, and quantify compounds, including proteins. HPLC utilizes a column that holds chromatographic packing material (stationary phase), a pump that moves the mobile phase(s) through the column, and a detector that shows the retention times of the molecules. Retention time varies depending on the interactions between the stationary phase, the molecules being analyzed, and the solvent(s) used.

3.10.8 SHOTGUN PROTEOMICS

Shotgun proteomics is a method of identifying proteins in complex mixtures using a combination of high-performance liquid chromatography combined with mass spectrometry. The name is derived from shotgun sequencing of DNA, which is itself named by analogy with the rapidly expanding, quasi-random firing pattern of a shotgun. In shotgun proteomics, the proteins in the mixture are digested and the resulting peptides are separated by liquid chromatography. Tandem mass spectrometry is then used to identify the peptides.

3.10.9 TOP-DOWN PROTEOMICS

Top-down proteomics is a method of protein identification that uses an ion-trapping mass spectrometer to store an isolated protein ion for mass measurement and tandem mass spectrometry analysis. The name is derived from the similar approach to DNA sequencing. Proteins are typically ionized by ESI and trapped in a Fourier transform ion cyclotron resonance (Penning trap) or quadruple ion trap (Paul trap) mass spectrometer. Fragmentation for tandem mass spectrometry is accomplished by electron capture dissociation or electron transfer dissociation.

PROBLEMS

Section A: Descriptive Type

Q1. Explain the function of proteins as enzymes.
Q2. Explain the role of proteins in cell signaling.
Q3. What are the different kinds of nonessential amino acids?
Q4. What is protein biosynthesis?
Q5. Discuss transcription in prokaryotes and eukaryotes.
Q6. Describe the tools for studying the structure of proteins.
Q7. What is protein folding?
Q8. Describe the regulation of gene expression for protein synthesis.
Q9. What is the operon model for gene regulation?

Section B: Multiple Choice

Q1. About how many reactions are known to be catalyzed by enzymes?
   a. 3000
   b. 4500
   c. 4000
   d. 5000
Q2. Cystine and aspartic acid are the nonessential amino acids. True/False
Q3. Protein synthesis starts with translation of proteins. True/False
Q4. Make the correct pair
   a. Upstream process 3′UTR
   b. Downstream process 5′UTR
Q5. What is the most common type of core promoter in eukaryotes?
   a. GATA box
   b. TATA box
   c. BATA box
Q6. In bacteria, transcription begins with the binding of RNA polymerase to the promoter in DNA. True/False
Q7. Which of the following can transcribe RNA into DNA?
   a. Bacteria
   b. Fungi
   c. Viruses
Q8. How many stages are there in the translation phase of protein synthesis?
   a. 3
   b. 4
   c. 5
Q9. Among the following diseases, which one is not caused by incorrect protein folding?
   a. Creutzfeldt–Jakob disease
   b. Alzheimer's disease
   c. Epilepsy
   d. Amyloid cardiomyopathy
Q10. Protein molecules are continuously synthesized and degraded in all living organisms. True/False
Q11. Where does protein breakdown take place?
   a. Nucleus
   b. Microtubules
   c. Mitochondria
   d. Golgi bodies

Q12. During protein synthesis, the operon model is used for . . .
   a. Protein transport
   b. Gene regulation
   c. Amino acid regulation

Q13. Which of the following is not native gel?
   a. BN-PAGE
   b. CN-PAGE
   c. SDS-PAGE
   d. QPNC-PAGE

Q14. What does ELISA mean?
   a. Enzyme assay
   b. Enzyme-linked immunoassay
   c. Enzyme-linked immunosorbent assay
   d. Enzyme-linked sorbent assay

Q15. What does HPLC mean?
   a. High protein liquid chromatography
   b. High-performance liquid chromatography
   c. High purified liquid chromatography

Section C: Critical Thinking

Q1. How can one distinguish normal protein from abnormal protein using proteomics tools?
Q2. Why do proteins undergo folding and refolding phases? Explain using examples.
Q3. What method would you employ to identify the structure of a newly synthesized protein?

ASSIGNMENT

Make a chart of the essential and nonessential amino acids showing their beneficial attributes. Discuss your chart with your classmates.

REFERENCES AND FURTHER READING

Alberts, B., Johnson, A., Lewis, J., Raff, M., Roberts, K., and Walters, P. The shape and structure of proteins. In: *Molecular Biology of the Cell*, 4th edn. Garland Science, New York, 2002.

Alexander, P.A.Y., He, Y., Chen, J., Orban, P.N., and Bryan, P.N. The design and characterization of two proteins with 88% sequence identity but different structure and function. *Proc. Natl. Acad. Sci. USA* 104(29): 11963–11968, 2007.

Anfinsen, C. The formation and stabilization of protein structure. *Biochem. J.* 128: 737–749, 1972.

Belay, E. Transmissible spongiform encephalopathies in humans. *Annu. Rev. Microbiol.* 53: 283–314, 1999.

Bu, Z., Cook, J., and Callaway, D.J.E. Dynamic regimes and correlated structural dynamics in native and denatured alpha-lactalbumin. *J. Mol. Biol.* 312: 865–873, 2001.

Büeler, H.A., Aguzzi, A., Sailer, R., Greiner, P., Autenried, M., Aguet, C., and Weissmann, C. Mice devoid of PrP are resistant to scrapie. *Cell* 73: 1339–1347, 1993.

Collinge, J. Prion diseases of humans and animals: Their causes and molecular basis. *Annu. Rev. Neurosci.* 24: 519–550, 2001.

Deechongkit, S., Nguyen, H., Dawson, P.E., Gruebele, M., and Kelly, J.W. Context dependent contributions of backbone H-bonding to β-sheet folding energetics. *Nature* 403: 101–105, 2004.

Fürst, P., and Stehle, P. What are the essential elements needed for the determination of amino acid requirements in humans? *J. Nutr.* 134: 1558–1565, 2004.

Gilch, S. et al. Intracellular re-routing of prion protein prevents propagation of PrPSc and delays onset of prion disease. *The EMBO J.* 20: 3957–3966, 2001.

Imura, K., and Okada, A. Amino acid metabolism in pediatric patients. *Nutrition* 14: 143–148, 1998.

Ironside, J.W. Variant Creutzfeldt-Jakob disease: Risk of transmission by blood transfusion and blood therapies. *Hemophilia* 12: 8–15, 2006.

Kim, P.S., and Baldwin, R.L. Intermediates in the folding reactions of small proteins. *Annu. Rev. Biochem.* 59: 631–660, 1990.

Kmiecik, S., and Kolinski, A. Characterization of protein-folding pathways by reduced-space modeling. *Proc. Natl. Acad. Sci. USA* 104: 12330–12335, 2007.

Kubelka, J., Hofrichter, J., and Eaton, W.A. The protein folding speed limit. *Curr. Opin. Struct. Biol.* 14: 76–88, 2004.

Lee, S., and Tsai, F. Molecular chaperones in protein quality control. *J. Biochem. Mol. Biol.* 38(3): 259–265, 2005.

Levinthal, C. Are there pathways for protein folding? *J. Chim. Phys.* 65: 44–45, 1968.

Pace, C., Shirley, B., McNutt, M., and Gajiwala, K. Forces contributing to the conformational stability of proteins. *FASEB J.* 10: 75–83, 1996.

Reeds, P.J. Dispensable and indispensable amino acids for humans. *J. Nutr.* 130: 1835–1840, 2000.

Rose, G., Fleming, P., Banavar, J., and Maritan, A. A backbone-based theory of protein folding. *Proc. Natl. Acad. Sci. USA* 103: 16623–16633, 2006.

Shortle, D. The denatured state (the other half of the folding equation) and its role in protein stability. *FASEB J.* 10: 27–34, 1996.

Telling, G., Scott, M., Mastrianni, J., Gabizon, R., Torchia, M., Cohen, F., DeArmond, S., and Prusiner, S. Prion propagation in mice expressing human and chimeric PrP transgenes implicates the interaction of cellular PrP with another protein. *Cell* 83: 93, 1995.

Van den Berg, B., Wain, R., Dobson, C.M., and Ellis, R.J. Macromolecular crowding perturbs protein refolding kinetics. Implications for folding inside the cell. *Embo J.* 19(15): 3870–3935, 2000.

Young, V.R. Adult amino acid requirements: The case for a major revision in current recommendations. *J. Nutr.* 124: 1517S–1523S, 1994.

4 Recombinant DNA Technology

LEARNING OBJECTIVES
- Define recombinant DNA (rDNA) technology
- Discuss the significance of rDNA technology
- Explain the steps involved in making rDNA products
- Discuss the role of restriction enzymes in rDNA technology
- Explain the significance of vectors and their characteristics
- Discuss applications of rDNA technology

4.1 INTRODUCTION

After learning the structure and function of genes and proteins, the next question that comes to mind is if it is possible to edit or manipulate genes? If answer is yes, then how can we edit or manipulate genes to make customized proteins? The use of technology to manipulate genes is called *genetic engineering* or *recombinant DNA technology* (rDNA technology). rDNA technology is a field of molecular biology in which scientists manipulate DNA to form new synthetic molecules, called *chimeras*. The practice of cutting, pasting, and copying DNA is based on Arthur Kornberg's successful replication of viral DNA in a breakthrough that served as a proof-of-concept for cloning. This was followed by the Swiss biochemist Werner Arber's discovery of restriction enzymes in bacteria that degrade foreign viral DNA molecules while sparing their own DNA. Arber effectively showed how to "cut" DNA molecules. Soon to follow was the understanding that ligase could be used to "glue" them together. These two achievements were the main reasons for the launching of rDNA technology research. The most common recombinant process involves combining the DNA of two different organisms. The rDNA technique was first proposed by Peter Lobban, a graduate student with A. Dale Kaiser at Stanford University, Department of Biochemistry. The technique was then realized by Lobban and his group in 1972–1974. rDNA technology was made possible by the discovery, isolation, and application of restriction endonucleases by Werner Arber, Daniel Nathans, and Hamilton Smith, for which they received the 1978 Nobel Prize in Medicine.

4.2 MAKING RECOMBINANT DNA

The making of rDNA is a multistep process, which includes isolation and insertion of the gene of interest in a specific vector (Figure 4.1). The steps are described briefly in the following sections.

4.2.1 STEPS IN MAKING A RECOMBINANT DNA PRODUCT

The following are the steps involved in making an rDNA product:

Step 1: rDNA technology begins with the isolation of the gene of interest (foreign DNA). The gene is then inserted into a vector and cloned. A *vector* is a piece of DNA that is capable of independent growth. The commonly used vectors are bacterial plasmids and viral phages. The gene of interest is integrated into the plasmid (A plasmid is a small

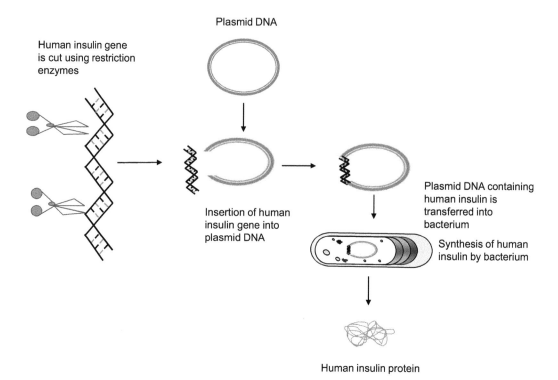

Human insulin gene
is cut using restriction
enzymes

Plasmid DNA

Insertion of human
insulin gene into
plasmid DNA

Plasmid DNA containing
human insulin is
transferred into
bacterium

Synthesis of human
insulin by bacterium

Human insulin protein

FIGURE 4.1   Recombinant DNA technology.

DNA molecule within a cell that is physically separated from a chromosomal DNA and can replicate independently. They are most commonly found in bacteria as small, circular, double-stranded DNA molecules; however, plasmids are sometimes present in archaea and eukaryotic organisms) or phage (A phage or bacteriophage is a virus that infects and replicates within a bacterium. Bacteriophages are composed of proteins that encapsulate a DNA or RNA genome and may have relatively simple or elaborate structures. Their genomes may encode as few as four genes, and as many as hundreds of genes. Phages replicate within the bacterium following the injection of their genome into its cytoplasm) which is referred to as *rDNA*.

Step 2: Before introducing the vector containing the foreign DNA into host cells to express the protein, it must be cloned. Cloning is necessary to produce numerous copies of the DNA because the initial supply is inadequate to insert into host cells.

Step 3: Once the vector is isolated in large quantities, it can be introduced into the desired host cells such as mammalian, yeast, or special bacterial cells. The host cells will then synthesize the foreign protein from the rDNA. When the cells are grown in vast quantities, the foreign or recombinant protein can be isolated and purified in large amounts (Figure 4.1).

4.2.2   Methods Involved in Making Recombinant DNA

There are three methods involved in making rDNA: *transformation*, *phage introduction*, and *nonbacterial transformation*.

4.2.2.1   Transformation

The first step in transformation is to select a piece of DNA to be inserted into a vector. The next step is to cut that DNA with a restriction enzyme and then ligate the DNA insert into the vector with

DNA ligase. The insert contains a selectable marker that allows for identification of recombinant molecules. An antibiotic marker is often used so that a host cell without a vector dies when exposed to a certain antibiotic, and the host with the vector will live because it is resistant. The vector is inserted into a host cell by a process called *transformation*. One example of a possible host cell is *E. coli*. The host cells must be specially prepared to take up the foreign DNA. Selectable markers can be for antibiotic resistance, color changes, or any other characteristic that can distinguish transformed hosts from untransformed hosts. Different vectors have different properties to make them suitable for different applications. Some properties can include symmetrical cloning sites, size, and high copy number.

4.2.2.2 Nonbacterial Transformation

Nonbacterial transformation is a process very similar to transformation. The only difference is that nonbacterial transformation does not use bacteria such as *E. coli* for the host. In microinjection, the DNA is injected directly into the nucleus of the cell being transformed. In biolistics, the host cells are bombarded with high velocity microprojectiles, such as particles of gold or tungsten that have been coated with DNA.

4.2.2.3 Phage Introduction

Phage introduction is a process of transfection that is very similar to transformation, except a phage is used instead of a bacterium. *In vitro* packaging of a vector is also used. This uses lambda or MI3 phages to produce phage plaques that contain recombinants. The recombinants that are created can be identified by differences in the recombinants and nonrecombinants by using various selection methods.

4.3 DEVELOPMENT OF RECOMBINANT DNA TECHNOLOGY

Over the past few years, rDNA technology has been gaining importance especially now that genetic diseases have become more prevalent. rDNA technology has a great impact on growing better crops (drought- and heat-resistant crops), making recombinant vaccines (such as for Hepatitis B), prevention and cure of sickle-cell anemia and cystic fibrosis, production of clotting factors, insulin and recombinant pharmaceuticals, plants that produce their own insecticides, and germ line and somatic gene therapy. The rDNA works when the host cell expresses protein from the recombinant genes. The host will only produce significant amounts of recombinant protein if expression factors are added. Protein expression depends on the gene being surrounded by a collection of signals that provide instructions for the transcription and translation of the gene by the cell. These signals include the promoter, the ribosome binding site, and the terminator. Expression vectors in which the foreign DNA is inserted contain these signals. Signals are species specific. In the case of *E. coli*, these signals must be *E. coli* signals as *E. coli* is unlikely to understand the signals of human promoters and terminators. Problems are encountered if the gene contains introns or signals that act as terminators to a bacterial host. This results in premature termination, and the recombinant protein may not be processed correctly, may be folded incorrectly, or may even be degraded. Production of recombinant proteins in eukaryotic systems generally takes place in yeast and filamentous fungi. The use of animal cells is a challenging approach because many need a solid support surface, unlike bacteria, and have complex growth needs. However, some proteins are too complex to be produced in a bacterium, so eukaryotic cells must be used. In addition, mammalian cells can also be used in rDNA proteins production but have many limitations.

4.4 SIGNIFICANCE OF RESTRICTION ENZYMES IN RECOMBINANT DNA TECHNOLOGY

We know that restriction enzyme is critical for rDNA technology, and it has been shown that *restriction enzyme* or *restriction endonuclease* cuts double-stranded or single-stranded DNA at specific

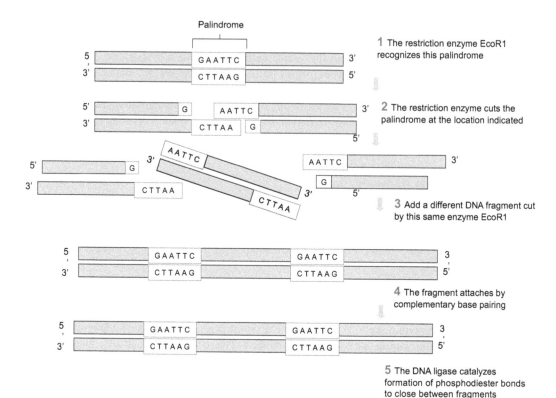

FIGURE 4.2　Restriction enzyme to cut a DNA fragment.

recognition nucleotide sequences, known as *restriction sites*. In fact, without restriction enzymes, it is not possible to make rDNA products. Restriction enzymes, which are found in bacteria and archaea, are thought to have evolved to provide a defense mechanism against invading viruses. Inside a bacterial host, the restriction enzymes selectively cut up foreign DNA in a process called *restriction*. Host DNA is then methylated by a modification enzyme (a methylase) to protect it from the restriction enzyme's activity. Collectively, these two processes form the restriction modification system. To cut the DNA, a restriction enzyme makes two incisions, once through each sugar-phosphate backbone (i.e., each strand) of the DNA double helix.

Owing to their benefits, more than 3000 restriction enzymes have been studied in detail; more than 600 of these are available commercially and are routinely used for DNA modification and manipulation in laboratories. Since their discovery in the 1970s, more than 100 different restriction enzymes have been identified in different bacteria. Each enzyme is named after the bacterium from which it was isolated using a naming system based on bacterial genus, species, and strain (Figure 4.2).

4.4.1　Types of Restriction Enzymes

Restriction enzymes are categorized into three general groups (types I, II, and III) based on their composition and enzyme cofactor requirements, the nature of their target sequence, and the position of their DNA cleavage site relative to the target sequence.

4.4.1.1　Type I Restriction Enzymes

Type I restriction enzymes were the first to be identified and are characteristic of two different strains of *E. coli* (K-12 and B). These enzymes cut at a site that differs and is some distance (at least

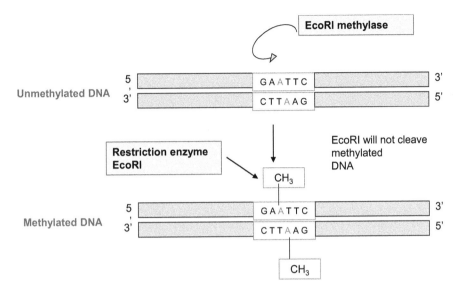

FIGURE 4.3 Restriction enzyme Type I.

1000 bp) away from their recognition site. The recognition site is asymmetrical and is composed of two portions: one containing 3–4 nucleotides, and another containing 4–5 nucleotides separated by a spacer of approximately 6–8 nucleotides. Several enzyme cofactors—including *S*-adenosyl-methionine (AdoMet), hydrolyzed ATP, and magnesium (Mg^{2+}) ions—are required for their activity. Type I restriction enzymes possess three subunits: HsdR, HsdM, and HsdS. The subunit HsdR is required for restriction, HsdM for adding methyl groups to host DNA (methyltransferase activity), and HsdS for specificity of cut site recognition in addition to its methyltransferase activity (Figure 4.3).

4.4.1.2 Type II Restriction Enzymes

Type II restriction enzymes are the most commonly available and widely used restriction enzymes. They differ from type I restriction enzymes in several ways: they are composed of only one subunit; their recognition sites are usually undivided and palindromic and measure 4–8 nucleotides in length; they recognize and cleave DNA at the same site; and they do not use ATP or AdoMet for their activity, because they usually require only Mg^{2+} as a cofactor. In the 1990s and early 2000s, new enzymes from this family were discovered that did not follow all the classical criteria of this enzyme class; consequently, new subfamily nomenclature was developed to divide this large family into subcategories based on deviations from typical characteristics of type II enzymes. These subgroups are defined using a letter suffix.

Type IIB restriction enzymes (e.g., BcgI and BplI) are multimers, containing more than one subunit. They cleave DNA on both sides of their recognition to cut out the recognition site. They require both AdoMet and Mg^{2+} cofactors. *Type IIE restriction enzymes* (e.g., NaeI) cleave DNA following interaction with two copies of their recognition sequence. One recognition site acts as the target for cleavage, while the other acts as an allosteric effector that speeds up or improves the efficiency of enzyme cleavage. Like type IIE enzymes, *type IIF restriction enzymes* (e.g., NgoMIV) interact with two copies of their recognition sequence but cleave both sequences at the same time. *Type IIG restriction enzymes* (e.g., Eco57I) do have a single subunit, like classical Type II restriction enzymes, but require the cofactor AdoMet to be active. *Type IIM restriction enzymes* (e.g., DpnI) can recognize and cut methylated DNA. *Type IIS restriction enzymes* (e.g., FokI) cleave DNA at a defined distance from their nonpalindromic asymmetric recognition sites. These enzymes may function as dimers. Similarly, *Type IIT restriction enzymes* (e.g., Bpu10I and BslI) are composed of

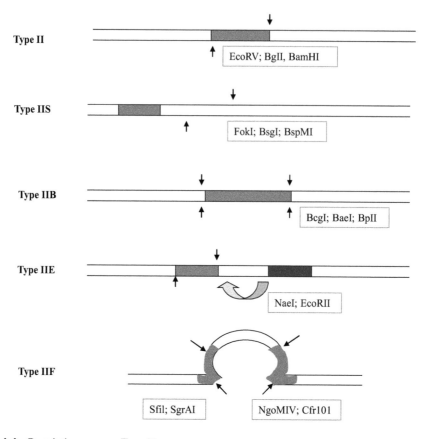

FIGURE 4.4 Restriction enzyme Type II

two different subunits. Some recognize palindromic sequences, but others have asymmetric recognition sites (Figure 4.4).

4.4.1.3 Type III Restriction Enzymes

Type III restriction enzymes (e.g., EcoP15) recognize two separate nonpalindromic sequences that are inversely oriented. They cut DNA approximately 20–30 bp after the recognition site. These enzymes contain more than one subunit and require AdoMet and ATP cofactors for their roles in DNA methylation and restriction, respectively.

4.5 STEPS IN GENE CLONING

After learning the importance of restriction enzymes, we can now focus on the process of gene cloning. The process of gene cloning is briefly described in the following five steps and diagrammatically depicted in Figure 4.5.

1. Identification and isolation of the desired gene or DNA fragment to be cloned.
2. Insertion of the isolated gene into a suitable vector.
3. Introduction of this vector into a suitable organism/cell, called the host (transformation).
4. Selection of the transformed host cells.
5. Multiplication/expression/integration followed by expression of the introduced gene in the host.

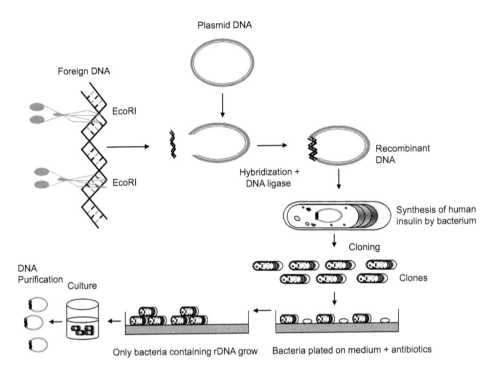

FIGURE 4.5   Gene cloning.

4.6   PCR AND GENE CLONING

PCR has many exciting and varied applications. It can be used to amplify a specific gene present in different individuals of a species and even in different somatic cells or gametes, such as human sperm. These copies can be used for cloning. Alternatively, they can be sequenced to obtain information on the mutational changes in the genes of different individuals, cells, or gametes. Such data can be used in disease diagnosis, population genetics, estimation of recombination frequencies, etc. PCR has been used to study DNA polymorphism in the genome using known sequences as primers. Synthetic nucleotides of any sequence can be used as random primers to amplify polymorphic DNAs having sequences specific to the primers used. Such an application of PCR generates random amplified polymorphic DNA (RAPD, pronounced "rapid"), which is detected as bands after electrophoresis. RAPD bands of different strains or species can be compared. They can be used to construct RAPD maps, like restriction fragment length polymorphism (RFLP) maps. PCR can be used to detect the presence of a gene transferred into an organism (transgene) by using the end sequences of the transgene for amplification of DNA from the putative transgenic organism. Amplification will occur only when the transgene is present in the organism. The amplified DNA is detected as a band on the electrophoretic gel. Microdissected segments of chromosomes, such as salivary gland chromosomes of Drosophila, can be used for PCR amplification to determine the physical location of specific genes in chromosomes. PCR can be used to determine the sex of embryos. Thus, the sex of *in vitro* fertilized cattle embryos could be determined using Y-chromosome-specific primers before their implantation in the uterus.

4.6.1   WHAT ARE THE DIFFERENCES BETWEEN PCR AND CLONING?

PCR has become a standard tool in forensic science because it can multiply very small samples of DNA for multiple crime lab testing. PCR has also become useful for archaeologists studying the evolutionary biology of different animal species, including samples thousands of years old. Cloning technology has made it relatively easy to isolate DNA fragments that contain genes to study gene function. Scientists believe that reliable cloning can be used to make farming more productive by replicating the best animals and crops and to make medical testing more accurate by providing test animals that all react the same way to the same drug. PCR is a revolutionary technology and is more efficient than gene cloning as it needs much less of the desired DNA (a single copy is enough). It is not difficult to store and does not require costly restriction enzymes, ligase, and vector DNA, thus drastically reducing the experimental cost. It needs far less work, time, and skill, and has many more applications than gene cloning. Typically, gene cloning experiments take 2–4 days, whereas PCR takes only up to 4–5 h. In addition, PCR is fully automated but gene cloning is not. Nevertheless, for PCR, one does need sequence information for construction of primers and a thermal cycler or PCR machine. It is expected that PCR will eventually take over most of the applications of gene cloning and will find many novel applications as well.

4.7 SIGNIFICANCE OF VECTORS IN RECOMBINANT DNA TECHNOLOGY

Recall that in molecular biology, a *vector* is a DNA molecule used as a vehicle to transfer foreign genetic material into another cell. Viral vectors are tools commonly used by molecular biologists to deliver genetic material into cells. This process can be performed inside a living organism (*in vivo*) or in a cell culture (*in vitro*). Viruses have evolved specialized molecular mechanisms to efficiently transport their genomes inside the cells they infect. Delivery of genes by a virus is termed *transduction* and the infected cells are called *transduced*. Molecular biologists first harnessed this machinery in the 1970s. Paul Berg used a modified SV40 virus containing DNA from the bacteriophage lambda to infect monkey kidney cells maintained in culture. The four major types of vectors are plasmids, bacteriophages and other viruses, cosmids, and artificial chromosomes. Common to all engineered vectors are an origin of replication, a multi-cloning site, and a selectable marker. The vector itself is generally a DNA sequence that consists of an insert (transgene) and a larger sequence that serves as the "backbone" of the vector. The purpose of a vector, which transfers genetic information to another cell, is typically to isolate, multiply, or express the insert in the target cell. Vectors called *expression vectors* (or *expression constructs*) are specifically for the expression of the transgene in the target cell and generally have a promoter sequence that drives expression of the transgene. Simpler vectors called *transcription vectors* are only capable of being transcribed but not translated. They can be replicated in a target cell but not expressed, unlike expression vectors. Transcription vectors are used to amplify their insert. Insertion of a vector into the target cell is generally called *transfection* (Figure 4.6).

4.7.1 PROPERTIES OF GOOD VECTORS

Vectors should be able to replicate autonomously. When the objective of cloning is to obtain many copies of the DNA insert, the vector replication must be under relaxed control so that it can generate multiple copies of itself in a single host cell. It should also be easy to isolate, purify, and introduce into the host cells. Transformation of the host with the vector should be easy. The vector should have suitable marker genes that allow easy detection and/or selection of the transformed host cells. When the objective is gene transfer, it should have the ability to integrate either itself or the DNA insert it carries into the genome of the host cell. The cells transformed with the vector containing the DNA insert (rDNA) should be easily identifiable and selectable from those transformed by the unaltered vector. A vector should contain unique target sites for as many restriction enzymes as possible into which the DNA insert can be integrated. When the expression of the DNA insert is desired, the vector should at least contain suitable control elements such as promoter, operator, and ribosome binding sites. It should be kept in mind that the DNA molecules used as vectors have coevolved with their specific natural host species and hence are adapted to function well in them and in their closely

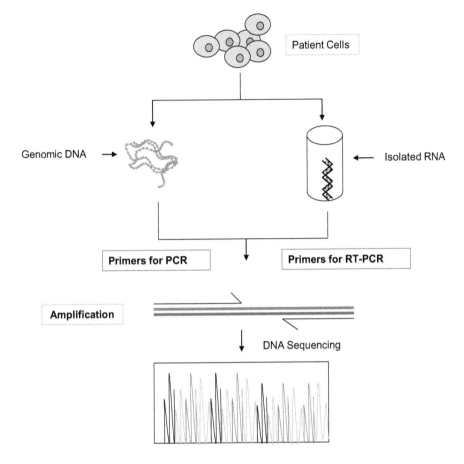

FIGURE 4.6   DNA sequencing.

related species. Therefore, the choice of vector largely depends on the host species into which the DNA insert of a gene is to be cloned. In addition, most naturally occurring vectors do not have all the required functions. Therefore, useful vectors have been created by joining together segments performing specific functions (called *modules*) from two or more natural entities.

4.7.2   Cloning and Expression Vectors

All vectors used for propagation of DNA inserts in a suitable host are called *cloning vectors*. However, when a vector is designed for the expression or production of the protein specified by the DNA insert, it is called an *expression vector*. As a rule, such vectors contain at least the regulatory sequences such as promoters, operators, and ribosomal binding sites having an optimum function in the chosen host. When a eukaryotic gene is to be expressed in a prokaryote, the eukaryotic coding sequence should be placed over the prokaryotic promoter and the ribosome building site since the regulatory sequences of eukaryotes are not recognized in prokaryotes. In addition, eukaryotic genes, as a rule, contain introns (noncoding regions) present within their coding regions. These introns must be removed from the DNA insert to enable the proper expression of eukaryotic genes, since prokaryotes lack the machinery needed for their removal from the RNA transcripts. When eukaryotic genes are isolated as cDNA, they are intron-free and therefore suitable for expression in prokaryotes. Expression vectors can be constructed by allowing the synthesis of fusion proteins—which are composed of amino acids encoded by a sequence—in the vector and those encoded by the DNA insert (translational fusion). Another way to construct expression vectors is by permitting the synthesis of pure proteins encoded exclusively by the DNA inserts (transcriptional fusion). Some examples of the first strategy, which produces fusion proteins, are the expression of rat insulin, a rat growth hormone, structural protein VP1 of foot and mouth disease virus, and human growth hormone. On the other hand, some examples of the second strategy, which produces unique proteins, are the rabbit β-globin, small t-antigen of SV40, human fibroblast interferon, and human IGF-I protein. It may be pointed out that in the case of translational fusion the undesired amino acids encoded by the vector sequence must be removed from the fusion proteins by a suitable chemical cleavage. Several other problems arise when eukaryotic genes are expressed in a prokaryotic system, such as removal of signal sequences from precursor proteins to obtain active mature protein molecules. Various strategies are being rapidly devised to effectively overcome these problems.

4.7.3 Applications of Viral Vectors

Vectors have been extensively used in molecular biology-based researches because of their advantage as the most desirable transfection vehicle. Viral vectors, especially retroviruses, stably expressing marker genes such as green fluorescent protein (GFP) are widely used to permanently label cells to track them and their progeny, such as in xenotransplantation experiments when infected cells are implanted into a host animal. One of the main applications of viral vectors is to develop vaccines to protect humans from various pathogens. Viruses expressing pathogen proteins are currently being developed as vaccines against those pathogens, based on the same rationale as DNA vaccines. T-lymphocytes recognize cells infected with intracellular parasites based on the foreign proteins produced within the cell. T-cell immunity is crucial for protection against viral infections and diseases such as malaria. A viral vaccine induces expression of pathogen proteins within host cells like the Sabin polio vaccine and other attenuated vaccines. However, because viral vaccines contain only a small fraction of pathogen genes, they are much safer, and sporadic infection by the pathogen is impossible. Adenoviruses are being actively developed as vaccines.

4.7.4 Classification of Vectors

There are two different classifications of vectors that have been extensively used in rDNA technology. In this section, we will describe these vector classifications in detail.

4.7.4.1 Bacterial Vectors

Bacterial vectors are broadly classified into *E. coli* vectors and plasmid vectors as described subsequently.

4.7.4.1.1 E. coli *Vectors*

Bacteria are the hosts of choice for DNA cloning. Among them, *E. coli* occupies a prominent position as cloning and isolating DNA inserts for structural analysis is easiest when using this host. Therefore, the initial cloning experiments are generally carried out using *E. coli*. The *E. coli* strain K12 is the most commonly used strain. It has several substrains such as C600, RRI, and HB101, each of which has some specific features important in cloning. For example, the substrain RRI has, in addition to certain other features, the mutation HsdR, which inactivates the restriction enzyme endogenous to *E. coli* K12. This minimizes the degradation of rDNA introduced into it.

4.7.4.1.2 Plasmid Vectors

Plasmid vectors are shortened linear lambda genomes containing DNA replication and lytic functions plus the cohesive ends of the phage. Their middle nonessential segment is replaced by a linearized plasmid with an intact replication module. In practice, a plasmid vector contains several tandem copies of the plasmid to make it longer than 38 kb (kilobase pairs), the minimum size needed for packaging in lambda. particles. During construction of the rDNA, one or more copies of the plasmid are deleted from the vector and the DNA insert is integrated into it, but generally one copy of the plasmid is retained in the rDNA. Plasmids, both recombinant and unaltered, are packaged in lambda. particles *in vitro* and used for infection of appropriate *E. coli* cells. If a plasmid lacks the lambda. gene cI, which produces the lysis repressor, it multiplies like a phage and produces plaques on a bacterial lawn. However, if the cI gene is present, the plasmid replicates like a plasmid. Furthermore, a plasmid may contain a mutant cI gene, which produces a temperature-sensitive CI protein (inactive at higher temperatures). Such vectors replicate as plasmids at lower temperatures but behave like phages at higher temperatures. This feature is quite useful in some experiments.

It is possible for plasmids of different types to coexist in a single cell. Several different plasmids have been found in *E. coli*. However, related plasmids are often incompatible in the sense that only one of them survives in the cell line owing to the regulation of vital plasmid functions. Therefore, plasmids can be assigned into compatibility groups.

Another way to classify plasmids is by function. There are five main groups of plasmids classified according to function:

Fertility F-plasmids, which contain transfer operon (tra genes), are capable of conjugation (transfer of genetic material between bacteria that are touching).

Resistance (R) plasmids contain genes that can build a resistance against antibiotics or poisons and help bacteria to produce pili. Before the nature of plasmids was understood, resistance plasmids were historically known as *R-factors*.

Col-plasmids contain genes that code for bacteriocins or proteins that can kill other bacteria.

Degradative plasmids enable the digestion of unusual substances such as toluene and salicylic acid.

Virulence plasmids turn a bacterium into a pathogen (one that causes disease).

Plasmids can belong to more than one of these functional groups. Plasmids that exist only as one or a few copies in each bacterium are in danger of being lost in one of the segregating bacteria during cell division. Such single-copy plasmids have systems that attempt to actively distribute a copy to both daughter cells. Some plasmids include an *addiction system* or "post-segregation killing system (PSK)," such as the host killing/suppressor of killing (hok/sok) system of plasmid R1 in *E. coli*. They produce both a long-lived poison and a short-lived antidote. Daughter cells that retain a copy of the plasmid survive, while a daughter cell that fails to inherit the plasmid dies or suffers a reduced growth rate because of the lingering poison from the parent cell.

4.7.4.2 Viral Vectors

In Section 4.8, we learned how viral vectors are important in rDNA technology. We will now study the unique characteristics that are tailored to their specific applications. Although viral vectors are occasionally created from pathogenic viruses, they are modified so as to minimize the risk of handling them. This usually involves the deletion of a part of the viral genome critical for viral replication. Such a virus can efficiently infect cells but, once the infection has taken place, requires a helper virus to provide the missing proteins to produce new versions. The viral vector should have a minimal effect on the physiology of the cell it infects. Another issue related to viral vectors is that some viruses are genetically unstable and can rapidly rearrange their genomes. Most viral vectors are engineered to infect as wide a range of cell types as possible. However, sometimes the opposite is preferred. The viral receptor can be modified to target the virus to a specific kind of cell. Viral vectors are being used to produce vaccines as diagrammatically depicted in Figure 4.7.

4.7.4.3 Retroviruses

Retroviruses are one of the mainstays of current gene therapy approaches. Recombinant retroviruses such as the Moloney murine leukemia virus can integrate into the host genome in a stable

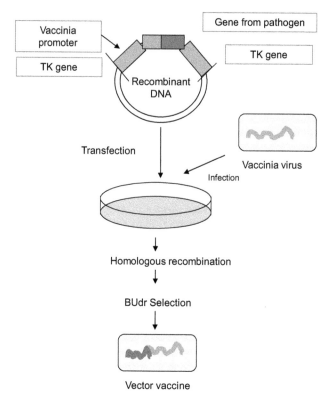

FIGURE 4.7 Viral vector in vaccine development.

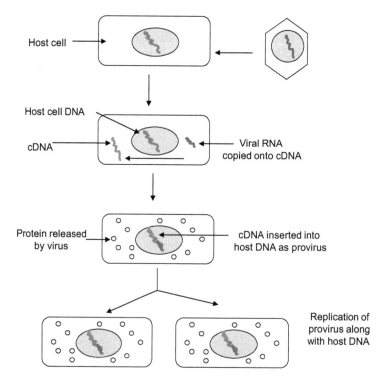

FIGURE 4.8 Replication of retrovirus.

fashion. They contain a reverse transcriptase that allows integration into the host genome. They have been used in several FDA-approved clinical trials, such as the SCID-X1 trial. Retroviral vectors can be either replication-competent or replication-defective. Replication-defective vectors are the most common choice in studies because the viruses have had the coding regions for the genes necessary for additional rounds of virion replication and packaging replaced with other genes or deleted. These viruses can infect their target cells and deliver their viral payload, but then fail to continue the typical lytic pathway, which would typically result in cell lysis and death. Conversely, replication-competent viral vectors contain all the necessary genes for virion synthesis and will continue to propagate themselves once infection occurs. Because the viral genome for these vectors is much lengthier, the length of the actual inserted gene of interest is limited compared to the possible length of the insert for replication-defective vectors (Figure 4.8).

Depending on the viral vector, the typical maximum length of an allowable DNA insert in a replication-defective viral vector is usually about 8–10 kb. Although this limits the introduction of many genomic sequences, most cDNA sequences can still be accommodated. The primary drawback to the use of retroviruses such as the Moloney retrovirus involves the requirement for cells to be actively dividing for transduction. As a result, cells such as neurons are very resistant to infection and transduction by retroviruses. There is a concern for insertional mutagenesis because of the integration into the host genome, which can lead to cancer or leukemia.

4.7.4.4   Lentiviruses

Lentiviruses are basically a subclass of retroviruses. They have recently been adapted as vectors because of their ability to integrate into the genome of nondividing cells; this is a unique feature of lentiviruses, as other retroviruses can only infect dividing cells. The viral genome in the form of RNA is reverse-transcribed when the virus enters the cell to produce DNA, which is then inserted into the genome at a random position by the viral integrase enzyme. The vector, now called a *provirus*, remains in the genome and is passed on to the progeny of the cell when it divides. The site of integration is unpredictable, which can pose a problem. The provirus can disturb the function of cellular genes and lead to activation of oncogenes that promote the development of cancer, which raises concerns for possible applications of lentiviruses in gene therapy. However, studies have shown that lentivirus vectors have a lower tendency than gamma-retroviral vectors to integrate in places that potentially cause cancer. More specifically, one study found that lentiviral vectors did not cause either an increase in tumor incidence or an earlier onset of tumors in a mouse strain with a much higher incidence of tumors. Moreover, clinical trials that utilized lentiviral vectors to deliver gene therapy for the treatment of HIV showed no increase in mutagenic or oncogenic activities in the patients. For safety reasons, lentiviral vectors never carry the genes required for their replication. To produce a lentivirus, several plasmids are transfected into a so-called *packaging cell line*, commonly HEK-293. One or more plasmids, generally referred to as *packaging plasmids*, encode the virion proteins, such as the capsid and the reverse transcriptase. Another plasmid contains the genetic material to be delivered by the vector. It is transcribed to produce the single-stranded RNA viral genome and is marked by the presence of the ψ (psi) sequence. This sequence is used to package the genome into the virion (Figure 4.9).

4.7.4.5   Adenoviruses

Unlike lentiviruses, adenoviruses do not integrate into the genome and do not replicate during cell division. This limits their use in basic research, although adenoviral vectors are occasionally used in *in vitro* experiments. Their primary applications are in gene therapy and vaccination. Since humans commonly encounter adenoviruses, which cause respiratory, gastrointestinal, and eye infections, they trigger a rapid immune response with potentially dangerous consequences. To overcome this problem, scientists are currently investigating adenoviruses to which humans do not have immunity (Figure 4.10).

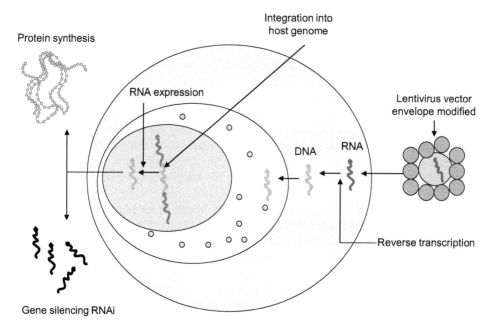

FIGURE 4.9 Lentivirus vector.

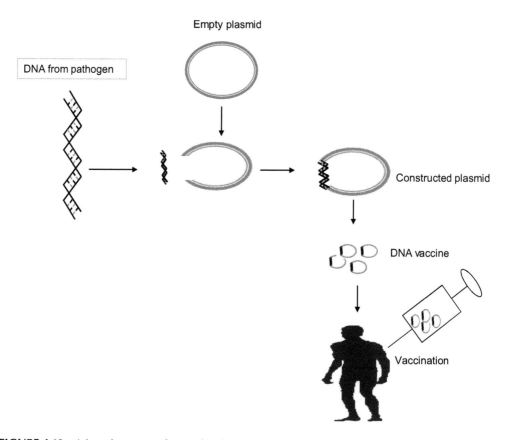

FIGURE 4.10 Adenovirus vector for vaccine development.

4.7.4.6   Adeno-Associated Viruses

An adeno-associated virus (AAV) is a small virus that infects humans and some other primate species. AAV is not currently known to cause disease, and consequently the virus causes a very mild immune response. AAV can infect both dividing and nondividing cells and may incorporate its genome into that of the host cell. These features make AAV a very attractive candidate for creating viral vectors for gene therapy.

4.7.4.7   Autonomously Replicating Sequence Vectors

In yeast chromosomes, the origin of replication is specified by about 100 bp sequences called *Autonomously Replicating Sequences* (ARS). All ARS have an 11 bp consensus sequence, which is essential for their function. Other functional but variable sequences are also present in ARS. Any DNA double helix containing an ARS can serve as a yeast vector. Such a vector will be maintained in yeast cells only if it is essential for their survival—for example, if it is the only source of an essential gene (such as TRP1) in yeast cells mutant for that gene (in this case, Trp1–yeast cells). In the absence of such a selection pressure, ARS vectors are rapidly lost.

4.7.4.8   Minichromosome Vectors

These shuttle vectors behave like very small chromosomes, in that they replicate only once during each cell division and are distributed to daughter cells like true chromosomes. A typical minichromosome vector contains the following functions:

FROM YEAST

- An ARS (provides replication origin).
- A centromere sequence (CEN sequence) (regulates replication and distribution during cell division like those of chromosomes).
- A functional gene, such as LEU2, that serves as a selectable marker in appropriate yeast strains (in this case LEU2 cells).

FROM A BACTERIAL PLASMID

- The origin of replication.
- A selectable marker (e.g., ampr). CEN sequences are of about 500 bp and come from the centromeric regions of yeast chromosomes. They are responsible for centromeric functions during cell division.

4.7.4.9   Yeast Artificial Chromosome Vectors

Linear vectors that behave like a yeast chromosome are called *yeast artificial chromosomes* (YACs). A typical YAC such as pYAC3 contains the following functional elements from yeast:

An ARS sequence for replication.
A CEN4 sequence for centromeric function.
Telomeric sequences at the two ends for protection from exonuclease action.
One or two selectable marker genes such as TRPJ and URA3 (strategy like other vectors).
SUP4, a selectable marker into which the DNA insert is integrated.
The necessary sequences from *E. coli* plasmid for selection and propagation in *E. coli*.
The telomeric sequence in yeast chromosomes is a 20–70 tandem repeat of the six base sequence 5'CCCCAA3' (its complementary sequence, 5TIGGGG3', occurs in the other strand). Theirs is a hairpin loop formation at the terminus, which makes the DNA duplex resistant to exonuclease action.

Vector pYAC3 is essentially a pBR322 plasmid into which the above described yeast sequences have been integrated. Subsequently, several YAC vectors have been constructed on the basic scheme of pYAC3. The YAC vector itself is propagated in *E. coli*, but cloning is done in yeast.

For cloning, the vector is restricted with a combination of BamHI and SnaBI. BamBI cleaves the vector at the junctions of the two TEL sequences with the fragment that is used to circularize the vector for propagation in *E. coli*; this fragment is discarded. The enzyme SnaBI recognizes the single sequence 5′T ACGT A3′ located in SUP4 and produces blunt-ended cleavage, thereby generating two arms of the YAC, each ending in a TEL sequence. The DNA insert, therefore, must have blunt ends. It is integrated within SUP4 to generate the linear YAC. The recombinant YAC is introduced into TRP1⁻ URA3⁻ yeast cells by protoplast transformation. Transformed cells are selected by plating them onto the minimal medium. Only those cells that correctly constructed YAC, containing one left and one right arm of each chromosome, can grow on this medium. Recombinant clones are identified owing to the insertional inactivation of SUP4, detected by a simple color test: recombinant colonies are white, whereas nonrecombinant ones are red. The TEL sequence of the vector is not the complete telomeric sequence, but it contains enough of this sequence to be able to support the creation of the complete telomere once the YAC is inside a yeast cell. Thus, a YAC is a shuttle vector that is propagated in circular form in *E. coli* and is used for cloning in yeast in a linear form. When a YAC is less than approximately 20 kb, the centromeric function is unable to control the copy number during mitosis, so that several copies of YAC accumulate per yeast cell.

The centromeric function improves in YACs of 50 kb or more. YACs of 150 kb or more behave like regular yeast chromosomes. YACs are the predominant vector system used for cloning of very large (up to 100–1400 kb) DNA segments for mapping of complex eukaryotic chromosomes. YACs are reported to suffer from many problems, including chimerism, tedious steps in YAC library construction, and low yields of YAC insert DNA. The yeast genes present in different yeast vectors can become integrated into the host genome. This is called *permanent transformation*. It generally occurs through homologous recombination between the gene present in a vector (e.g., LEU2) and that present in the yeast chromosomes (e.g., LEU2). Rarely, the gene may become inserted at a random chromosome site. The homologous recombination may occur by regular crossing over or it may involve gene conversion (a nonreciprocal recombination).

4.7.4.10 Vectors for Animals

Animal cells such as *in vitro* cultured cell lines, Xenopus oocytes, and early embryos obtained from transgenic animals may be used for transfection or transformation. DNA fragments of animals are generally cloned in *E. coli* to obtain sufficient copies for structural analysis. However, expression of the DNA inserts has to be studied in animal cells. Therefore, many of the vectors are shuttle vectors for *E. coli* and animal cells. The different vectors used in animal cells are derived from viral genomes and are either virus-like (rDNAs produce virions, which are virus particles) or plasmid-like vectors (these replicate but do not produce virions). Some other vectors are designed to be unable to replicate. Another way of looking at animal vectors relates to their ability to replicate in animal cells. Some vectors are capable of replication (replicating vectors), whereas others are not (non-replicating vectors). Furthermore, any DNA fragment, or a mixture of fragments, can be used to transfect animal cells and this DNA segment becomes integrated into the cellular genome and expresses itself. In animal cells, vectors that remain and replicate in the extra-chromosomal state are quickly lost in a few days, even when selection conditions favoring their retention are imposed. This is called *transient transfection*. However, some animal vectors do remain stable in the extra-chromosomal state, such as those derived from bovine papilloma-virus (BPV) and those that contain the origin of DNA replication from the herpes virus Epstein-Barr. However, permanent or stable transfection is most often because of the integration of transfecting DNA into the genome of animal cells. This is also known as *insertional transfection*. The DNA integration is usually by nonhomologous recombination, hence in random locations in the genome. This is in contrast to the situation in yeast, where recombination is homologous and, as a result, site-specific. The first animal vector was devised from the primate papova virus, simian virus 40 (SV40). Subsequently, vectors have been developed from many other viruses such as papillomavirus, adenoviruses, the Epstein-Barr herpes virus, vaccinia viruses (all for mammals), and baculoviruses (for insects).

4.7.4.11   SV40 Vectors

The SV40 vector is a spherical virus with a circular, double-stranded 5243 bp chromosome, which encodes five proteins: small-T, large-T (both early proteins), and the virion proteins (VPs) VP1, VP2, and VP3. It has an origin of replication (approximately 80 bp) and is complexed with histones to form chromatin. Large-T is essential for viral replication, while VP1, VP2, and VP3 form the viral capsid. In the laboratory, it is multiplied in cultured kidney cells of the African green monkey. The SV40 genome has mainly been used to develop three types of vectors: transducing vectors, plasmid vectors, and transforming vectors.

4.7.4.12   Bovine Papillomavirus Vectors

Bovine papillomavirus (BPV) belongs to the papovaviral class and causes warts. It has a circular 8 kb genome organized in nucleosomes. The BPV vector replicates as a stable plasmid in rodent and many bovine cells, and the cells are not killed. The viral genome transforms cells that behave like tumor cells and form piled up colonies of cells instead of the typical monolayer. (Transformation describes the conversion of normal cells into timorous cells.) The transformed state is because of the genes present in the "transforming region" (approximately 5500 bp) of the virus genome. The virus genome is generally used to produce shuttle vectors by using its transforming region. Eukaryotic DNA segments are first cloned in *E. coli* to select rDNAs. Then the *E. coli* plasmid, such as pBR322, is deleted from the vector and the linear rDNA is introduced into animal cells. The rDNA now becomes circular and replicates as a plasmid. The *E. coli* neo gene may be included within the vector. This allows easy selection of transfected cells by culturing them on a medium containing the aminoglycoside G-418.

4.8   INTEGRATION OF THE DNA INSERT INTO THE VECTOR

Once the DNA fragments to be cloned are prepared and the appropriate vector is selected, the DNA segments will have to be integrated into the vector at an appropriate site. The vector is cut open with a restriction enzyme that has a unique (single) target site located in the sequence where the DNA insert is to be integrated. There are five possible situations with respect to the vector and the DNA insert.

4.8.1   Both Ends Cohesive and Compatible

The simplest strategy concerns the presence of compatible cohesive ends in both the vector and the DNA insert. This happens when the vector is cut open by the same restriction enzyme that was used to isolate the DNA insert. The opened-up vector and the DNA insert are mixed under annealing conditions that allow pairing between the compatible cohesive ends of the vector and the DNA insert. The nicks (the broken covalent bond between the 5'-phosphate and the 3'-OH of two neighboring nucleotides within a DNA strand) remaining after annealing are sealed, joined, or ligated by DNA ligase. Initially, DNA ligase from *E. coli* was used, but T4 (an *E. coli* phage) ligase is currently preferred. DNA inserts joined in this manner can be easily and precisely isolated from the rDNA using the same restriction enzyme that was used to generate the inserts and to open the vector. In the rDNAs produced in this way, the DNA insert may be present in either of the two orientations relative to the sequences of the vector. If the two cohesive ends of the vector are marked as 1 and 2 and those of the DNA insert as 1' and 2', the insert may join the vector to either yield 1–1' and 2–2' junctions, or 1–2' and 2–1' junctions.

The orientation of DNA insert within the vector is not important when only copies of the insert are to be obtained. However, it is extremely important when expression of the DNA insert is desired. In addition to the formation of rDNA, the cohesive ends of the vector itself will pair together to produce unaltered vector molecules. Similarly, the two ends of the DNA insert will also join to yield a circular DNA insert molecule. A circularized insert is not a problem as it lacks an origin of replication and, as a result, is diluted out of the transformed cells. The formation of an unchanged circularized vector can be prevented by treating the opened-up vector with alkaline phosphatase, which removes the 5'-phosphate present as a monoester at the vector ends. When such a vector is mixed, annealed, and ligated with the DNA insert, two nicks remain in the rDNA owing to a lack of phosphate at the 5'-ends of the vector. These nicks are readily repaired once the rDNA is introduced into appropriate host cells.

4.8.2   Both Ends Cohesive and Separately Matched

One protruding end of the vector may be compatible with one end of the DNA insert, while the other end of vector is compatible with the second end of the insert. This situation arises when one end of the vector as well as that of the DNA insert is generated by one restriction enzyme, while their other end is cut by a different enzyme. As earlier, the opened vector and the DNA insert are mixed under annealing conditions to allow pairing between the compatible ends of the vector and the DNA insert. T4 ligase is then used to seal the nicks to yield rDNA. In such a situation, only the rDNA is circularized (since the two ends of the vector are unmatched as are the two ends of the DNA insert) and the DNA insert is integrated in only one orientation or direction.

4.8.3   Both Ends Cohesive and Unmatched

Often, the DNA insert is prepared using one restriction enzyme, while the vector is opened with another enzyme. This generates cohesive ends in the vector and the DNA insert, which are unmatched. In this situation, the following approaches are available. The protruding ends are converted into blunt ends by removing the protruding ends either by digestion or by extending the recessed ends, using Klenow fragments or reverse transcriptase. The blunt ends can then be joined together by T4 ligase. The blunt ends so produced can again be changed into protruding ends by 3'-tailing. A poly-T tail may be added to the 3'-ends of the vector, while poly-A tails are added to the DNA insert. The two now have matched protruding ends, which pair together under annealing conditions to yield rDNA. Often, the tails are of different sizes. Therefore, the gaps remaining in the rDNA are filled with DNA polymerase I prior to ligation. Alternatively, linkers can be attached to the blunt ends of the vector and the DNA insert. The linkers are cleaved with the appropriate restriction enzyme to generate protruding ends that are compatible. The vector and the insert are now joined together. In addition, appropriate adaptors may be joined to the protruding ends of the vector and/or the DNA insert to generate completely matched cohesive ends.

4.8.4   Both Ends Flush/Blunt

Flush or blunt-ended vector and DNA insert can be joined together by T4 DNA ligase. However, high concentrations of both the enzyme and the vector and the insert DNA are required. The DNA insert gets ligated in either orientation, and it can be easily and precisely separated from the vector only if the two were cleaved by the same restriction enzyme (producing blunt ends). Alternatively, the blunt ends can be converted into cohesive ends, as outlined previously.

4.8.5   One End Cohesive and Compatible, the Other End Blunt

If the vector and the DNA insert have one compatible and cohesive end and one flush end, their cohesive end pairs together when they are mixed under annealing conditions. The T4 ligase seals the nick at the cohesive end and joins the blunt ends as well. It should be noted that under such a situation, only rDNAs are produced, because vector and DNA insert molecules cannot circularize owing to their one cohesive end and one blunt end.

4.9   INTRODUCTION OF THE RECOMBINANT DNA INTO THE SUITABLE HOST

rDNA is constructed *in vitro*. It is then generally introduced into *E. coli* to select the rDNA from the unchanged vector, to obtain many copies of the rDNA, or to express the DNA insert in *E. coli* itself. Purified rDNA may subsequently be introduced into another bacterium (such as *Bacillus subtilis* and Streptomyces), yeast, and higher plants or animals. The various approaches for introducing rDNAs into bacteria, especially *E. coli*, are briefly summarized here.

4.10   INCREASED COMPETENCE OF *E. COLI* BY CACL₂ TREATMENT

E. coli cells are generally poorly accessible to DNA molecules, but treatment with $CaCl_2$ makes them permeable to DNA. The rDNA is then added. Efficient transformation takes only a few minutes, and the cells are plated on a suitable medium for the selection of transformed clones. The frequency of transformed cells is 106–107/μg of plasmid DNA. This is approximately one transformation per 10,000 plasmid molecules. This frequency can be further improved by using special *E. coli* strains such as SK1590, SK1592, and X1766, and by some specific conditions during transformation. These may raise the frequency to 5×10^8 transformed cells/μg of plasmid DNA. The transformed cells are suitably diluted and spread thinly on a suitable medium, so that each cell is well separated and produces a separate colony. Generally, the medium is designed to permit only the transformed cells to divide and produce colonies.

4.11   INFECTION BY RECOMBINANT DNAS PACKAGED AS VIRIONS

Alternatively, those rDNA that have the λ phage cos sequences such as those derived from cosmids, phasmids, and λ vectors, are generally packaged *in vitro* into specially produced empty λ phage heads and complete λ particles are constituted. These phage particles are used to infect *E. coli* cells. This process is often called transfection. These rDNAs can also be used to transform *E. coli* cells directly as naked DNA, using the $CaCl_2$ technique. Generally, transfection is far more efficient than direct transformation. For example, the frequency of transfection by recombinant λ phage DNAs packaged in phage particles is up to 10^8 plaques/μg of DNA, whereas it is $<10^3$ plaques/μg DNA when the rDNA is used for transformation by the $CaCl_2$ technique. The infected/transformed bacterial cells are spread on a lawn of susceptible cells, where clear areas or plaques develop in the lawn. Plaques containing the rDNA (λ vector and phasmids) are identified, and the phage particles collected from such plaques provide the purified vector/rDNA.

4.12   SELECTION OF RECOMBINANT CLONES

When rDNA is constructed and used for transformation of *E. coli*, cells following the types of bacterial cells are obtained. Most of the cells are nontransformed. A proportion of the transformed cells contain an unaltered vector, while the remaining cells have rDNA. The first objective of cloning experiments is to identify and isolate the small number of cells that contain the rDNA from among a very large number of nontransformed cells. Because the DNA inserts are generally mixtures, particularly when cDNA preparations and genomic DNA fragments are used, the various transformed clones would contain a variety of different DNA inserts. The next step, therefore, is to identify the clone having the desired DNA insert from among the large number of clones containing the rDNAs. Suitable selection strategies have been devised to achieve these two critical objectives. This is the most important step in DNA cloning.

4.13 IDENTIFICATION OF CLONES HAVING RECOMBINANT DNAS

The next step consists of identification and isolation of those clones that are transformed by the rDNAs from among those that contain the unaltered vector. This may be achieved in one of several ways listed in the following:

In case the vector has two selectable markers such as pBR322, the DNA insert may be placed within one of these markers (for instance, the ampT gene). The other marker (in this case, tetr) is used for elimination of the nontransformed cells. The transformed clones are then replica plated on an ampicillin-containing medium. The clones containing the rDNAs will be sensitive to ampicillin owing to inactivation of the gene ampT by insertion of the DNA fragment. Such clones are identified and isolated from the master plate.

Some vectors contain a gene, or sometimes only part of a gene, which complements a function missing in their host cells, such as gene lacZα in the pUC vectors, which complements such lacZ–E. coli strains in which lacZα is deleted. The same combination is used for some λ vectors and M13 phage vectors. In all such cases, the DNA insert is so placed that it disrupts the expression of lacZα.

Therefore, E. coli cells containing the rDNA are deficient in β-galactosidase and produce white colonies or plaques on a medium containing X-gal and IPTG. On the other hand, cells having the unchanged vector produce active β-galactosidase and give rise to blue colonies or plaques on the same medium. This allows an easy identification of the clones containing the rDNAs.

When the DNA insert codes for a gene product that is defective in the auxotrophic host cells, a direct selection for the rDNA is possible. The host cells are grown on a medium lacking the compound needed by the auxotrophic host. Only those cells that contain the rDNA can grow and form colonies. Obviously, this approach is limited in application.

Similarly, selection by suppression of nonsense mutations present in the host also permits a direct selection for the rDNA.

Some λ vectors, such as λgt10, retain the lysogenic function as well. In such vectors, the DNA insert may be placed within the lysis repressor gene cI-, so that the vector becomes cI. As a result, cells transfected by the rDNA will give rise to clear plaques, whereas those infected by the unaltered vector will yield cloudy or turbid plaques. Thus, the rDNAs are readily identified and isolated.

Some vectors such as λ replacement vectors and cosmids are much shorter than the minimum genome length needed for their packaging within virus particles. In such cases, the length of the DNA insert can be adjusted to allow the packaging of only the rDNA. This provides an efficient selection strategy for rDNA.

4.14 SELECTION OF CLONES CONTAINING A SPECIFIC DNA INSERT

Once we obtain recombinant clones, the next step is to identify clones that have the DNA inserts of interest. The technique used for identification must be highly precise and extremely sensitive to allow the accurate detection of a single clone from among the thousands obtained from a cloning experiment. The various strategies used for the purpose are briefly outlined in the following.

4.14.1 COLONY HYBRIDIZATION

The most efficient and rapid strategy for identification of a clone having the desired insert uses the technique of colony hybridization. The bacterial colonies are replica plated or phage plaques directly lifted on nitrocellulose filters. The cells are lysed and their DNA is denatured. The filter is incubated with the specific radioactive (^{32}P-labelled) probe under annealing conditions. After some time, the probe is washed out leaving only those probe molecules that have hybridized with the denatured DNA from bacterial cells or phage particles. The colonies/plaques with whose DNA the probe has hybridized are identified by autoradiography. These contain the desired DNA insert. These colonies/plaques are isolated from the master plate used for replica plating. A very large number of colonies or plaques (up to 10,000 plaques) can be lifted onto a single 10 cm diameter filter. However, it is essential that a specific probe for the DNA insert is available.

A *probe* is a polynucleotide molecule (DNA or RNA and usually small molecules of as few as 15 bases, but more often of 2530 bases) of a specific base sequence, which is used to detect DNA molecules having the same base sequence by complementary base pairing. Generally, the probes are labeled with ^{32}P to enable autoradiography for easy identification of the DNA samples that base pair with the probe. It is desirable that the probes are single-stranded to avoid pairing between the two strands of the probe itself. Either DNA or RNA can be used as a probe. There are several approaches for developing specific probes.

4.14.2   OTHER APPROACHES FOR DEVELOPING SPECIFIC PROBES

When specific probes are not available, many indirect approaches may be used for identifying clones having the desired DNA insert. These approaches or procedures are not generally convenient for screening of many clones. Two such procedures, called *hybrid arrested translation* (HART) and *hybrid selection*, use *in vitro* translation systems followed by identification of the resulting polypeptide(s). It is therefore necessary that the protein product of the DNA insert being searched should be known, at least in terms of its electrophoretic mobility.

4.14.3   COMPLEMENTATION

The cloned DNA insert may express itself in the bacterial cells. This is possible for prokaryotic genes, for some yeast genes, and for eukaryotic cDNAs cloned in suitable expression vectors. Eukaryotic sequences isolated from genomic DNA must be expressed in appropriate eukaryotic hosts, such as yeast cells or animal cells. If the protein produced by the desired DNA insert is deficient in the host cells, this insert will correct the deficiency of the cells transformed by it, which means it will complement the deficiency of the host cells. This can be stated in general terms as follows. The host cells are deficient in a protein A, which means they are A–. These cells can be used to isolate the DNA fragment coding for protein A from a mixture of DNA fragments. Expression of rDNAs is prepared from the DNA fragments and A– host cells are transformed. These cells are now cultured under selective conditions that require functional A product. Only those host cells that contain the DNA insert encoding protein A will be able to multiply under the selective conditions (as the DNA insert will provide functional protein A). This strategy is limited in application by the availability of appropriate host cells.

4.14.4   UNIQUE GENE PRODUCTS

Alternatively, the protein product of the DNA insert can be identified by its unique function, that is, a function not performed by the proteins of nontransformed host cells. Such functions may relate to enzyme activities or hormone effects for which appropriate assays exist.

4.14.5   ANTIBODIES SPECIFIC TO THE PROTEIN PRODUCT

Finally, if the protein lacks a recognizable and measurable function, it can be detected by using specific antibodies. A practical approach is to divide the large number of recombinant clones into a convenient number of groups and to assay for the presence of the protein. The positive group is again divided into subgroups and assayed. In this manner, the positive groups are subdivided again and again, until a single positive clone is identified. This approach is applicable to the previous strategy as well. The identification of proteins using antibodies may be achieved by western blotting, precipitation and electrophoresis, or ELISA.

4.14.6 COLONY SCREENING WITH ANTIBODIES

An efficient and rapid screening using antibodies is as follows. The antibody specific to the concerned gene product (i.e., protein) is spread uniformly over a solid support such as a plastic or paper disc, which is placed in contact with an agar layer containing lysed bacterial colonies or phage plaques. If any clone is producing the protein in question, it will bind to the antibody molecules present on the disc. The disc is removed from the agar and is treated with a second radiolabeled (generally with ^{125}I) antibody, which is also specific to the same protein but in a region different from that recognized by the first antibody. These antibodies will therefore also bind to the protein molecule held by the first antibody. The location of the radioactivity on the disc is determined by autoradiography. The colonies/plaques producing the protein are then identified and isolated from the master plate. This technique is analogous to colony hybridization and can screen large numbers of clones rather rapidly. However, for this technique, two different antibodies that bind to two distinct domains of the desired protein are required. This protein must not be produced by the nontransformed host cells.

4.14.7 FLUORESCENCE-ACTIVATED CELL SORTER

In the case of animal cells, an automated system, called a *fluorescence-activated cell sorter* (FACS), can be used for very rapid (up to 1000 cells/s) sorting of transformed cells. This is applicable to all the genes whose products become arranged on the cell surface and are available for binding of specific antibodies. Therefore, these proteins must not be produced by the nontransformed host cells. The antibody molecules are attached to a fluorescent molecule, and the transformed cells are treated with this antibody specific for the desired protein. The cells containing the protein in question on their surface will interact with the fluorescent antibodies. Cells are then passed one by one in a stream between a laser and a fluorescence detector. The cells that fluoresce are deflected into a microculture tray, while the nonfluorescing cells are drawn away by an aspirator. This approach is also applicable to the genes encoding receptor proteins present on the cell surface. In such cases, fluorescent ligands (the concerned molecule to which the receptor binds) are used in the place of fluorescent antibodies.

4.15 RECOMBINANT DNA PRODUCTS

In this section, we will learn about various recombinant DNA products produced by using rDNA technology.

4.15.1 GENETICALLY MODIFIED ORGANISMS

In this subsection, we will concentrate on how genetically modified organisms (GMOs) have revolutionized the research, development, and industrial sectors. A *genetically modified organism* or *genetically engineered organism* (GEO) is an organism whose genetic material has been altered using rDNA technology. The rDNA technology uses DNA molecules from different sources, which are combined into one molecule to create a new set of genes. This DNA is then transferred into an organism, giving it modified or novel genes. Transgenic organisms, a subset of GMOs, are organisms that have inserted DNA that originated in a different species.

4.15.1.1 Transgenic Microbes

A bacterium having a simple genetic makeup was the first organism to be modified in the laboratory. Bacteria are now used for several purposes and are particularly important in producing large amounts of therapeutic proteins for treating various ailments and diseases, such as the genetically modified (GM) bacteria used to produce the protein insulin to treat diabetes. Similar bacteria have been used to produce clotting factors to treat hemophilia and human growth hormone to treat various forms of dwarfism. In addition to the use of GM bacteria to make therapeutic proteins, GM bacteria are also being used to treat dental disease. For example, tooth decay is caused by the bacteria *Streptococcus mutans*; these bacteria consume leftover sugars in the mouth, producing lactic acid that corrodes tooth enamel and ultimately causes cavities. Scientists have recently modified *Streptococcus mutans* so that they do not produce lactic acid. These transgenic bacteria, if properly colonized in a person's mouth, could reduce the formation of cavities. In recent research, transgenic microbes have also been used to kill or hinder tumors. GM bacteria are also used in some soils to facilitate crop growth and to produce chemicals that are toxic to crop pests.

4.15.1.2 Transgenic Animals

Transgenic animals are used as experimental models to perform phenotypic tests with genes whose function is unknown. Genetic modification can also produce animals that are susceptible to certain compounds or stresses for testing in biomedical research. In biological research, transgenic fruit flies (*Drosophila melanogaster*) are model organisms used to study the effects of genetic changes on development. Fruit flies are often preferred over other animals because of their short life cycle, low maintenance requirements, and relatively simple genome compared with many vertebrates. Transgenic mice are often used to study cellular and tissue-specific responses to disease. This is possible because mice can be created with the same mutations that occur in human genetic disorders. The production of the human disease in these mice allows treatments to be tested. In 2009, scientists in Japan announced that they had successfully transferred a gene into a primate species (marmosets) and produced a stable line of breeding transgenic primates for the first time. It is hoped that this will aid research into human diseases that cannot be studied in mice, such as Huntington's disease and strokes.

Besides mammalians and flies, rDNA technology has been used to create transgenic fish and cnidarians such as *Hydra* to study the evolution of immunity. For analytical purposes, an important technical breakthrough was the development of a transgenic procedure for generation of stably transgenic hydras by embryo microinjection. The creation of transgenic fish to produce higher levels of growth hormone has resulted in dramatic growth enhancement in several species, including salmonids, carps, and tilapias. These fish have been created for use in the aquaculture industry to increase the speed of development and to potentially reduce fishing pressure on wild stocks.

4.15.1.3 Transgenic Plants

rDNA technology has been widely used in generating plants with desirable traits, including resistance to pests, herbicides, or harsh environmental conditions; improved product shelf-life; and increased nutritional value. In 1996, transgenic plants were cultivated at the commercial level. Thereafter, there was a tremendous increase in the generation of transgenic plants that are not only tolerant to the herbicides glufosinate and glyphosate but are also resistant to viral damage—as in Ringspot virus-resistant GM papaya grown in Hawaii—and produce the Bt toxin, a potent insecticide.

4.15.1.4 Cisgenic Plants

In some GMOs, cells do not contain DNA from other species and are therefore not transgenic but are what is called *cisgenic*. GM sweet potatoes have been enhanced with protein and other nutrient values, whereas golden rice, developed by the International Rice Research Institute (IRRI), is a good source of vitamin A. As vitamin A deficiency causes deformities in children, eating cisgenic golden rice is highly recommended for children with a vitamin A deficiency. In January 2008, scientists altered a carrot so that it would produce calcium and become a possible cure for osteoporosis.

However, people would need to eat 1.5 kg of carrot/day to reach the required amount of calcium. The coexistence of GM plants with conventional and organic crops has raised significant concern in many European countries. As there is separate legislation for GM crops and high demand from consumers for the freedom of choice between GM and non-GM foods, measures are required to separate foods and feed produced from GMO plants and from conventional and organic foods. European research programs such as Co-Extra, Transcontainer, and SIGMEA are investigating appropriate tools and rules. At the field level, biological containment methods include isolation distances and pollen barriers.

4.16   ADVANCED TECHNIQUES

Microarray: A microarray is a tool used to detect/identify the expression of thousands of genes at the same time. DNA microarrays are microscopic slides that are printed with thousands of tiny spots in defined positions, with each spot containing a known DNA sequence or gene (Figure 4.11).

DNA Chip: A DNA microarray is also commonly known as a DNA chip or biochip, which is a collection of microscopic DNA spots attached to a solid surface. Researchers use DNA microarrays to measure the expression levels of large numbers of genes simultaneously or to genotype multiple regions of a genome (Figure 4.12).

cDNA Library: A cDNA library is a combination of cloned cDNA (complementary DNA) fragments inserted into a collection of host cells, which constitute some portion of the transcriptome of the organism and are stored as a library (Figure 4.13).

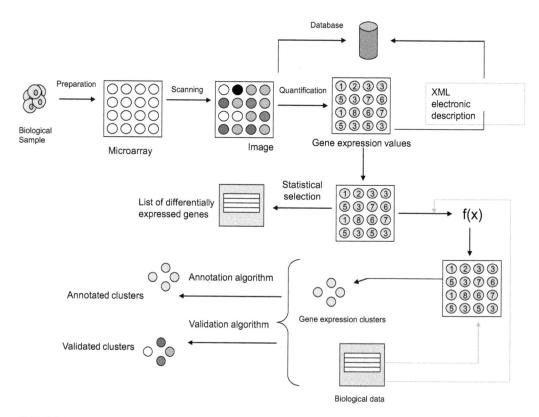

FIGURE 4.11   Microarray technique.

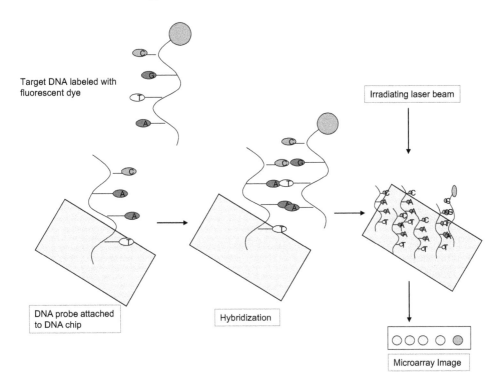

FIGURE 4.12 DNA chip.

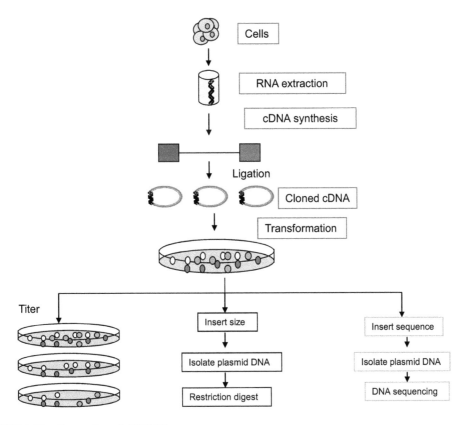

FIGURE 4.13 Generation of cDNA library.

PROBLEMS

Section A: Descriptive Type

Q1. What is rDNA technology?

Q2. What are the steps involved in making rDNA?

Q3. Describe nonbacterial transformation.

Q4. Why are restriction enzymes so important in rDNA technology?

Q5. How can we amplify genes using PCR?

Q6. Differentiate between PCR and gene cloning.

Q7. What is a vector? Describe its role in rDNA technology.

Section B: Multiple Choice

Q1. Which of the following methods is not used in making rDNA product?
   a. Translation
   b. Transformation
   c. Phage introduction
   d. Nonbacterial transformation

Q2. Restriction enzyme cuts double-stranded or single-stranded DNA at a specific site. True/False

Q3. Type IIM restriction endonucleases such as DpnI can recognize and cut methylated DNA. True/False

Q4. All vectors used for propagation of DNA inserts in a suitable host are called cloning vectors. True/False

Q5. Which of the following is *not* an *E. coli* strain?
   a. K12
   b. C600
   c. HB101
   d. B250

Q6. Resistance plasmids contain genes that do not build resistance against antibiotics or poisons. True/False

Q7. What is the spherical virus with a circular, double-stranded 5243 bp chromosome?
   a. SV40 vector
   b. YAC vector
   c. ARS vector
   d. Adenovirus

Q8. What does FACS mean?
   a. Fluorescence acquired cell sorter
   b. Fluorescence activated cell sorter
   c. Fluorescence auto cell sorter

Q9. What do you call GM plants that do not contain the DNA of other species?
   a. Cisgenic plants
   b. Transgenic plants
   c. Autogenic plants
   d. Heterogenic plants

Section C: Critical Thinking

Q1. What problems may arise if, instead of microbes, mammalian cells are used as vectors for producing rDNA products?

Q2. Is it possible to cut DNA fragments without restriction enzyme? Explain why.

Q3. Explain why 30–45 cycles are usually carried out in most PCR reactions.

ASSIGNMENTS

Make a poster based on the various rDNA products such as insulin, growth hormones, or erythro-poietin. Display it in the classroom and discuss it with your colleagues.

REFERENCES AND FURTHER READING

Berg, P., Baltimore, D., Brenner, S., Roblin III, R.O., and Singer, M.F. Summary statement of the Asilomar conference on recombinant DNA molecules. *Proc. Natl. Acad. Sci. USA* 72(6): 1981–1984, 1975.

Braslavsky, I., Hebert, B., Kartalov, E., and Quake, S.R. Sequence information can be obtained from single DNA molecules. *Proc. Natl. Acad. Sci. USA* 100: 3960–3964, 2003.

Church, G.M. Genomes for all. *Sci. Am.* 294: 46–54, 2006.

Cohen, S.N., Chang, A.C.Y., Boyer, H.W., and Helling, R.B. Construction of biologically functional bacterial plasmids in vitro. *Proc. Natl. Acad. Sci. USA* 70(11): 3240–3244, 1973.

Colowick, S.P., and Kapian, O.N. *Methods in Enzymology—Volume 68; Recombinant DNA*. Academic Press, Burlington, MA, 1980.

Ewing, B., and Green, P. Base-calling of automated sequencer traces using phred. II: Error probabilities. *Genome Res.* 8: 186–194, 1998.

Fiers, W., Contreras, R., Duerinck, F. et al. Complete nucleotide sequence of bacteriophage MS2 RNA: Primary and secondary structure of the replicase gene. *Nature* 260: 500–507, 1976.

Garret, R.H., and Grisham, C.M. *Biochemistry*. Saunders College Publishers, Philadelphia, PA, 2000.

Genentech. 1978. The insulin synthesis is the first laboratory production DNA technology. Press release. Archived from the original on May 9, 2006. http://web.archive.org/web/20060509151511/www.gene.com/gene/news/press-releases/display.do?method=detail&id=4160 (retrieved on January 7, 2009).

Gilbert, W. DNA sequencing and gene structure. Nobel Lecture. December 8, 1980.

Gilbert, W., and Maxam, A. The nucleotide sequence of the lac operator. *Proc. Natl. Acad. Sci. USA* 70: 3581–3584, 1973.

Hall, N. Advanced sequencing technologies and their wider impact in microbiology. *J. Exp. Biol.* 210: 1518–1525, 2007.

Hanna, G.J., Johnson, V.A., Kuritzkes, D.R. et al. Comparison of sequencing by hybridization and cycle sequencing for genotyping of human immunodeficiency virus type 1 reverse transcriptase. *J. Clin. Microbiol.* 38: 2715–2721, 2000.

Inoue, N., Takeuchi, H., Ohashi, M., and Suzuki, T. The production of recombinant human erythropoietin. *Biotechnol. Ann. Rev.* 1: 297–300, 1995.

Johnston, S.A., and Tang, D.C. Gene gun transfection of animal cells and genetic immunization. *Method. Cell Biol.* 43(Pt A): 353–365, 1994.

Ju, J., Ruan, C., Fuller, C.W., Glazer, A.N., and Mathies, R.A. Fluorescence energy transfer dye-labeled primers for DNA sequencing and analysis. *Proc. Natl. Acad. Sci. USA* 92: 4347–4351, 1995.

Kruzer, H., and Massay, A. *Recombinant DNA and Biotechnology: A Guide for Student*. ASM Press, Washington, DC, 2001.

Leader, B., Baca, Q.J., and Golan, D.E. Protein therapeutics: A summary and pharmacological classification. *Nat. Rev. Drug Discov.* (A guide to drug discovery) 7: 21–39, 2008.

Lee, L.Y., and Gelvin, S.B. T-DNA binary vectors and systems. *Plant Physiol.* 146(2): 325–332, 2008.

Margulies, M., Egholm, M., Altman, W.E. et al. Genome sequencing in microfabricated high-density picolitre reactors. *Nature* 437: 376–380, 2005.

Maxam, A.M., and Gilbert, W. A new method for sequencing DNA. *Proc. Natl. Acad. Sci. USA* 74: 560–564, 1977.

Min Jou, W., Haegeman, G., Ysebaert, M., and Fiers, W. Nucleotide sequence of the gene coding for the bacteriophage MS2 coat protein. *Nature* 237: 82–88, 1972.

Park, F. Lentiviral vectors: Are they the future of animal transgenesis? *Physiol. Genomics* 31(2): 159–173, 2007.

Pipe, S.W. Recombinant clotting factors. *Thromb. Haemost.* 99: 840–850, 2008.

Ronaghi, M., Karamohamed, S., Pettersson, B., Uhlen, M., and Nyren, P. Real-time DNA sequencing using detection of pyrophosphate release. *Anal. Biochem.* 242: 84–89, 1996.

Sanger, F. Determination of nucleotide sequences in DNA. Nobel Lecture. December 8, 1980.

Sanger, F., and Coulson, A.R. A rapid method for determining sequences in DNA by primed synthesis with DNA polymerase. *J. Mol. Biol.* 94: 441–448, 1975.

Sanger, F., Nicklen, S., and Coulson, A.R. DNA sequencing with chain-terminating inhibitors. *Proc. Natl. Acad. Sci. USA* 74: 5463–5467, 1977.

Shendure, J. Accurate multiplex polony sequencing of an evolved bacterial genome. *Science* 309: 1728–1732, 2005.

Shreeve, J. Secretes of the gene. *Natl. Geogr.* 1966: 42–75, 1999.

Smith, L.M., Fung, S., Hunkapiller, M.W., Hunkapiller, T.J., and Hood, L.E. The synthesis of oligonucleotides containing an aliphatic amino group at the 5' terminus: Synthesis of fluorescent DNA primers for use in DNA sequence analysis. *Nucleic Acids Res.* 13: 2399–2412, 1985.

Smith, L.M., Sanders, J.Z., Kaiser, R.J. et al. Fluorescence detection in automated DNA sequence analysis. *Nature* 321: 674–679, 1986.

Walsh, G. Therapeutic insulins and their large-scale manufacture. *Appl. Microbiol. Biotechnol.* 67: 151–159, 2005.

Watson, J., and Tooze, J. *The DNA Story: A Documentary History of Gene Cloning.* W.H. Freeman & Co., San Francisco, CA, 1981.

Weiner, D., and Kenendy, R. Genetic vaccines. *Sci. Am.* 281: 50–57, 1999.

5 Microbial Biotechnology

LEARNING OBJECTIVES

- Define microbes and explain their various attributes
- Discuss structural and functional characteristics of microbes
- Explain the growth and culture of microbes
- Discuss microbial genetics and genetic transformations
- Discuss the role of microbes in food, medical, agricultural, and environmental biotechnology

5.1 INTRODUCTION

Microbes are single-celled organisms, and they are the oldest form of life on earth. Microbe fossils date back more than 3.5 billion years, to a time when the earth was covered with oceans that regularly reached boiling point, hundreds of millions of years before dinosaurs roamed the earth. Without microbes, we could not eat or breathe. Without us, they would probably be just fine. Understanding microbes is vital to understanding our own past and future and that of our planet. Microbes are everywhere. There are more of them on a person's hand than there are people on the entire planet! Microbes are in the air we breathe, the ground we walk on, the food we eat—they are even inside us! We could not digest food without them, and neither could animals. Without microbes, plants could not grow, garbage would not decay, and there would be a lot less oxygen to breathe in. In fact, without these invisible companions, our planet as we know it would not survive!

With a view to understanding their significance in our daily life, it is very important to have historical information pertaining to microorganisms. It is believed that the ancestors of modern bacteria were single-celled microorganisms that were the first forms of life to develop on earth, approximately 4 billion years ago. For about 3 billion years, all organisms were microscopic, and bacteria and archaea were the dominant forms of life. Although bacterial fossils such as stromatolites exist, their lack of distinctive morphology prevents them from being used to examine the history of bacterial evolution or to date the time of origin of a bacterial species. However, gene sequences can be used to reconstruct bacterial phylogeny, and these studies indicate that bacteria diverged first from the archaeal/eukaryotic lineage. The most recent common ancestor of bacteria and archaea was a hyperthermophile that lived approximately 2.5–3.2 billion years ago.

Bacteria were first observed in 1676 by Antonie van Leeuwenhoek using a single-lens microscope of his own design. He called them "animalcules" and published his observations in a series of letters to the Royal Society. The name *bacterium* was introduced much later, in 1838, by Christian Gottfried Ehrenberg. In 1859, Louis Pasteur demonstrated that the fermentation process is caused by the growth of microorganisms, and that this growth is not because of spontaneous generation. (Yeasts and molds that are commonly associated with fermentation are not bacteria, but rather fungi.) Along with his contemporary, Robert Koch, Pasteur was an early advocate of the germ theory of disease. Robert Koch was a pioneer in medical microbiology and worked on cholera, anthrax, and tuberculosis. In his research into tuberculosis, Koch finally proved the germ theory, for which he was awarded a Nobel Prize in 1905. In *Koch's postulates*, he set out criteria to test if an organism is the cause of a disease. These postulates are still used today. Though it was known in the nineteenth century that bacteria are the cause of many diseases, no effective antibacterial treatments were available. In 1910, Paul Ehrlich developed the first antibiotic, by changing dyes that selectively stained *Treponema pallidum*—the spirochete that causes syphilis—into compounds that selectively killed the pathogen. In 1908, Ehrlich was awarded a Nobel Prize for his work in immunology and pioneering the use of stains to detect and identify bacteria, with his work being the basis of the Gram stain and the Ziehl–Neelsen stain. A major step forward in the study of bacteria was the recognition, in 1977, by Carl Woese that archaea have a separate line of evolutionary descent from bacteria. This new phylogenetic taxonomy was based on the sequencing of 16S ribosomal RNA and divided prokaryotes into two evolutionary domains as part of the three-domain system.

5.2 ANATOMY OF MICROBES

5.2.1 STRUCTURE

Bacteria display a wide diversity of shapes and sizes, called *morphologies*. Bacterial cells are approximately one-tenth the size of eukaryotic cells and are typically 0.5–5.0 μm in length. However, a few species, such as *Thiomargarita namibiensis* and *Epulopiscium fishelsoni*, are up to half a millimeter long and are visible to the unaided eye. Among the smallest bacteria are members of the genus *Mycoplasma*, which measure only 0.3 μm—as small as the largest viruses. Some bacteria may be even smaller, but these ultramicrobacteria are not well studied. Most bacterial species are either spherical, called *cocci*, or rod-shaped, called *bacilli*. Some rod-shaped bacteria, called *vibrio*, are slightly curved or comma-shaped. Others can be spiral-shaped, called *spirilla*, or tightly coiled, called *spirochetes*. A small number of species even have tetrahedral or cuboidal shapes. More recently, bacteria that grow as long rods with a star-shaped cross section were discovered deep under the earth's crust. The large surface area-to-volume ratio of this morphology may give these bacteria an advantage in nutrient-poor environments. This wide variety of shapes is determined by the bacterial cell wall and cytoskeleton and is important because it can influence the ability of the bacteria to acquire nutrients, attach to surfaces, swim through liquids, and escape predators. Many bacterial species exist simply as single cells, while others associate in characteristic patterns: *Neisseria* form diploids (pairs), *Streptococci* form chains, and *Staphylococcus* group together in "bunch of grapes" clusters. Bacteria such as *Actinobacteria* can also be elongated to form filaments.

Filamentous bacteria are often surrounded by a sheath that contains many individual cells. Certain types, such as species of the genus *Nocardia*, even form complex, branched filaments, similar in appearance to fungal mycelia. Bacteria often attach to surfaces and form dense aggregations called *biofilms* or *bacterial mats*. These films can range from a few micrometers in thickness to up to half a meter in depth, and may contain multiple species of bacteria, protists, and achaea. Bacteria living in biofilms display a complex arrangement of cells and extracellular components, forming secondary structures such as microcolonies through which there are networks of channels to enable better diffusion of nutrients. In natural environments, such as soil or the surfaces of plants, most bacteria are bound to surfaces in biofilms. Biofilms are also important in medicine, as these structures are often present during chronic bacterial infections or in infections of implanted medical devices, and bacteria protected within biofilms are much harder to kill than individual isolated bacteria. Even more complex morphological changes are sometimes possible. For example, when starved of amino acids, *Mycobacterium* detect surrounding cells in a process known as *quorum sensing*, migrate toward each other, and aggregate to form fruiting bodies up to 500 μm long and containing approximately 100,000 bacterial cells. In these fruiting bodies, the bacteria perform separate tasks. This type of cooperation is a simple type of multicellular organization. For example, approximately 1 in 10 cells migrate to the top of these fruiting bodies and differentiate into a specialized dormant state called *myxospores*, which are more resistant to drying and other adverse environmental conditions than are ordinary cells (Figure 5.1).

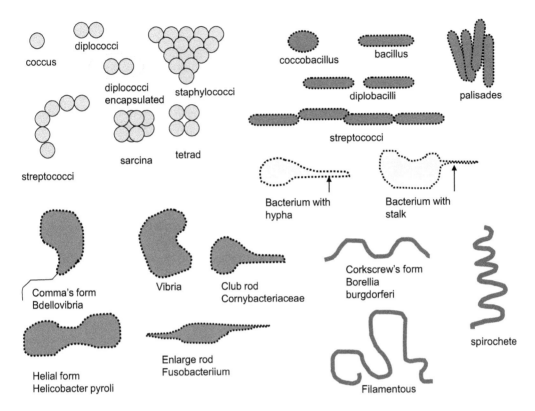

coccus

diplococci

diplococci encapsulated

staphylococci

sarcina

tetrad

streptococci

coccobacillus

bacillus

diplobacilli

palisades

streptococci

Bacterium with hypha

Bacterium with stalk

Comma's form
Bdellovibria

Vibria

Club rod
Cornybacteriaceae

Corkscrew's form
Borellia
burgdorferi

Helial form
Helicobacter pyroli

Enlarge rod
Fusobacteriium

Filamentous

spirochete

FIGURE 5.1   Family of microorganisms.

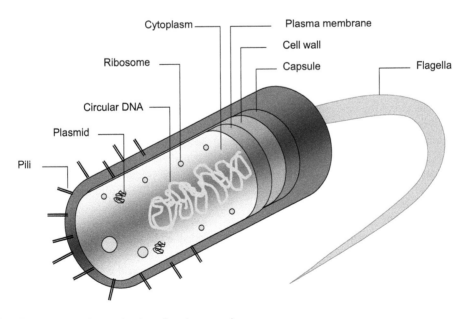

Cytoplasm

Plasma membrane

Cell wall

Ribosome

Capsule

Flagella

Circular DNA

Plasmid

Pili

FIGURE 5.2   Internal organization of a microorganism.

5.2.2 INTRACELLULAR ORGANIZATION

The bacterial cell is surrounded by a lipid membrane, or *cell membrane*, which encloses the contents of the cell and acts as a barrier to hold nutrients, proteins, and other essential components of the cytoplasm within the cell. As they are prokaryotes, bacteria do not tend to have membrane-bound organelles in their cytoplasm and thus contain few large intracellular structures (Figure 5.2). They consequently lack a nucleus, mitochondria, chloroplasts, and the other organelles present in eukaryotic cells, such as the Golgi apparatus and the endoplasmic reticulum (Figure 5.3). Bacteria were once seen as simple bags of cytoplasm, but elements such as a prokaryotic cytoskeleton and the localization of proteins to specific locations within the cytoplasm have been found to show levels of complexity. These subcellular compartments have been called *bacterial hyperstructures*. Microcompartments such as carboxysome provide a further level of organization; microcompartments are compartments within bacteria that are surrounded by polyhedral protein shells rather than by lipid membranes. These polyhedral organelles localize and compartmentalize bacterial metabolism, a function performed by the membrane-bound organelles in eukaryotes. Many important biochemical reactions, such as energy generation, occur by concentration gradients across membranes, a potential difference also found in a battery. The general lack of internal membranes in bacteria means reactions such as electron transport occur across the cell membrane between the cytoplasm and the periplasmic space. However, in many photosynthetic bacteria the plasma membrane is highly folded and fills most of the cell with layers of light-gathering membrane. These light-gathering complexes may even form lipid-enclosed structures called *chlorosomes* in green sulfur bacteria. Other proteins import nutrients across the cell membrane or expel undesired molecules from the cytoplasm.

Bacteria do not have a membrane-bound nucleus, and their genetic material is typically a single circular chromosome located in the cytoplasm in an irregularly shaped body called the *nucleoid*. The nucleoid contains the chromosome with associated proteins and RNA. The order *Planctomycetes* are an exception to the general absence of internal membranes in bacteria because they have a membrane around their nucleoid and contain other membrane-bound cellular structures. Like all living organisms, bacteria contain ribosomes to produce proteins, but the structure of the bacterial ribosome is different from those of eukaryotes and archaea. Some bacteria produce intracellular nutrient storage granules such as glycogen, polyphosphate, or sulfur. These granules enable bacteria to store compounds for later use. Certain bacterial species, such as the photosynthetic Cyanobacteria, produce internal gas vesicles, which they use to regulate their buoyancy, allowing them to move up or down into water layers with different light intensities and nutrient levels.

5.2.3 EXTRACELLULAR ORGANIZATION

Around the outside of the cell membrane is the bacterial *cell wall*. Bacterial cell walls are made of peptidoglycan, called *murein* in older sources, which is made from polysaccharide chains cross-linked by unusual peptides containing D-amino acids. Bacterial cell walls are different from the cell walls of plants and fungi, which are made of cellulose and chitin, respectively. The cell walls of bacteria are also distinct from that of archaea, which do not contain peptidoglycan. The cell wall is essential to the survival of many bacteria, and the antibiotic penicillin can kill bacteria by inhibiting a step in the synthesis of peptidoglycan. Broadly speaking, there are two different types of bacterial cell walls, called *Gram-positive* and *Gram-negative*. The names originate from the reaction of cells to the *Gram stain*, a test long employed for the classification of bacterial species. Gram-positive bacteria possess a thick cell wall containing many layers of peptidoglycan and teichoic acids. In contrast, Gram-negative bacteria have a relatively thin cell wall consisting of a few layers of peptidoglycan surrounded by a second lipid membrane containing lipopolysaccharides and lipoproteins. Most bacteria have the Gram-negative cell wall, and only the *Firmicutes* and *Actinobacteria* (previously known as the *low G+C Gram-positive bacteria* and *high G+C Gram-positive bacteria*, respectively) have the alternative Gram-positive arrangement (Figure 5.4). These

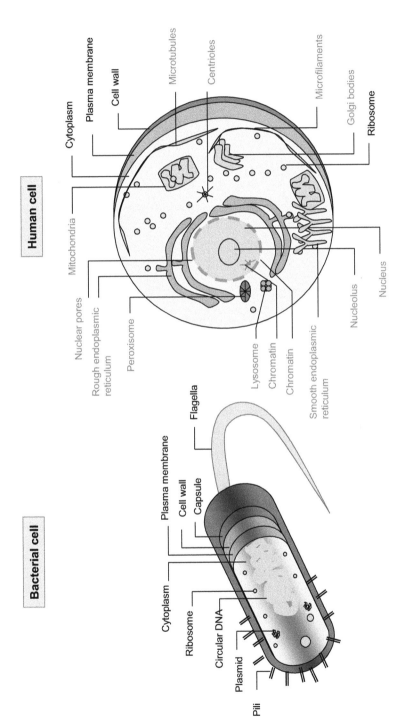

FIGURE 5.3 Similarity and difference between a bacterial cell and a human cell.

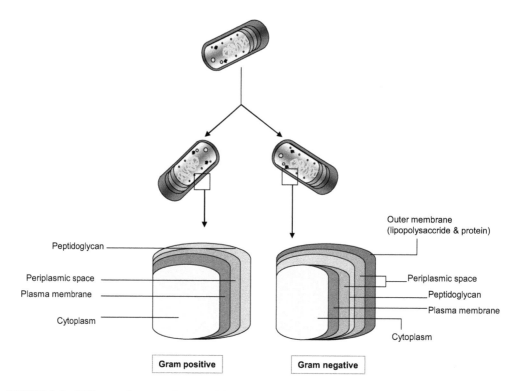

FIGURE 5.4   Difference between Gram positive and Gram-negative microbes.

differences in structure can produce differences in antibiotic susceptibility. For instance, vancomycin can kill only Gram-positive bacteria and is ineffective against Gram-negative pathogens such as *Haemophilus influenzae* or *Pseudomonas aeruginosa*.

In many bacteria, an S-layer of rigidly arrayed protein molecules covers the outside of the cell. This layer provides chemical and physical protection for the cell surface and can act as a macromolecular diffusion barrier. S-layers have diverse but mostly poorly understood functions, although they are known to act as virulence factors in *Campylobacter* and contain surface enzymes in *Bacillus stearothermophilus*. *Flagella* are rigid protein structures, approximately 20 nm in diameter and up to 20 μm in length, that are used for motility. Flagella are driven by the energy released by the transfer of ions down an electrochemical gradient across the cell membrane. *Fimbriae* are fine filaments of protein, just 2–10 nm in diameter and up to several micrometers in length. They are distributed over the surface of the cell and resemble fine hairs when seen under the electron microscope. Fimbriae are believed to be involved in attachment to solid surfaces or to other cells and are essential for the virulence of some bacterial pathogens. *Pili* (pilus in the singular) are cellular appendages slightly larger than fimbriae, which can transfer genetic material between bacterial cells in a process called *conjugation*.

5.3   MICROBIAL METABOLISM

Microbial metabolism is the process through which microorganisms obtain the energy and nutrients to live and reproduce. Microbes use many different types of metabolic strategies, and microbes can often be differentiated from each other based on metabolic characteristics. The specific metabolic properties of a microbe are the major factors in determining their usefulness in industrial applications.

5.3.1 Heterotrophic Microbial Metabolism

Most microbes are *heterotrophic* (more precisely, chemoorganoheterotrophic), which means they use organic compounds as both carbon and energy sources. Heterotrophic microbes live off nutrients that they scavenge from living hosts (e.g., commensals or parasites) or find in dead organic matter of all kinds (Figure 5.5). Microbial metabolism is the main contributor to the bodily decay of all organisms after death. Many eukaryotic microorganisms are heterotrophic by predation or parasitism, properties also found in some bacteria, such as *Bdellovibrio* (an intracellular parasite of other bacteria, causing the death of its victims) and Myxobacteria such as *Myxococcus* (predators of other bacteria that are killed and lysed by cooperating swarms of many single cells of Myxobacteria). Most pathogenic bacteria can be viewed as heterotrophic parasites of humans or of the other eukaryotic species they affect. Heterotrophic microbes are extremely abundant in nature and are responsible for the breakdown of large organic polymers such as cellulose, chitin, or lignin, which are generally indigestible to larger animals. Generally, the breakdown of large polymers to carbon dioxide (mineralization) requires several different organisms, with one breaking down the polymer into its constituent monomers, one able to use the monomers and excrete simpler waste compounds as byproducts, and one able to use the excreted wastes. There are many variations on this theme, as different organisms can degrade different polymers and secrete different waste products. Some organisms are even able to degrade more recalcitrant compounds such as petroleum compounds or pesticides, making them useful in bioremediation.

Biochemically, prokaryotic heterotrophic metabolism is much more versatile than that of eukaryotic organisms, although many prokaryotes share the most basic metabolic models with eukaryotes, such as using glycolysis (also called the Embden-Meyerhof-Parnas pathway) for sugar metabolism and the citric acid cycle to degrade acetate, producing energy in the form of ATP and reducing power in the form of nicotinamide adenine dinucleotide (NADH) or quinols. These basic pathways are well conserved, because they are also involved in biosynthesis of many conserved building blocks needed for cell growth (sometimes in the reverse direction). However, many bacteria and archaea utilize alternative metabolic pathways other than glycolysis and the citric acid cycle. A well-studied example is sugar metabolism via the keto-deoxy-phosphogluconate pathway (also called the Entner-Doudoroff pathway) in *Pseudomonas*. Moreover, there is a third alternative sugar-catabolic pathway used by some bacteria, the pentose-phosphate pathway. The metabolic diversity

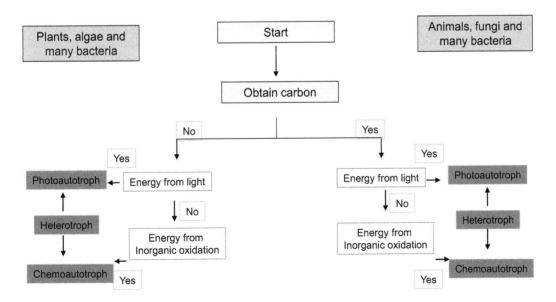

FIGURE 5.5   Heterotrophic metabolism in bacteria.

and ability of prokaryotes to use a large variety of organic compounds arises from the much deeper evolutionary history and diversity of prokaryotes as compared to eukaryotes. It is also noteworthy that the mitochondrion, the small membrane-bound intracellular organelle that is the site of eukaryotic energy metabolism, arose from the endosymbiosis of a bacterium related to obligate intracellular *Rickettsia* and to plant-associated *Rhizobium* or *Agrobacterium*. Therefore, it is not surprising that all mitrochondriate eukaryotes share metabolic properties with these *Proteobacteria*. Most microbes respire (use an electron transport chain), although oxygen is not the only terminal electron acceptor that may be used. As discussed subsequently, the use of terminal electron acceptors other than oxygen has important biogeochemical consequences.

5.3.2 Fermentation

Fermentation is a specific type of heterotrophic metabolism that uses organic carbon instead of oxygen as a terminal electron acceptor. This means that these organisms do not use an electron transport chain to oxidize NADH to NAD+, and therefore must have an alternative method of reducing and maintaining a supply of NAD+ for the proper functioning of normal metabolic pathways (e.g., glycolysis). As oxygen is not required, fermentative organisms are anaerobic. Many organisms can use fermentation under anaerobic conditions and anaerobic respiration when oxygen is not present. These organisms are facultative anaerobes. To avoid the overproduction of NADH, obligately fermentative organisms usually do not have a complete citric acid cycle. Instead of using an adenosine triphosphatase (ATPase) as in respiration, ATP in fermentative organisms is produced by substrate-level phosphorylation in which a phosphate group is transferred from a high-energy organic compound to adenosine diphosphate (ADP) to form ATP. Because of the need to produce high-energy, phosphate-containing organic compounds (generally in the form of CoA-esters), fermentative organisms use NADH and other cofactors to produce many different reduced metabolic by-products, often including hydrogen gas (H_2) (Figure 5.6). These reduced organic compounds are generally small organic acids and alcohols derived from pyruvate, the end product of glycolysis. Examples include ethanol, acetate, lactate, and butyrate. Fermentative organisms are very important industrially and are used to make many different types of food products. The different metabolic end products produced by each specific bacterial species are responsible for the different tastes and properties of each food.

Not all fermentative organisms use substrate-level phosphorylation. Instead, some organisms can couple the oxidation of low-energy organic compounds directly to the formation of a proton (or sodium) motive force, and therefore ATP synthesis. Examples of these unusual forms of fermentation include succinate fermentation by *Propionigenium modestum* and oxalate fermentation by *Oxalobacter formigenes*. These are extremely low-energy-yielding reactions. Humans and other higher animals also use fermentation to produce lactate from excess NADH, although this is not the major form of metabolism as it is in fermentative microorganisms.

5.3.3 Anaerobic Respiration

Although aerobic organisms use oxygen as a terminal electron acceptor during respiration, anaerobic organisms use other electron acceptors. These inorganic compounds have a lower reduction potential than oxygen, which means respiration is less efficient in these organisms and leads to slower growth rates than those of aerobes. Many facultative anaerobes can use either oxygen or alternative terminal electron acceptors for respiration, depending on the environmental conditions. Most respiring anaerobes are heterotrophs, although some do live autotrophically. All the processes described following are dissimilative, which means that they are used during energy production and not to provide nutrients for the cell (assimilative). Assimilative pathways for many forms of anaerobic respiration are also known.

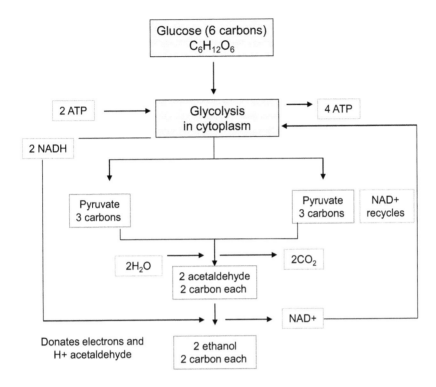

FIGURE 5.6 Bacterial metabolism through fermentation.

5.3.4 DENITRIFICATION

Denitrification is the utilization of nitrate (NO_3^-) as a terminal electron acceptor (Figure 5.7). It is a widespread process that is used by many members of Proteobacteria. Many facultative anaerobes use denitrification because nitrate, like oxygen, has a high reduction potential. Many denitrifying bacteria can also use ferric iron (Fe^{3+}) and some organic electron acceptors. Denitrification involves the stepwise reduction of nitrate to nitrite (NO_2^-), nitric oxide (NO), nitrous oxide (N_2O), and dinitrogen (N_2) by the enzymes nitrate reductase, nitrite reductase, nitric oxide reductase, and nitrous oxide reductase, respectively. Protons are transported across the membrane by the initial NADH reductase, quinones, and nitrous oxide reductase to produce the electrochemical gradient critical for respiration. Some organisms (such as *E. coli*) only produce nitrate reductase and therefore can accomplish only the first reduction, leading to the accumulation of nitrite. Others (such as *Paracoccus denitrificans* and *Pseudomonas stutzeri*) reduce nitrate completely. Complete denitrification is an environmentally significant process because some intermediates of denitrification (nitric oxide and nitrous oxide) are important greenhouse gases (GHGs) that react with sunlight and ozone to produce nitric acid, a component of acid rain. Denitrification is also important in biological wastewater treatment, where it is used to reduce the amount of nitrogen released into the environment.

5.3.5 NITROGEN FIXATION

Nitrogen is an element required for growth by all biological systems. Although extremely common in the atmosphere (80% by volume), dinitrogen gas (N_2) is generally biologically inaccessible because of its high activation energy. Throughout all of nature, only specialized bacteria and

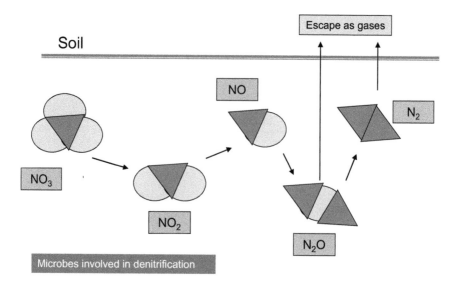

FIGURE 5.7   Denitrification cycle.

archaea are capable of nitrogen fixation, converting dinitrogen gas into ammonia (NH_3), which is easily assimilated by all organisms. These prokaryotes are therefore very important ecologically and are often essential for the survival of entire ecosystems. This is especially true in the ocean, where nitrogen-fixing cyanobacteria are often the only sources of fixed nitrogen, and in soils, where specialized symbioses exist between legumes and their nitrogen-fixing partners to provide the nitrogen needed by these plants for growth.

Nitrogen fixation can be found distributed throughout nearly all bacterial lineages and physiological classes but is not a universal property.

5.4   MICROBIAL CELL CULTURE AND GROWTH

Bacterial growth is the division of one bacterium into two daughter cells in a process called *binary fission*. Providing no mutational event occurs, the resulting daughter cells are genetically identical to the original cell. Hence, "local doubling" of the bacterial population occurs. Both daughter cells from the division do not necessarily survive. However, if the number of surviving cells exceeds unity on average, the bacterial population undergoes exponential growth. The measurement of an exponential bacterial growth curve in batch culture was traditionally a part of the training of all microbiologists. The basic means requires bacterial enumeration (cell counting) by direct and individual (microscopic, flow cytometry), direct and bulk (biomass), indirect and individual (colony counting), or indirect and bulk (most probable number, turbidity, nutrient uptake) methods. Models reconcile theory with the measurements.

5.4.1   PHASES OF MICROBIAL GROWTH

The bacterial growth in batch culture can be modeled with four different phases: lag phase, exponential or log phase, stationary phase, and death phase (Figure 5.8).

During the *lag phase*, bacteria adapt themselves to growth conditions. It is the period in which the individual bacteria are maturing and not yet able to divide. During the lag phase of the bacterial growth cycle, synthesis of RNA, enzymes, and other molecules occurs.

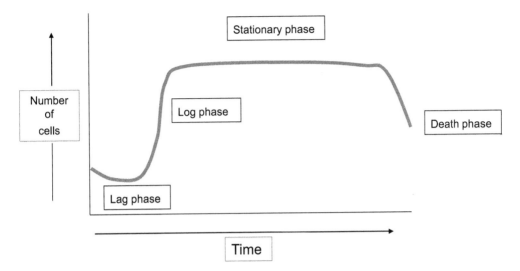

FIGURE 5.8   Phases of microbial growth.

The *exponential phase* (sometimes called the *log phase*) is a period characterized by cell doubling. The number of new bacteria appearing per unit time is proportional to the present population. If growth is not limited, doubling will continue at a constant rate so that both the number of cells and the rate of population increase doubles with each consecutive time. For this type of exponential growth, plotting the natural logarithm of cell number against time produces a straight line. The slope of this line is the specific growth rate of the organism, which is a measure of the number of divisions per cell per unit time. The actual rate of this growth (i.e., the slope of the line in the figure) depends upon the growth conditions, which affect the frequency of cell division events and the probability of both daughter cells surviving. Exponential growth cannot continue indefinitely, however, because the medium is soon depleted of nutrients and enriched with wastes.

During the *stationary phase*, the growth rate slows because of nutrient depletion and accumulation of toxic products. This phase is reached as the bacteria begin to exhaust the resources that are available to them. This phase is a constant value, as the rate of bacterial growth is equal to the rate of bacterial death.

At the *death phase*, bacteria run out of nutrients and die.

5.4.2   Factors That Influence Microbial Growth

Microorganisms can grow under many different conditions and there are six main factors that can affect the growth of microorganisms:

Food: Microorganisms grow best in foods that are high in protein or carbohydrates, such as meat, poultry, seafood, milk, rice, and eggs.

pH (Acid): The measure of acidity or alkalinity of a food also affects the growth of microorganisms. Most disease-causing bacteria multiply best at a pH of 5–8, which is near the neutral pH of 7. Fresh foods such as meat, seafood, and milk tend to have a pH near 7 (neutral).

Temperature: Food-poisoning microorganisms can multiply rapidly at temperatures between 4°C (40°F) and 60°C (140°F). This is known as the food temperature danger zone. Hazardous foods should spend as little time as possible in the danger zone. Hot foods should be kept hot (above 60°C or 140°F) and cold foods cold (below 4°C or 40°F).

Time: Microorganisms often need time to grow in the food and they can double in number every 20 min under ideal conditions.

Oxygen: Some microorganisms will only grow when there is oxygen present in the food or environment (aerobic organisms). On the other hand, some microorganisms will only grow when there is **no** oxygen present in the food or environment (anaerobic organisms).

Moisture: Microorganisms need water to grow and multiply. However, some microorganisms can survive when there is little water, although they will not be able to grow very well.

5.5 MICROBIAL GENETICS

Most bacteria have a single circular chromosome that can range in size from only 159,662 bp in the endosymbiotic bacteria *Candidatus Carsonella ruddii* to 13,033,779 bp in the soil-dwelling bacteria *Sorangium cellulosum* (Figure 5.9). Spirochetes of the genus *Borrelia* are a notable exception to this arrangement, with bacteria such as *Borrelia burgdorferi*, the cause of Lyme disease, containing a single linear chromosome. The genes in bacterial genomes are usually a single continuous stretch of DNA, and although several different types of introns do exist in bacteria, these are much rarer than in eukaryotes. Its length is found to be 1100 μm and its molecular weight is 2.6×10^9 Da. In addition to this major chromosome, an *E. coli* cell often possesses one or more minor chromosomes, each called a plasmid, which may contain 0.5%—2% of the DNA of the cell. Plasmids usually maintain a distinct existence from the main chromosome and replicate independently of it. Thus, in *E. coli*, the transmission of more than one genetic element from parent to offspring must frequently be followed. Finally, though bacteria reproduce chiefly by asexual reproductive means, there are several different avenues by which DNA from one bacterial cell can undergo genetic exchange with the DNA from another bacterial cell.

5.5.1 MUTATIONS

Mutation is a natural phenomenon resulting in variations within any population of cells. This is a change in the DNA sequence of a gene and is said to lead to a change in the genotype of the organism. There are various types of mutants found in microorganisms (Figure 5.10).

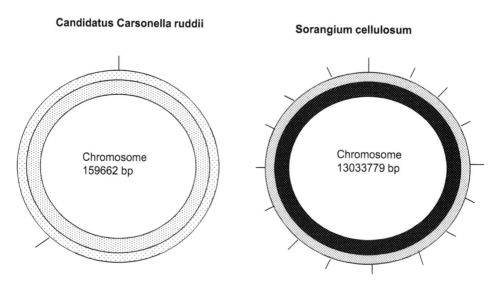

FIGURE 5.9 Chromosomes in different microbes.

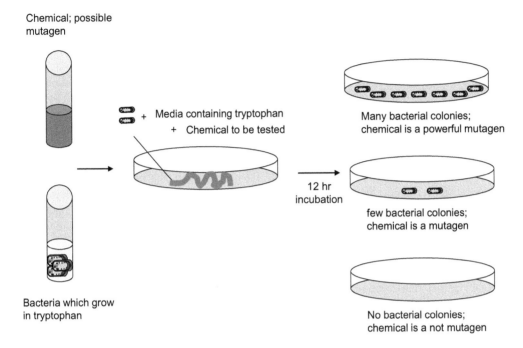

Chemical; possible mutagen

Media containing tryptophan
+ Chemical to be tested

Many bacterial colonies; chemical is a powerful mutagen

12 hr incubation

few bacterial colonies; chemical is a mutagen

Bacteria which grow in tryptophan

No bacterial colonies; chemical is a not mutagen

FIGURE 5.10   Mutagenic test using microbes.

5.6   GENETIC RECOMBINATION IN BACTERIA

Genetic changes owing to mutations can result in the acquisition of new biological characteristics and thereby allow evolutionary change. However, the evolution of the fittest organism in an environment can be enhanced if transfer of genes between organisms is made possible by genetic recombination. As compared with eukaryotes, where sexual recombination is of an ordered nature, the process is less well-developed in prokaryotes. It does not involve a true fusion of male and female gametes to produce a diploid zygote; instead, there is transfer of only some genes from the donor cell to produce a partial diploid. This is followed by recombination to restore the haploid state. There are three mechanisms by which these DNA fragments can pass from a donor to a recipient cell: transformation, transduction, and conjugation.

5.6.1   BACTERIAL TRANSFORMATION

In the 1940s, it was recognized that inheritance in bacteria was basically governed by the same mechanisms as those in higher eukaryotic organisms. It was also realized that bacteria represent a useful tool to understand the mechanism of heredity and genetic transfer and were therefore being increasingly used in genetic studies. The first observation that bacterial properties can be changed using heat-inactivated cell material was, however, discovered in 1928 by Frederick Griffith. Griffith found that in *Streptococcus pneumoniae* (earlier called *Pneumococcus*), virulence to mice was related to the presence of a capsular material, and loss of the ability to produce the capsule made the bacteria virulent.

Mutants lacking the capsular material were designated as rough (R), because colonies formed by these on solid media appeared rough, as opposed to the colonies formed by the virulent capsule-forming strains that were smooth and shining (S). Griffith's experiments involved the infection of mice with heat-killed and living preparations from two different strains of *Streptococcus pneumoniae*. When he injected the mice with either the dead S cells or a small number of living R cells, no death occurred. However, when the mice were injected with a mixture of dead S cells and a small number of live R cells, the mice died (Figure 5.11).

From these experiments, he concluded that the dead S cells that contained the capsule contributed to the killing effect by the R cells, as neither of the preparations was effective by itself. Although these observations were not well understood at that time, the term "transformation" was used to describe this phenomenon whereby one type of cell was converted by contact with the dead cells of a second strain. The material responsible for causing transformation was thought to be the capsular polysaccharide.

However, in 1943, the material responsible for bringing about this change was identified. It was left to Avery, McLeod, and McCarty in 1944 to identify the transforming principle in capsulated cells as the DNA. Their studies with purified DNA from the smooth cells of *S. pneumoniae* and its ability to transform rough cells in a test tube explained the observations made by Griffith in 1928. It was then possible to conclude that the heat-killed encapsulated cells carried the information for the synthesis of the capsule that was transferred to the live non-capsulated cells. Consequently, cells that received the genetic material for capsule formation became encapsulated and virulent.

At that time, this remarkable finding did not receive as much attention as it should have, because most believed that proteins rather than nucleic acids were the genetic elements and those proteins in the DNA preparations were responsible for bringing about transformation. Since then, however, using highly purified DNA preparations and other genetic markers, it has been shown beyond doubt that the transforming principle is DNA and not protein. The process of transformation has now also been demonstrated in several other bacteria, such as *Bacillus subtilis*, *Haemophilus influenzae*, *Rhizobium*, *E. coli*, *Streptococcus*, and *Streptomyces*. The process of transformation in all these organisms has two common features: (i) the purified donor DNA is first transported across the cell membrane into the recipient "competent cells" (cells that can take up DNA), and (ii) the DNA then undergoes recombination with the recipient DNA and is then expressed (Figure 5.12).

The uptake process is apparently not very specific as it has been found that even calf thymus DNA can be taken in by bacterial cells, but the subsequent process of integration is highly specific. Although the double-stranded DNA is necessary for transformation, single-stranded DNA can also

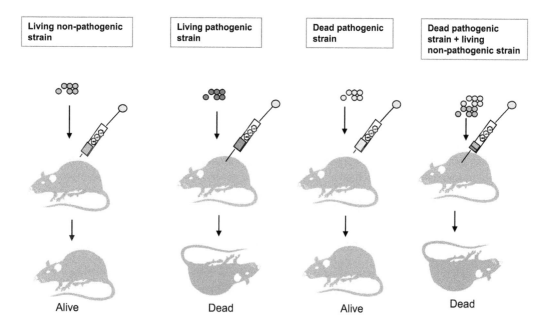

FIGURE 5.11 Genetic transformation experiment by Griffith.

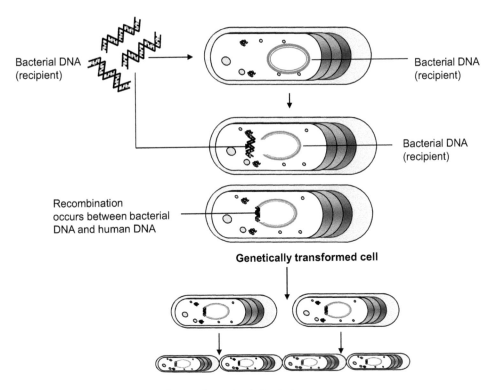

Bacterial DNA (recipient)

Bacterial DNA (recipient)

Bacterial DNA (recipient)

Recombination occurs between bacterial DNA and human DNA

Genetically transformed cell

FIGURE 5.12   Mechanism of genetic transformation in bacteria.

penetrate bacterial cells. Following uptake by the recipient cells, the transforming DNA immediately undergoes modifications and an "eclipse" period, lasting for a few minutes, is seen.

5.6.2   BACTERIAL TRANSDUCTION

Bacterial transduction is a process by which the genetic material in bacteria is transferred from one cell to another through the mediation of bacterial viruses (Figure 5.13). This process was first discovered by Norton Zinder and Joshua Lederberg in 1952 during their experiments to see whether the process of conjugation existed in *Salmonella*. In performing the "D" tube experiments, they found that the recombinants appeared only in one arm of the tube without cell contact. Also, cell-free filtrates from one culture could yield recombinants when mixed with the other. The active factor in the filtrate was, however, resistant to DNase, and this ruled out transformation involving DNA.

Bacteriophages are viruses that parasitize bacteria and use their machinery for their own replication. During the process of replication inside the host bacteria, the bacterial chromosome or plasmid is erroneously packaged into the bacteriophage capsid. Thus, newer progeny of phages may contain fragments of the host chromosome along with their own DNA or entirely contain the host chromosome. When such a phage infects another bacterium, the bacterial chromosome in the phage also gets transferred to the new bacterium. This fragment may undergo recombination with the host chromosome and confer a new property to the bacterium. The life cycle of bacteriophages may be either lytic or lysogenic. In the former, the parasitized bacterial cell is killed with the release of mature phages, whereas in the latter, DNA gets incorporated into the bacterial chromosome as a prophage.

It was subsequently proved that the active component was a bacteriophage that was carried by one of the strains in the prophage condition. Some bacteria can carry phage DNA within their own DNA, and such bacteria are known as *lysogenic bacteria*. In lysogenic bacteria, the prophage

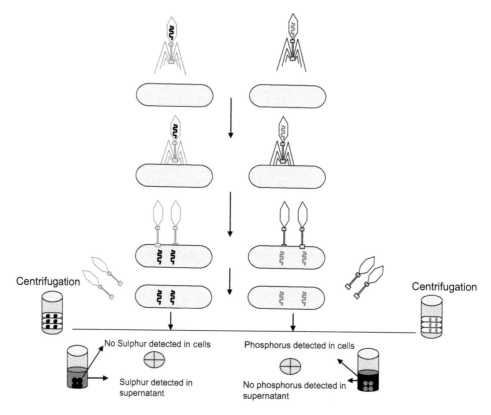

FIGURE 5.13   Genetic transformation by virus transduction.

becomes active under certain conditions. It multiplies and destroys the host cell with the release of several phage particles. The phage particles released from a small number of bacterial cells attack sensitive cells, multiply, and release more phage particles. The lysogenic strains, however, are resistant to the same phage that they carry. Sometimes, when the prophage is released as the vegetative phage, it also carries a small fragment of the host DNA in addition to its own DNA. These phages can infect other bacteria and carry the bacterial DNA to the recipient cells. Such phages are called *transducing phages* and act as carriers of bacterial DNA from one cell to another. The size of the DNA transferred by transduction is small when compared to either transformation or conjugation, and the amount of DNA is generally <1% of the bacterial genome. This technique is therefore useful only in determining the relative positions of very closely located markers and mapping regions within a gene.

5.6.2.1   Stages of Transduction

The following are the stages of transduction:

1. A lytic bacteriophage adsorbs to a susceptible bacterium.
2. The bacteriophage genome enters the bacterium. The phage DNA directs the bacterium's metabolic machinery to manufacture bacteriophage components and enzymes.
3. Occasionally during maturation, a bacteriophage capsid erroneously incorporates a fragment of the donor bacterium's chromosome or a plasmid instead of a phage genome.
4. The bacteriophages are released with the lysis of the bacterium.
5. The bacteriophage carrying the donor bacterium's DNA adsorbs to another recipient bacterium.

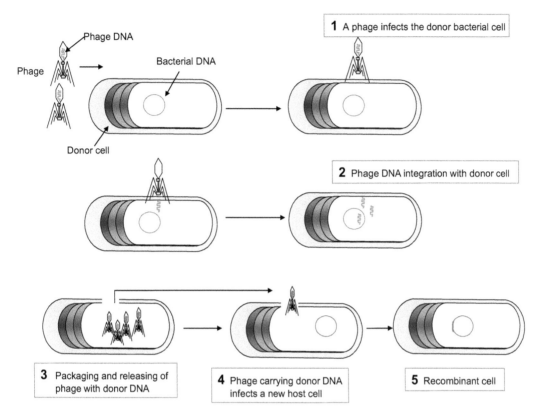

FIGURE 5.14 Genetic transformation transduction process.

6. The bacteriophage inserts the donor bacterium's DNA that it is carrying into the recipient bacterium.

7. The donor bacterium's DNA is exchanged by recombination for some of the recipient's DNA (Figure 5.14).

5.6.3 Conjugation Mechanism in Gene Recombination

Literature in bacterial morphology contains many descriptions of microscopic observations of cell pairs that were identified as indicators of mating and sexuality in bacteria. However, no confirmatory genetic evidence was available till the discovery of conjugation in *E. coli* by Lederberg and Tatum in 1946, in a study in which they mixed auxotrophic mutants and selected rare recombinants. In their initial experiments, Lederberg and Tatum plated *E. coli* mutants having triple and complementary nutritional requirements (abcDEF × ABCdef) on minimal agar and obtained prototrophic bacteria (ABCDEF).

These recombinants were stable, could be propagated, and arose at a frequency of 10^{-6}, 10^{-7}. Further evidence to show that the development of prototrophic colonies required the cooperation of intact bacteria of both types was obtained by the "U" tube experiments. Neither the culture filtrates nor the cell-free culture extracts were productive, suggesting that actual cell contact was necessary. Lederberg also examined many of the prototrophic colonies to learn whether the process was reciprocal. He found that most colonies contained only one class of recombinants, suggesting that recombination in bacteria may be of an unorthodox kind. In addition, detailed analysis of prototrophs showed an initial heterozygous nature that later was converted to haploids.

These studies by Lederberg and his colleagues proved that bacteria possess sex, which made them amenable to formal genetic analysis and also revealed the existence of genetic material in a

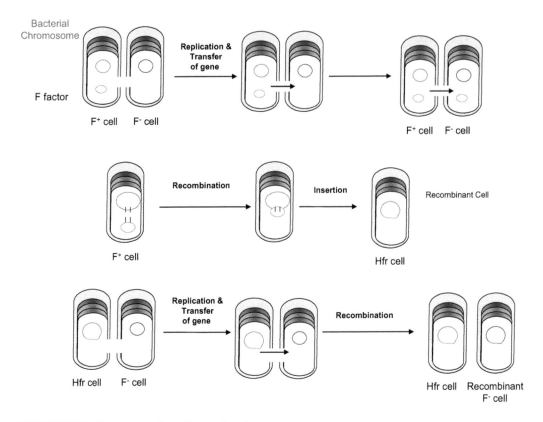

FIGURE 5.15   Genetic transformation conjugation process.

chromosomal organization. Subsequent studies carried out to determine the size of the DNA fragment involved, by detecting the number of genetic markers transferred, suggested that more than
one marker could be transferred at a time and, interestingly, the linkage between certain markers
was always seen. In this process of conjugation, it was concluded that (i) large fragments of DNA
were transferred from one bacterium to another in a nonreciprocal manner, and (ii) transfer always
occurred from a given point. It was also found that the size of the DNA transferred from one cell to
another was much larger than in transformation, and this technique appeared to be a more useful
technique for gene mapping in bacteria. The bacteria that transfer DNA are called *donor bacteria*,
whereas those that receive the DNA are called *recipient bacteria* (Figure 5.15).

5.7   TRANSPOSABLE GENETIC ELEMENTS

Transposable genetic elements are segments of DNA that have the capacity to move from one
location to another (i.e., jumping genes). Transposable genetic elements can move from any DNA
molecule to any DNA of another molecule, or even to another location on the same molecule.
The movement is not totally random. There are preferred sites in a DNA molecule at which the
transposable genetic element will insert. The transposable genetic elements do not exist autonomously; thus, to be replicated, they must be a part of some other replicon. Transposition requires
little or no homology between the current location and the new site. The transposition event
is mediated by an enzyme, transposase, which is coded by the transposable genetic element.
Recombination that does not require homology between the recombining molecules is called *ille-
gitimate* or *nonhomologous recombination*. In many instances, transposition of the transposable

genetic element results in the removal of the element from the original site and insertion at a new site. However, in some cases, the transposition event is accompanied by the duplication of the transposable genetic element. One copy remains at the original site and the other is transposed to the new site.

5.7.1 Types of Transposable Genetic Elements

5.7.1.1 Insertion Sequences

Insertion sequences are transposable genetic elements that carry no known genes except those that are required for transposition. Insertion sequences are small stretches of DNA that have repeated sequences, which are involved in transposition, at their ends. In between the terminal repeated sequences there are genes involved in transposition and sequences that can control the expression of the genes, but no other nonessential genes are present. The introduction of an insertion sequence into a bacterial gene will result in the inactivation of the gene. The sites at which plasmids insert into the bacterial chromosome are at or near the insertion sequence in the chromosome. In *Salmonella*, there are two genes that code for two antigenically different flagellar antigens. The expression of these genes is regulated by an insertion sequence.

5.7.1.2 Transposons

Transposons are transposable genetic elements that carry one or more other genes in addition to those that are essential for transposition. The structure of a transposon is like that of an insertion sequence; the extra genes are located between the terminal repeated sequences. Many antibiotic resistance genes are located on transposons. Because transposons can jump from one DNA molecule to another, these antibiotic-resistant transposons are a major factor in the development of plasmids, which can confer multiple drug resistance on a bacterium harboring such a plasmid. These multiple drug resistance plasmids have become a major medical problem.

5.8   USE OF *E. COLI* IN MICROBIAL CLONING

The microorganism *Escherichia coli* has a long history of use in the biotechnology industry and is still the microorganism of choice for most gene cloning experiments. Although *E. coli* is known to the general population for the infectious nature of one strain (0157:H7), few people are aware of how versatile and useful *E. coli* is to genetic research. There are several reasons why *E. coli* became so widely used and is still a common host for recombinant DNA.

5.8.1 Genetic Simplicity

Bacteria make useful tools for genetic research because of their relatively small genome size compared to eukaryotes. *E. coli* cells only have about 4,400 genes, whereas the human genome project has determined that humans contain approximately 30,000 genes. Also, bacteria, including *E. coli*, live their entire lifetime in a haploid state with no second allele to mask the effects of mutations during protein engineering experiments.

5.8.2 Growth Rate

Bacteria typically grow much faster than more complex organisms. *E. coli* grows rapidly, at a rate of one generation per 20 min under typical growth conditions. This allows for preparation of log-phase (mid-way to maximum density) cultures overnight and genetic experimental results in mere hours instead of several days, months, or years. Faster growth also means better production rates when cultures are used in scaled-up fermentation processes.

5.8.3 Safety

E. coli is naturally found in the intestinal tracts of humans and animals, where it helps provide nutrients (vitamins K and B12) to its host. There are many different strains of *E. coli* that may produce toxins or cause varying levels of infection if ingested or allowed to invade other parts of the body. Despite the bad reputation of one particularly toxic strain (O157:H7), *E. coli* is generally relatively innocuous if handled with reasonable hygiene.

5.8.4 Conjugation and the Genome Sequence

The *E. coli* genome was the first to be completely sequenced. Genetic mapping in *E. coli* was made possible by the discovery of conjugation. *E. coli* is the most highly studied microorganism, and an advanced knowledge of its protein expression mechanisms makes it simpler to use in experiments where expression of foreign proteins and selection of recombinants is essential.

5.8.5 Ability to Host Foreign DNA

Most gene cloning techniques were developed using *E. coli*, and these techniques are still more successful and effective when using *E. coli* rather than other microorganisms. *E. coli* is readily transformed with plasmids and other vectors, easily undergoes transduction, and preparation of competent cells (cells that will take up foreign DNA) is not complicated. Transformations with other microorganisms are often less successful.

5.9 PATHOGENIC BACTERIA

If bacteria form a parasitic association with other organisms, they are classed as *pathogens*. Pathogenic bacteria are a major cause of human death and disease and cause infections such as tetanus, typhoid fever, diphtheria, syphilis, cholera, foodborne illness, leprosy, and tuberculosis. A pathogenic cause for a known medical disease may only be discovered after many years, as was the case with *Helicobacter pylori* and peptic ulcer disease. Bacterial diseases are also important in agriculture, with bacteria causing leaf spot, fire blight, and wilts in plants, as well as John's disease, mastitis, salmonellosis, and anthrax in farm animals.

Each species of pathogen has a characteristic spectrum of interactions with its human hosts. Some organisms, such as *Staphylococcus* or *Streptococcus*, can cause skin infections, pneumonia, meningitis, and even overwhelming sepsis, a systemic inflammatory response producing shock, massive vasodilation, and death. Yet these organisms are also part of the normal human flora and usually exist on the skin or in the nose without causing any disease at all. Other organisms invariably cause disease in humans, such as *Rickettsia*, which are obligate intracellular parasites able to grow and reproduce only within the cells of other organisms. One species of *Rickettsia* causes typhus and another causes Rocky Mountain spotted fever. Chlamydiae, another phylum of obligate intracellular parasites, contains species that can cause pneumonia or urinary tract infection and may be involved in coronary heart disease. Finally, some species such as *Pseudomonas aeruginosa*, *Burkholderia cenocepacia*, and *Mycobacterium avium* are opportunistic pathogens and cause disease mainly in people suffering from immunosuppression or cystic fibrosis.

Bacterial infections may be treated with antibiotics, which are classified as bactericidal if they kill bacteria or bacteriostatic if they just prevent bacterial growth. There are many types of antibiotics, and each class inhibits a process that is different in the pathogen from that found in the host. Examples of how antibiotics produce selective toxicity are chloramphenicol and puromycin, which inhibit the bacterial ribosome but not the structurally different eukaryotic ribosome. Antibiotics are used both in treating human disease and in intensive farming to promote animal growth, where they may be contributing to the rapid development of antibiotic resistance in bacterial populations.

Infections can be prevented by antiseptic measures, such as sterilization of the skin prior to piercing it with the needle of a syringe, and by proper care of indwelling catheters. Surgical and dental instruments are also sterilized to prevent contamination by bacteria. Disinfectants such as bleach are used to kill bacteria or other pathogens on surfaces to prevent contamination and further reduce the risk of infection.

5.10   APPLICATION OF MICROBES

The economic importance of bacteria derives from the fact that bacteria are exploited by humans in several beneficial ways. Even though some bacteria play harmful roles, such as causing disease and spoiling food, the economic importance of bacteria includes both their useful and harmful aspects. Applications of microbes in industry are well known. Various microorganisms are used for commercial production of alcohols, acids, fermented foods, vitamins, medicines, enzymes, etc. One recent development in industrial microbiology has been the production of immobilized enzymes and cells for production of these chemicals at enhanced rates, with simultaneous recovery of the enzyme(s) involved in such processes. New strains of microbes have also been developed through recombinant DNA technology for overproduction of metabolites. Immobilized enzymes and cells could have their maximum application in industrial microbiology. Immobilized enzymes have also been utilized in medicine. In view of their various applications, microbes have become the most sought-after entities for various applications.

5.10.1   AGRICULTURE

Besides being important in the biogeochemical cycling of nutrients, microbes play a vital role in the maintenance of soil fertility and in crop protection. Microbes are being exploited in two important ways: as biofertilizers and for creating new nitrogen-fixing organisms. The potential of *Rhizobium, Azotobacter, Beijerinckia, Azospirillum,* and Cyanobacteria—such as species of *Aulosira, Anabaena, Nostoc, Plectonema, Scytonema, Tolypothrix,* and *Azalia*—as biofertilizers has been exploited so that these could serve as alternatives to chemical fertilizers. Many brands of rhizobial inoculants are already used in Indian market today, and several organizations and manufacturers are producing huge quantities of *Rhizobium* culture. These include Micro Bac India, Shyam Nagar, Parganas; Bacifil Inoculants, Lucknow; Govt. of Tamil Nadu; Nitro Fix Industries, Calcutta (W. Bengal); and Indian Organic Chemical Ltd., Bombay. In some other states, units are being prepared for an increase in the production of *Rhizobium*. Much progress has also been made with cyanobacteria in this direction. Mycorrhizae, both ecto and endomycorrhiza, help in the uptake of N, P, K, and Ca. They particularly help in phosphorous nutrition.

5.10.2   NITROGEN FIXERS

Through recombinant DNA technology, efforts have been made to introduce nitrogen-fixing (nif) genes into wheat, corn, rice, etc. Plasmids of the bacterium *E. coli* and yeast are being worked on for such a possibility. Hybrid *E. coli* plasmids cloned with nif genes of a nitrogen-fixing bacterium *Klebsiella pneumoniae* and hybrid yeast plasmids are then integrated.

5.10.3   BIOPESTICIDES AND BIOWEEDICIDES

Several microbes are being developed as suitable biopesticides for management of insect and nematode pests. Some fungi have good potential for their use as bionematicides to control nematode pests of vegetables, fruit, and cereal crops. Some bacterial and fungal products are also in use to control diseases of roots and shoots of plants. Several fungi have been found to be very useful in the control of troublesome weeds in crop fields. Registered products are available for use in the market in several countries. The bacteria are used as a Lepidopteran-specific insecticide under trade names such as Dipel and Thuricide. Because of their specificity, these pesticides are regarded as environmentally friendly with little or no effect on humans, wildlife, pollinators, and most other beneficial insects. The bacterium *Bacillus thuringiensis* has been the most successful bioinsecticide so far. Several registered products of different strains of this microbe, such as Thuricide, are available for the control of insects, including mosquitoes—the carriers of malaria.

5.10.4 Acetone Butanol Fermentations

The *acetone butanol fermentation* is one of the oldest types of fermentation known. The fermentation is based on culturing various strains of clostridia in carbohydrate-rich media under anaerobic conditions to yield butanol and acetone. *Clostridium acetobutylicum* is the organism of choice in the production of these organic solvents. These fermentations were out of favor until very recently because of the availability of acetone and butanol from the petroleum industry. Today, there is a considerable amount of interest in these fermentations. However, the concentration of end products in these fermentations is quite small, and the fermentations are a type of mixed fermentation yielding a mixture of compounds such as butyric acid, butanol, acetone, etc. Attempts to increase yields by using genetically altered strains or by changing the fermentation conditions have been partially successful.

5.10.5 Recovery of Metals and Petroleum

Recently, microbes have been found to be very useful in enhanced recovery of metals, including uranium from low-grade ores. Through bioleaching, these microbes are able to solubilize the metals from their ores. Microbes thus play an important role in mining and recovery of metals. For instance, *Thiobacillus thiooxidans* and *T. ferrooxidans* can be used in the recovery of copper. Microbes are used in tertiary recovery of petroleum. For instance, the bacterium *Xanthomonas campestris* is being exploited for this purpose. Some thiobacilli have also been found to have this potential.

5.10.6 Paper Industry

Mechanical pulping in the process of manufacturing paper from wood needs much energy and does not preserve the quality of the product. Therefore, the potential of some lignin-decomposing fungi (lignolytic fungi) has been exploited for this process. Biological pulping by use of these higher fungi (Basidiomycotina) could find application in the paper industry. *Phanerochaete chrysosporium* has been mostly studied for its use in the biopulping process. Other lignolytic fungi found suitable for this process are *Pholiota mutabilis*, *Tremetes versicolor*, and *Phlebia* spp.

5.10.7 Medicine

The production of antibiotics and other chemotherapeutic agents by a range of microbes is well-known. Recent development in this area has been the use of microbial biotechnology in steroid transformations and biotransformation of natural penicillin G to several semi-synthetic penicillins. Penicillin acylase, produced by *Saccharomyces cerevisiae* and *Kluyvera citrophila*, is used in biotransformation of penicillin G to semi-synthetic penicillins. Microbial transformation of steroids is very important in the pharmaceutical industry. *Rhizopus nigricans* hydroxylates progesterone, forming another steroid. *Cunninghamella blakesleeana* hydroxylates cortexolone to hydrocortisone.

Microorganisms are used to produce insulin, growth hormone, and antibodies. Diagnostic assays that use monoclonal antibodies, DNA probe technology, or real-time PCR are used as rapid tests for pathogenic organisms in the clinical laboratory. Microorganisms may also help in the treatment of diseases such as cancer. Research shows that clostridia can selectively target cancer cells. Various strains of nonpathogenic clostridia have been shown to infiltrate and replicate within solid tumors. Clostridia therefore have the potential to deliver therapeutic proteins to tumors. *Lactobacillus* spp. and other lactic acid bacteria possess numerous potential therapeutic properties, including anti-inflammatory and anticancer activities.

5.10.8 SYNTHETIC FUELS

Several microbes have been found to be helpful in solving the energy crisis. Some synthetic fuels produced by the activity of microbes include ethanol, methane, hydrogen, and hydrocarbons. Gasohol, a 9:1 blend of gasoline and ethanol, is a popular fuel in the United States. The most efficient microbes used for synthetic fuels are *Zymomonas mobilis* and *Thermoanaerobacter ethanolicus*. Methane is produced by methanogenic bacteria and biogas. A mixture of CH_4, CO_2, H_2, N_2, and O_2 is produced during fermentation of cattle dung by several bacteria, including methanogens.

5.10.9 ENVIRONMENTAL CLEANING

Microorganisms play several key roles in the environment. Besides their well-known activities in biogeochemical cycling, soil fertility maintenance, etc., several microbes may prove to be very helpful in the maintenance of environmental quality through biodegradation of wastes (urban, municipal, and industrial) into useful products and also in biodegradation of harmful pesticides used in crop protection and public health.

DOT, lindane heptachlor, chlordane, and malathion are biodegraded by several bacteria and fungi. These microbes are efficient purifiers of the environment. Aside from this, microbes can also remove toxic heavy metals from industrial waste. Some bacteria can also metabolize the hydrocarbons in petroleum and are thus very useful in removal of oil spills and grease from water bodies. In the United States, a strain of *Pseudomonas aeruginosa* has been developed that can produce a glycolipid emulsifier that reduces the surface tension of an oil–water interface, thus removing oil from water. For removal of grease deposits, a mixture of several bacteria is used. Some microorganisms can also be used in environmental monitoring and biomonitoring. Environmental pollutants can be detected by use of appropriate strains of microbes as biosensors. *Biosensor* is a biophysical device used to detect the presence and quantify the specific substances (sugars, proteins, hormones, pollutants) in a specific environment.

5.11 FOOD MICROBIOLOGY

Most foods are excellent media for rapid growth of microorganisms. There is abundant organic matter in foods, their water content is usually sufficient, and the pH is either neutral or slightly acidic. Foods consumed by humans and animals are ideal ecosystems in which bacteria and fungi can multiply. The mere presence of microorganisms in foods in small numbers need not be harmful, but their unrestricted growth may render the food unfit for consumption and can result in spoilage or deterioration. Some organisms grow and produce secondary metabolites that may affect food quality, which may be either desirable or undesirable. For example, the lactic fermentation of milk is a desired change and is not considered spoilage, whereas acidification of wine is an undesirable microbial spoilage. Some organisms may not only cause food spoilage but also produce metabolites that may be extremely toxic to man and animals. Some examples are the production of toxins by clostridia in proteinaceous foods and the production of aflatoxin by *Aspergillus* in feeds. Generally, foods carry a variety of organisms, most of which are saprophytic. Their presence cannot be avoided as these are mostly from the environment in which the food is prepared or processed. In addition, their complete elimination is difficult. However, it is possible to reduce their number or decrease their activities by altering environmental conditions. Knowledge of the factors that either favor or inhibit their growth is therefore essential in understanding the principles of food spoilage and preservation.

5.11.1 Microbes Associated with Food Spoilage

Fruits, vegetables, meat, poultry, seafood, dairy products, and various food products differ in their biochemical composition and are therefore subject to spoilage by different microbial populations. Such changes depend upon the nature of the microbes involved in the spoilage. Thus, degradation of apple juice by yeast gives an alcoholic taste to the juice. Yeasts convert the carbohydrate into ethanol. Bacteria that attack food proteins convert these into amino acids, which are broken down again into foul-smelling end products. Digestion of cysteine, for example, yields hydrogen sulfide, giving a rotten egg smell to food. Digestion of tryptophan yields indole and skatole, which give food a fecal odor. Two other products of the microbial metabolism of carbohydrates are (i) acid that causes foods to become sour and (ii) gases that cause sealed cans to swell. Digestion of fats, as in spoiled butter, yields fatty acids, giving a rancid odor or taste to food. Food may become slimy because of the production of capsules in bacteria. There may be pigment development, giving an odd color to foods.

5.11.2 Meat and Fish

Microorganisms that cause meat and fish spoilage are usually introduced during handling, processing, packaging, and storage. For example, if a piece of meat is ground, the surface organisms accumulate in the teeth of the grinder along with other dust-borne organisms. Bacteria from the hands of the preparer or from an errant sneeze may add more microbes. Processed meats may become contaminated during handling. For example, sausages, which are made from animal intestines, may contain residual bacteria, especially botulism spores. Organ meats such as liver and kidney spoil quickly and may contain many bacteria trapped in their filtering tissues. Greening on meat surfaces is usually because of the Gram-positive rod, *Lactobacillus*, or the Gram-positive coccus, *Leuconostoc*.

5.11.3 Poultry and Eggs

There are reports that suggest that *Salmonella* can cause diseases in humans following consumption of infected chickens and turkeys or poultry or eggs. Processed foods such as potpies, egg salad, and omelets may be sources of salmonellosis. Chicken products may become contaminated because of improper handling or from the water used for cleaning the product. Eggs may become contaminated by *Proteus* causing black rot (here, hydrogen sulfide accumulates because of digestion of the egg cysteine), by *Pseudomonas* causing green rot, and by *Serratia marcescens* causing red rot. The yolk is the main part of the egg prone to contamination (the white of an egg is inhibitory to Gram-positive bacteria because of the presence of the inhibitor enzyme lysozyme).

5.11.4 Breads and Bakery Products

The ingredients of bread products such as flour, egg, sugar, and salt are usually the sources of spoilage organisms. Some bacteria and molds are able to survive the baking temperatures. Some species of *Bacillus* give the bread a soft and cheesy texture, with long, stringy threads. This kind of bread is called *ropy*. Cream rolls, custards from whole eggs, and whipped cream are good media for growth of *Salmonella*, *Lactobacillus*, and *Streptococcus* species, which produce acids.

5.11.5 Other Foods

Many cereals, fruits, and vegetables are spoiled by microorganisms. The chief agents of spoilage are molds and bacteria, giving foods an unpleasant odor. Grains are spoiled by a mold, *Aspergillus flavors*, which is also present in peanut products and other foods. The mold forms a toxin called *aflatoxin*. *Claviceps purpurea* is also an agent of grain spoilage, causing ergot disease in rye, wheat, and barley grains. The mold toxin may induce convulsions and hallucinations. The drug LSD is derived from this toxin.

5.11.6 Importance of Microbes in Foods

Molds, yeasts, and bacteria play a significant part in food spoilage. It is true that molds are involved in the spoilage of many foods, but some molds are important in food manufacture, especially in mold-ripened cheeses and the preparation of oriental foods. Various fungi, such as species of *Aspergillus*, *Fusarium*, and *Penicillium*, have been found in foods, and some have been implicated in toxin production. Yeasts are both useful as well as problematic organisms, and this depends upon the food. For example, the production of wine is dependent on the growth and activity of the yeast *Saccharomyces cerevisiae*, but wine can be oxidized to CO_2 and water by wild yeast. Important yeasts in foods include species of *Saccharomyces*, which are useful in fermentations. On the other hand, species of *Zygosaccharomyces* that are osmophilic are involved in the spoilage of materials such as honey, whereas species of *Pichia* form pellicles in liquids such as beer and wines. A variety of bacteria are found in foods, and the important ones are grouped based on their biochemical properties.

5.11.7 Food Fermentation

Fermentation of food results in the production of organic acids, alcohols, and esters, which not only help in preserving the food but may also generate distinctive new food products. The fermentation may be by yeast, bacteria, molds, or by a combination of these organisms. Food products such as bread, beer, and wine are produced using yeast, whereas both yeast and bacteria are involved in the production of vinegar. Bacteria are also involved in the production of fermented milks, whereas molds are important in the production of oriental foods such as soy sauce and other soybean products. By virtue of their growth, these organisms bring about desirable changes in the food composition and, at the same time, bring about a certain degree of preservation. For example, in fermented pickles, the end product of fermentation, namely lactic acid, serves as a preservative. Fermented pickles, therefore, do not need the addition of any preservative.

In bread making, microorganisms are involved in gas production, which helps in the production of bread with a porous structure, and are also involved in the production of flavoring substances. During the preparation of dough, the yeast ferments the sugars to produce CO_2 and alcohol. Instead of bread yeast (baker's yeast), other gas-forming yeasts such as wild yeasts and heterolactic acid bacteria have also been used. Little growth occurs during the leavening process, but fermentation begins as soon as the dough is mixed and continues until the temperature of the oven inactivates the enzymes.

Addition of many yeast cells can hasten the fermentation process and discourage growth. During fermentation, conditioning of the dough takes place, which results from the action on gluten by proteolytic enzymes in the flour by the added yeast or from the malt and a reduction in pH. Sometimes dough conditioners are also added to stimulate yeast growth. The main objective of the baker during leavening is to have enough gas produced and to have the dough in such a condition that it will hold the gas. Yeasts are also reported to contribute to the flavor of bread through products such as alcohols, acids, aldehydes, esters, etc. that are released during the fermentation. If enough time is given for growth of bacteria before baking, they may also add to the flavor. During baking, the temperature inside the loaf does not reach 100°C, but the heat serves to kill the yeast, inactivate their enzymes, and allow expansion of the gas present to give the right structure to the bread.

Beer is the principal fermented malt beverage produced worldwide from malt, hops, and yeast and malt adjuncts. The production of malt wort and fermentation has been discussed earlier. In Japan, soya sauce is prepared using *Aspergillus oryzae*. In the initial stages, this organism produces a variety of enzymes that break down soy protein and starch. In the subsequent lactic fermentation, lactic bacteria produce lactic acid. More acid production by *Pediococcus halophilus* and alcoholic fermentation by *Saccharomyces rouxi* and *Zygosaccharomyces soyae* occurs. An Indonesian food called Temphe is also prepared from soybeans. Soybeans are soaked at 25°C, seed coats are then removed, and the split beans are cooked in water for 20 min. These are then cooled and inoculated with spores of *Rhizopus* sp. (*R. arrhizus* or *R. oryzae*). The mash is packed into plastic containers or rolled in banana leaves and incubated at 32°C for 20 h to allow mycelial growth. The product is then sliced, dipped in saltwater, and fried in fat before it is consumed. A variety of Japanese and Chinese foods such as miso, angkbak, and soybean cheese are also prepared using fungi.

5.12   MICROBIAL BIOTECHNOLOGY

Over the last few decades, microbial biotechnology has been employed for large-scale production of a variety of biochemicals, ranging from alcohols to antibiotics, and in the processing of foods and feeds. The use of microbes to obtain a product or service of economic importance constitutes industrial microbiology or microbial biotechnology. With respect to its scope, objectives, and activities, industrial microbiology is synonymous with "fermentation," as fermentation includes any process mediated by or involving microorganisms in which a product of economic value is obtained (Casida, Jr., 1968). The goods provided by microorganisms may be entire, live, or dead microbial cells; processed microbial biomass; components of microbial cells; intracellular or extracellular enzymes; or chemicals produced by the microbes utilizing the constituents of the medium or the substrate provided. Degradation of organic wastes, detoxification of industrial wastes and toxic compounds, and degradation of oil spills are some of the services rendered by microbes. Production of biocontrol agents and biofertilizers also comes under industrial microbiology. Any activity of industrial microbiology involves isolation of microorganisms from nature, their screening to establish product formation, improvement of product yields, and maintenance of cultures, mass culture using bioreactors, and, finally, recovery of products and their purification.

5.12.1   PRODUCTION OF ENZYMES BY MICROORGANISMS

Enzymes can be extracted from the tissues of higher organisms. It is easier and cheaper to get the enzymes from microorganisms. The first industrial production of enzymes from microorganisms dates to 1894, when fungal Taka-Diastase was produced in the United States to be employed as a pharmaceutical agent for digestive disorders. Enzymes have found a variety of applications in medicine, in the food industry, in the textiles and leather industries, and in analytical processes. The most commercialized microbial enzymes come from a small number of fungi and bacteria such as *Aspergillus*, *Fusarium, Trichoderma*, and *Humicola* (all belonging to ascomycetes). *Mucor* and *Rhyzomucor* (belonging to zygomycetes) are the fungi from which biocatalysts are retrieved. *Bacillus* and *Pseudomonas* are the bacterial strains employed to produce enzymes. The most important industrially produced microbial enzymes are alpha amylase, proteases, and lipases. Three different molecular techniques are adopted for the large-scale production of microbial enzymes: expression cloning, molecular screening, and protein engineering.

PROBLEMS

Section A: Descriptive Type

    Q1. Describe the intracellular organization of microbes.
    Q2. Write a brief note on pathogenic bacteria.
    Q3. How can microbes be used in the field of agriculture?
    Q4. Describe the significance of bacteria as material for genetic studies.

Q5. Describe mutations in bacteria.

Q6. What steps are involved in microbial transformation?

Q7. Explain the significance of *E. coli* in microbial cloning.

Section B: Multiple Choice

Q1. Microbe fossils were found to be present how many years ago?
   a. 3 billion
   b. 3.5 billion
   c. 5 million

Q2. Bacteria were first observed by Antonio van Leeuwenhoek in the year . . .
   a. 1600
   b. 1676
   c. 1650
   d. 1680

Q3. Bacteria do not have a membrane-bound nucleus. True/False

Q4. Cyanobacteria produce internal gas vesicles which they use to regulate . . .
   a. Metabolism
   b. Buoyancy
   c. Temperature
   d. Reproduction

Q5. Denitrification is the utilization of nitrate as a terminal electron acceptor. True/False

Q6. The process of bacterial growth is called . . .
   a. Primary fission
   b. Binary fission
   c. Tertiary fission

Q7. How many phases are there in industrial-level bacterial production?
   a. 3
   b. 4
   c. 2
   d. 5

Q8. Transduction involves the carrying over of DNA from one organism to another by an intermediate agent. True/False

Q9. This genome was the first to be completely sequenced.
   a. Virus
   b. *E. coli*
   c. *Salmonella*
   d. *Nocardia*

Q10. When was the first industrial production of enzymes from microorganism achieved?
   a. 1800
   b. 1850
   c. 1894

Section C: Critical Thinking

Q1. How can a bacterium be differentiated by Gram-positive and Gram-negative staining?

Q2. To get good bacterial growth in the bioreactor, what are the different parameters that can be regulated or modified?

Q3. Why is the lactic fermentation of milk not considered spoilage?

Q4. What would be the scenario of microbial growth in the absence of the log or exponential phase?

REFERENCES AND FURTHER READING

Barbara, A., Nelson, K.E., Pop, M., Creasy, H.H., Giglio, M.G., Huttenhower, C., Gevers, D., Petrosino, J.F. et al. The human microbiome project consortium, "A framework for human microbiome research". *Nature* 486(7402): 215–221, 2012. doi:10.1038/nature11209. PMC 3377744. PMID 22699610.

Demain, A.L., and Davies, J.E. *Manual of Industrial Microbiology and Biotechnology.* ASM Press, Washington, DC, 1999.

Gillor, O., Kirkup, B.C., and Riley, M.A. Colicins and microcins: The next generation antimicrobials. *Adv. Appl. Microbiol.* 54: 129–146, 2004.

Gillor, O., Nigro, L.M., and Riley, M.A. Genetically engineered bacteriocins and their potential as the next generation of antimicrobials. *Curr. Pharm. Des.* 11: 1067–1075, 2005.

Glazer, A.N., and Nikaido, H. *Microbial Biotechnology: Fundamentals of Applied Microbiology.* Cambridge University Press, Cambridge and New York, 2007.

Huttenhower, C., Gevers, D., Knight, R., Abubucker, S., Badger, J.H., Chinwalla, A.T., Creasy, H.H., Earl, A.M. et al. The human microbiome project consortium, "Structure, function and diversity of the healthy human microbiome". *Nature* 486(7402): 207–214, 2012. doi:10.1038/nature11234. PMC 3564958. PMID 22699609.

Kirkup, B.C. Bacteriocins as oral and gastrointestinal antibiotics: Theoretical considerations, applied research, and practical applications. *Curr. Med. Chem.* 13: 3335–3350, 2006.

Lee, L.K. *Microbial Biotechnology: Principles and Applications.* World Scientific Publishing Company, Singapore, 2007.

Marquez, B. Bacterial efflux systems and efflux pumps inhibitors. *Biochimie* 87: 1137–1147, 2005.

Marri, P.R., Paniscus, M., Weyand, N.J., Rendón, M.A.A., Calton, C.M., Hernández, D.R., Higashi, D.L., Sodergren, E., Weinstock, G.M., Rounsley, S.D., So, M., and Ahmed, Niyaz, ed. Genome sequencing reveals widespread virulence gene exchange among human neisseria species. *PLoS One* 5(7): e11835, 2010. doi:10.1371/journal.pone.0011835. PMC 2911385. PMID 20676376.

Metlay, J.P., Camargo, C.A., MacKenzie, T. et al. Cluster-randomized trial to improve antibiotic use for adults with acute respiratory infections treated in emergency departments. *Ann. Emerg. Med.* 50: 221–230, 2007.

Spurling, G., Del Mar, C., Dooley, L., and Foxlee, R. Delayed antibiotics for respiratory infections. *Cochrane Database of Systematic Reviews (Online)* (3): CD004417. doi:10.1002/14651858.CD004417.pub3. PMID 17636757, 2007.

Toratoa, G., Funke, B., and Case, C.L. *Microbiology: An Introduction.* Pearson Benjamin Cummings, San Francisco, CA, 2007.

Turnbaugh, P.J., Ley, R.E., Hamady, M., Fraser-Liggett, C.M., Knight, R., and Gordon, J.I. The human microbiome project. *Nature* 449(7164): 804–810, 2007.

6 Plant Biotechnology

LEARNING OBJECTIVES
- Explain the significance of plant biotechnology
- Discuss plant-breeding techniques by the classical approach
- Discuss plant-breeding techniques by the modern approach
- Discuss plant diseases and explain their causes
- Explain how genetically modified plants are created
- Discuss the benefits of genetically modified plants
- Provide examples of genetically modified plants

6.1 INTRODUCTION

With the growing world population, it is critical to have enough food grains and vegetables to feed them. The traditional methods to cultivate plant crops is not sufficient to produce enough food grains and vegetables, hence the use of modern technology is required. Plant biotechnology is an advanced technology that allows plant breeders or farmers to make precise genetic changes in plants to impart beneficial traits, which include size, yield, color, taste, and appearance. For centuries, farmers and plant breeders have labored to improve crop plants. Traditional breeding methods include selecting and sowing seeds of the strongest and most desirable plants to produce the next generation of crops.

6.2 PLANT BREEDING

Plant breeding has been in practice for thousands of years. It is now practiced worldwide by individuals such as gardeners and farmers, or by professional plant breeders employed by organizations such as government institutions, universities, crop-specific industry associations, or research centers. International development agencies believe that breeding new crops is important for ensuring food security by developing new varieties that are higher-yielding, resistant to pests and diseases, and drought-resistant or regionally adapted to different environments and growing conditions (Figure 6.1). The extensive use of plant breeding in certain situations may lead to domestication of wild plants. Domestication of plants is an artificial selection process conducted by humans to produce plants that have more desirable traits than wild plants. Many of the crops nowadays are the result of domestication in ancient times, about 5000 years ago in the Old World and 3000 years ago in the New World. In the Neolithic period, domestication took a minimum of 1000 years and a maximum of 7000 years. Almost all the domesticated plants used today for food and agriculture were domesticated in ancient times. A plant whose origin or selection is because of human activity is called a *cultigen*, and a cultivated crop species that has evolved from wild populations because of selective pressures from traditional farmers is called a *landrace*. Landraces, which can be the result of natural forces or domestication, are plants that are ideally suited to a particular region or environment. Some examples of landraces are rice, *Oryza sativa* subspecies *indica*, which was developed in South Asia, and *Oryza sativa* subspecies *japonica*, which was developed in China.

Plant breeding is defined as identifying and selecting desirable traits in plants and combining these into one individual plant. Since 1900, Mendel's laws of genetics have provided the scientific basis for plant breeding. As all traits of a plant are controlled by genes located on chromosomes, conventional plant breeding can be considered as the manipulation of the combination of chromosomes.

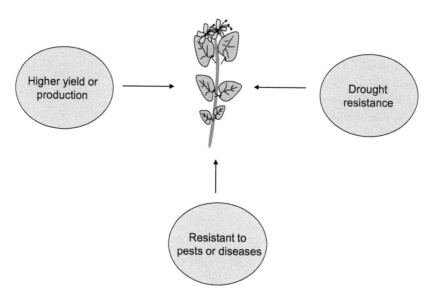

FIGURE 6.1 Major focus in plant breeding.

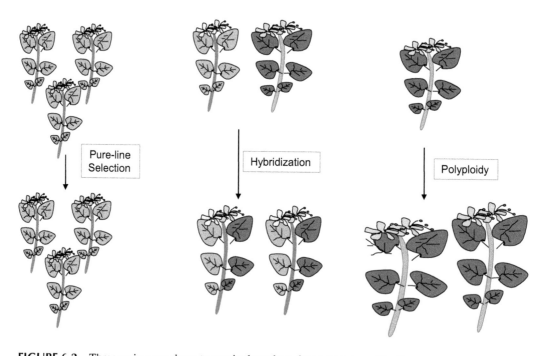

FIGURE 6.2 Three main procedures to manipulate plant chromosome combinations.

In general, there are three main procedures to manipulate plant chromosome combinations. First, plants of a given population, which show desired traits, can be selected and used for further breeding and cultivation, a process called *pure-line selection* (Figure 6.2). Second, desired traits found in different plant lines can be combined altogether to obtain plants that exhibit all the traits simultaneously, a method termed *hybridization* (Figure 6.2). *Heterosis*, a phenomenon of increased vigor, is obtained by hybridization of inbred lines. Third, *polyploidy* (increased number of chromosome sets) can contribute to crop improvement (Figure 6.2). Plant breeding may be classified as classical breeding or modern breeding.

6.2.1 CLASSICAL PLANT BREEDING

In classical plant breeding, the closely or distantly related plants are crossed to produce new crop varieties or lines with desirable properties. Plants are crossbred to introduce traits/genes from one variety or line into a new genetic background. For example, a mildew-resistant pea may be crossed with a high-yielding but susceptible pea. The goal of the crossbreeding is to introduce mildew resistance without losing the high-yield characteristic. Progeny from the cross would then be crossed with the high-yielding parent to ensure that the progeny was most like the high-yielding parent (backcrossing). The progeny from that cross would then be tested for yield and mildew resistance and high-yielding, mildew-resistant plants would be further developed. Plants may also be crossed with themselves to produce inbred varieties for breeding. Classical breeding relies largely on homologous recombination between chromosomes to generate genetic diversity. The classical plant breeder may also make use of several *in vitro* techniques such as protoplast fusion embryo rescue or mutagenesis to generate diversity and produce hybrid plants that would not exist in nature. For over 100 years, plant breeders have tried to incorporate various traits into crop plants to increase quality and yield; to increase tolerance to environmental pressures (such as salinity, extreme temperature, and drought); to enhance resistance against viruses, fungi and bacteria; and to increase tolerance to pests and herbicides.

6.2.1.1 Selection

In plant breeding, selection is the most ancient and basic procedure and it generally involves three distinct steps. First, many selections are made from the genetically variable original population. Second, progeny rows are grown from the individual plant selections for observational purposes. After obvious elimination, the selections are grown over several years to permit observations of performance under different environmental conditions for making further eliminations. Finally, the selected and inbred lines are compared with existing commercial varieties for their yielding performance and other aspects of agronomic importance.

6.2.1.2 Hybridization

Hybridization has been the most frequently employed plant-breeding technique. Its aim is to bring together desired traits found in different plant lines into one plant line via cross-pollination. The first step in this technique is to generate homozygous inbred lines. This is normally done by using self-pollinating plants in which pollen from male flowers pollinates the female flowers from the same plants. Once a pure line is generated, it is outcrossed or combined with another inbred line. Then the resulting progeny is selected for the combination of the desired traits. If a trait (such as resistance against diseases) from a wild relative of a crop species is to be brought into the genome of the crop, a large quantity of undesired traits (like low yield, bad taste, or low nutritional value) is transferred to the crop as well. These unfavorable traits must be removed by time-consuming backcrossing, that is, by repeated crossing with the crop parent. There are two types of hybrid plants: *interspecific* and *intergeneric hybrids*. Beyond this biological boundary, hybridization cannot be accomplished because of sexual incompatibility, which limits the possibilities of introducing desired traits into crop plants. *Heterosis* is an effect that is achieved by crossing highly inbred lines of crop plants. Inbreeding of most crops leads to a strong reduction of vigor and size in the first generation. After six or seven generations, no further reduction in vigor or size is found. When such highly inbred plants are crossed with other inbred varieties, very vigorous, large-sized, and large-fruited plants may result. The most notable and successful hybrid plant ever produced is the hybrid maize. By 1919, the first commercial hybrid maize was available in the United States. Two decades later, nearly all maize was hybrid, as it is today, although farmers must buy new hybrid seed every year because the heterosis effect is lost in the first generation after hybridization of the inbred parental lines.

6.2.1.3 Polyploidy

Plants that have three or more complete sets of chromosomes are commonly known as *polyploidy plants*. The chromosome numbers can be increased artificially by treating the plant cells with colchicine, which leads to a doubling of the chromosome number. Generally, the main effect of polyploidy is an increase in size and genetic variability. On the other hand, polyploid plants often have lower fertility and slower growth. Instead of relying only on the introduction of genetic variability from the wild species gene pool or from other cultivars, an alternative is the introduction of mutations induced by chemicals or radiation. The mutants obtained are tested and further selected for desired traits. The site of the mutation cannot be controlled when chemicals or radiation is used as agents of mutagenesis. Because the great majority of mutants carry undesirable traits, this method has not been widely used in breeding programs.

6.2.2 MODERN PLANT BREEDING

Modern plant-breeding techniques basically involve molecular or genetic engineering techniques to select or to insert desirable traits into plants. In recent years, biotechnology has developed rapidly as a practical means for accelerating success in plant breeding and improving economically important crops. Some of the modern plant breeding methods used today are described below.

6.2.2.1 *In Vitro* Cultivation

In this method, plants are cultivated using *in vitro* culture from cultured isolated somatic plant cells. The parts of plants can be cultured *in vitro* and are capable of proliferation and organization into tissues and eventually into complete plants (Figure 6.3). The process of regenerating whole plants

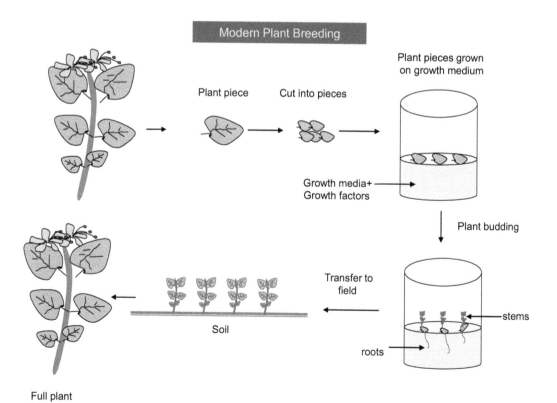

FIGURE 6.3 Plant cultivation by *in vitro* cell culture.

out of plant cells is called *in vitro regeneration*. Some factors affecting plant regeneration are genotype, explants source, culture conditions, culture medium, and environment. Different mixtures of plant hormones and other compounds in varying concentrations are used to achieve regeneration of plants from cultured cells and tissues. As the plant hormonal mechanisms are not yet completely understood, the development of *in vitro* cultivation and regeneration systems is still largely based on empirically testing variations of the above-mentioned factors.

6.2.2.2 *In Vitro* Selection and Somaclonal Variation

Plants regenerated from *in vitro* cell cultures may exhibit phenotypes differing from their parent plants, sometimes at quite high frequencies. If these are heritable and affect desirable agronomic traits, such "somaclonal variation" can be incorporated into breeding programs. However, finding specific valuable traits by this method is largely left to chance and hence is inefficient. Rather than relying on this undirected process, *in vitro* selection targets specific traits by subjecting large populations of cultured cells to the action of a selective agent in a Petri dish. For disease resistance, this selection can be provided by pathogens or isolated pathotoxins that are known to have a role in pathogenesis. The selection will only allow those cells that are resistant to the challenge to survive and proliferate. Selection of cells also plays an important role in genetic engineering, in which special marker genes are used to select for transgenic cells (Figure 6.4).

6.2.2.3 Somatic Hybrid Plants

Somatic hybrid plants are generated by fusion of somatic cells. Cell fusion was developed after the successful culture of many plant cells that were stripped of their cell walls. The resulting cells without walls are referred to as *protoplasts*. Because protoplasts from phylogenetically unrelated species can also be fused, attempts have been made to overcome sexual incompatibility using protoplast

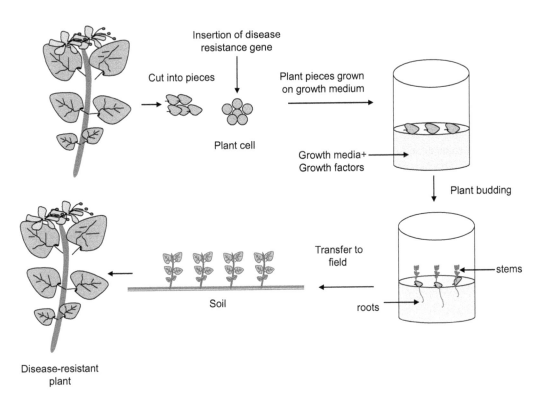

FIGURE 6.4 Plant cultivation by *in vitro* selection.

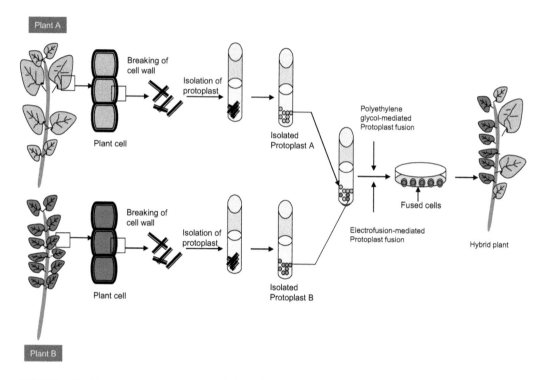

FIGURE 6.5   Plant breeding by somatic hybridization technique.

fusion. In most cases, these attempts failed because growth and division of the fused cells did not take place when only distantly related cells were fused. Although successful fusions between sexually incompatible petunia species and between potatoes and tomatoes did not lead to economically interesting products, important contributions to the understanding of cell wall regeneration and other mechanisms were achieved (Figure 6.5).

6.2.2.4   Breeding by Restriction Fragment Length Polymorphism

Traditional plant-breeding techniques are very time-consuming and sometimes a lot of undesired genes are introduced into the genome of a plant. The undesired genes must be "sorted out" by backcrossing. The use of restriction fragment length polymorphism (RFLP) greatly facilitates conventional plant breeding, because one can progress through a breeding program much faster, with smaller populations, and without relying entirely on testing for the desired phenotype. RFLP makes "use of restriction endonucleases and these enzymes recognize and cut specific nucleotide sequences in DNA. For example, the sequence GAATTC is cut by the endonuclease *EcoRI*. After treatment of a plant genome with endonucleases, the plant DNA is cut into pieces of different lengths, depending on the number of recognition sites on the DNA. These fragments can be separated according to their size by using gel electrophoresis and are made visible as bands on the gel by hybridizing the plant DNA fragments with radiolabeled or fluorescent DNA probes. As two genomes are not identical even within a given species because of mutations, the number of restriction sites and therefore the length and numbers of DNA fragments differ, resulting in a different banding pattern on the electrophoresis gel. This variability has been termed RFLP. The closer two organisms are related, the more the pattern of bands overlap. If a restriction site lies close to or even within an important gene, the existence of a particular band correlates with the particular trait of a plant, such as disease resistance. By looking at the banding pattern, breeders can identify individuals that have inherited resistance genes, and resistant plants can be selected for further breeding. The use of this technique not only considerably accelerates progress in plant breeding, but also facilitates the identification of resistance genes, thereby opening new possibilities in plant breeding.

6.2.2.5 Plant Breeding by Gene Transfer

In conventional breeding, the pool of available genes and the traits they code for is limited because of sexual incompatibility with other lines of the crops. This restriction can be overcome by using the methods of genetic engineering, which in principle allow introduction of valuable traits coded for by specific genes of any organism (other plants, bacteria, fungi, animals, viruses) into the genome of any plant. The first gene transfer experiments with plants took place in the early 1980s. Normally, transgenes are inserted into the nuclear genome of a plant cell. Recently, it has become possible to introduce genes into the genome of chloroplasts and other plastids (small organelles of plant cells that possess a separate genome). Transgenic plants have been obtained using *Agrobacterium*-mediated DNA transfer and direct DNA transfer, the latter including methods such as particle bombardment, electroporation, and polyethylene glycol permeabilization. Most plants have been transformed using *Agrobacterium*-mediated transformation.

6.2.2.6 *Agrobacterium*-Mediated Gene Transfer

The Agrobacterium-mediated technique involves the natural gene transfer system in the bacterial plant pathogens of the genus *Agrobacterium*. In nature, *Agrobacterium tumefaciens* and *Agrobacterium rhizogenes* are the causative agents of the crown gall and the hairy root diseases, respectively. The utility of *Agrobacterium* as a gene transfer system was first recognized when it was demonstrated that these plant diseases were produced because of the transfer and integration of genes from the bacteria into the genome of the plant. Both *Agrobacterium* species carry a large plasmid (small circular DNA molecule) called Ti in *A. tumefaciens* and Ri in *A. rhizogenes*. A segment of this plasmid designated T-(for transfer) DNA is transmitted by this organism into individual plant cells, usually within wounded tissue. The T-DNA segment penetrates the plant cell nucleus and integrates randomly into the genome where it is stably incorporated and inherited like any other plant gene in a predictable, dominant Mendelian fashion. Expression of the natural genes on the T-DNA results in the synthesis of gene products that direct the observed morphological changes such as tumor or hairy root formation. In genetic engineering, the tumor-inducing genes within the T-DNA, which cause the plant disease, are removed and replaced by foreign genes. These genes are then stably integrated into the genome of the plant after infection with the altered strain of *Agrobacterium*, just like the natural T-DNA. Because all tumor-inducing genes are removed, the gene transfer does not induce any disease symptoms. This reliable method of gene transfer is well suited for plants that are susceptible to infection by *Agrobacterium*. Unfortunately, many species, especially economically important legumes and monocotyledons such as cereals, do not respond positively to *Agrobacterium*-mediated transformation (Figure 6.6).

6.2.2.7 Particle Bombardment

Particle bombardment, also referred to as biolistic transformation (from biological ballistics), involves coating biologically active DNA onto small tungsten or gold particles and accelerating them into plant tissues at high velocity. The particles penetrate the plant cell wall and lodge themselves within the cell where the DNA is liberated, resulting in the transformation of the individual plant cell in an explant. This technique is generally less efficient than *Agrobacterium*-mediated transformation but has nevertheless been particularly useful in several plant species, most notably in cereal crops. The introduction of DNA into organized, morphogenic tissues such as seeds, embryos, or meristems has enabled the successful transformation and regeneration of rice, wheat, soybean, and maize, thus demonstrating the enormous potential of this method (Figure 6.7).

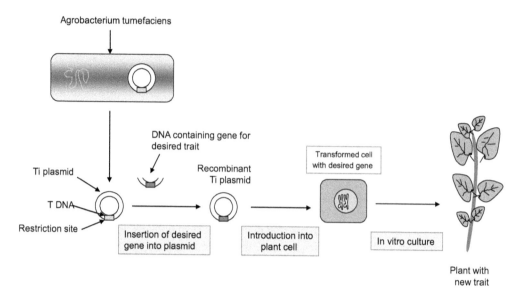

FIGURE 6.6 *Agrobacterium*-mediated gene transfer.

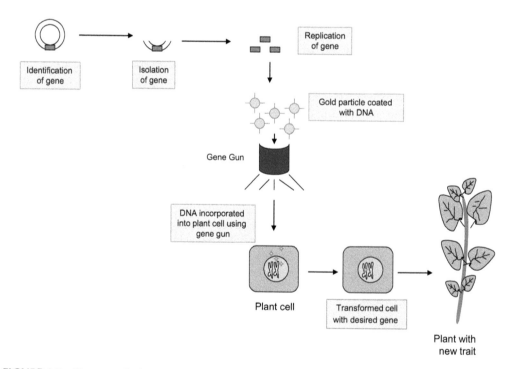

FIGURE 6.7 Gene transfer by particle bombardment.

6.2.2.8 Electroporation and Direct DNA Entry into Protoplasts

Electroporation is a process whereby very short pulses of electricity are used to reversibly permeabilize lipid bilayers of plant cell membranes. The electrical discharge enables the diffusion of DNA through an otherwise impermeable plasma membrane. Because the plant cell wall will not allow the efficient diffusion of many transgene constructs, protoplasts (cells without cell walls) must be prepared. DNA uptake by plant protoplasts can also be stimulated by phosphate or calcium/poly-ethylene glycol coprecipitation. However, all these methods suffer from the drawback that they use protoplasts as the recipient host, which often cannot be regenerated into whole plants.

6.2.2.9 Transgene Expression

The success of transgene expression is generally based on transcription of mRNA and then into translation leading to protein synthesis. *Promoter* is a sequence of nucleic acids where RNA poly-merase (a complex enzyme synthesizing the mRNA transcript) attaches to the DNA template. The nature of the promoter defines (together with other expression-regulating elements) under which conditions and intensity a gene will be transcribed. The promoter of the 35S gene of cauliflower mosaic virus (CaMV) is used very frequently in plant genetic engineering. This promoter confers high-level expression of exogenous genes in most cell types from virtually all species tested. As it is often advantageous to express a transgene only in certain tissues or quantities or at certain times, several other promoters can also be used such as promoters inducing gene expression after wound-ing or during fruit ripening only. Methods of gene transfer currently employed result in the random integration of foreign DNA throughout the genome of the recipient cells. The site of insertion may have a strong influence on the expression levels of the exogenous gene, resulting in different expres-sion levels of an introduced gene even if the same promoter/gene construct was used. The exact mechanism of this phenomenon is not yet fully understood (Figure 6.8).

6.2.2.10 Selection and Plant Regeneration

To select only cells that have incorporated the new genes, the genes coding for the desired trait are fused to a gene that allows selection of transformed cells, so-called *marker genes*. The expression of the marker gene enables the transgenic cells to grow in the presence of a selective agent, usually an antibiotic or an herbicide, while cells without the marker gene die. One of the most commonly used markers is the bacterial aminoglycoside-3′ phosphotransferase gene (APH(3′)II), also referred to as neomycin phosphotransferase II (NPTII). This gene codes for an enzyme that inactivates the antibiotics kanamycin, neomycin, and G418 through phosphorylation. In addition to NPTII, several other antibiotic resistance genes have been used as selective markers, such as hygromycin

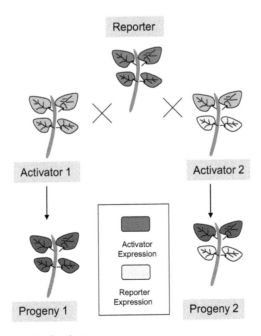

FIGURE 6.8 Transgene expression in plant.

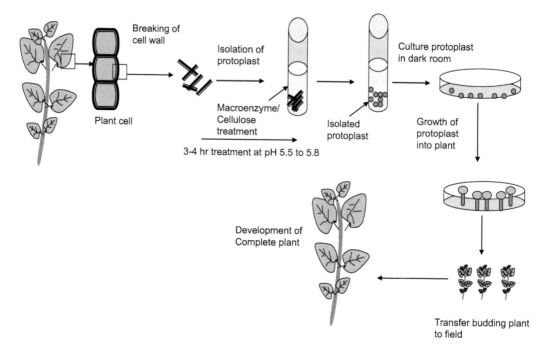

FIGURE 6.9 Plant regeneration by using protoplast.

phosphotransferase gene conferring resistance to hygromycin. Another group of selective markers is herbicide tolerance genes. Herbicide tolerance has been obtained through the incorporation and expression of a gene that either detoxifies the herbicide in a manner similar to that of the antibiotic resistance gene products or a gene that expresses a product that acts like the herbicide target but is not affected by the herbicide. Herbicide tolerance may not only serve as a useful trait for selection in the development of transgenic plants but also has some commercial interest. Transformations of plant protoplasts, cells, and tissues are usually only useful if they can be regenerated into whole plants. The rate of regeneration varies greatly not only among different species but also between cultivars of the same species. Besides the ability to introduce a gene into the genome of a plant species, regeneration of intact, fertile plants out of transformed cells or tissues is the most limiting step in developing transgenic plants (Figure 6.9).

6.2.2.11 Reverse Breeding and Doubled Haploids

Reverse breeding and doubled haploids (DH) is the most efficient method to produce homozygous plants from a heterozygous starting plant, which have all desirable traits. This starting plant is induced to produce DH from haploid cells, and later to create homozygous/DH plants of those cells. Although in natural offspring recombination occurs and traits can be unlinked from each other, in DH cells and in the resulting DH plants, recombination is no longer an issue. Here, a recombination between two corresponding chromosomes does not lead to un-linkage of alleles or traits, because it just leads to recombination with its identical copy. Thus, traits on one chromosome stay linked. Selecting those offspring that have the desired set of chromosomes and crossing them will result in a final F1 hybrid plant having the same set of chromosomes, genes, and traits as the starting hybrid plant. The homozygous parental lines can reconstitute the original heterozygous plant by crossing, if desired even in a large quantity. An individual heterozygous plant can be converted into a heterozygous variety (F1 hybrid) without the necessity of vegetative propagation, but as the result of the cross of two homozygous/DH lines derived from the originally selected plant.

6.2.2.12   Genetic Modification

Genetic modification of plants is achieved by adding a specific gene or genes to a plant or by knocking out a gene with the RNA interference (RNAi) technique to produce a desirable phenotype. The plants resulting from adding a gene are often referred to as *transgenic plants*. If for genetic modification, genes of the species or of a crossable plant are used under the control of their native promoter, then they are called *cisgenic plants*. Genetic modification can produce a plant with the desired trait or traits faster than classical breeding because most the plant's genome is not altered. To genetically modify a plant, a genetic construct must be designed so that the gene to be added or removed will be expressed by the plant. To do this, a promoter to drive transcription and a termination sequence to stop transcription of the new gene as well as the gene or genes of interest must be introduced into the plant. A marker for the selection of transformed plants is also included. In the laboratory, antibiotic resistance is a commonly used marker: plants that have been successfully transformed will grow on media containing antibiotics; plants that have not been transformed will die. In some instances, markers for selection are removed by backcrossing with the parent plant prior to commercial release. The construct can be inserted in the plant genome by genetic recombination using the bacteria *A. tumefaciens* or *A. rhizogenes* or by direct methods like the gene gun or microinjection. Using plant viruses to insert genetic constructs into plants is also a possibility, but the technique is limited by the host range of the virus. For example, CaMV only infects cauliflower and related species. Another limitation of viral vectors is that the virus is not usually passed on to the progeny, so every plant must be inoculated. The majority of commercially released transgenic plants are currently limited to plants that have introduced resistance to insect pests and herbicides. Insect resistance is achieved through incorporation of a gene from *B. thuringiensis* (Bt) that encodes a protein that is toxic to some insects. For example, when the cotton bollworm, a common cotton pest, feeds on Bt cotton it will ingest the toxin and die. Herbicides usually work by binding to certain plant enzymes and inhibiting their action. The enzymes that the herbicide inhibits are known as the herbicides *target site*. Herbicide resistance can be engineered into crops by expressing a version of *target site* protein that is not inhibited by the herbicide. This is the method used to produce glyphosate-resistant crop plants.

6.3   APPLICATIONS OF MOLECULAR AND GENETIC TOOLS IN AGRICULTURE

Recombinant DNA technology has not only enhanced the health of humans, but also contributed to exciting developments in agricultural biotechnology. Using rDNA methods, transgenic plants and animals with desirable properties such as resistance to diseases/herbicides have been developed. Flowers with exotic shapes and colors have been genetically engineered by transgenic expression of pigment genes. Recombinant growth hormones are now available for farm animals, resulting in leaner meat, improved milk yield, and more efficient feed utilization. In the future, transgenic plants and animals may serve as bioreactors to produce medicinal or protein pharmaceuticals.

6.3.1   Expression of Viral Coat Protein to Resist Infection in Agriculture

Viruses are a serious problem for many agricultural crops and animals. Infections can result in reduced growth, less yield, and low quality. Through a standard genetic trick termed cross-protection, infection of a plant/animal with a strain of virus that produces only mild effect protects the plants/animals against infection by more damaging strains. The same principle/mechanism when applied to animals is called *vaccination*. Although the mechanism of cross-protection is not entirely known, it is thought that a viral-encoded protein is responsible for the protective effect.

6.3.2   Expression of Bacterial Toxin in Agriculture Using Molecular Techniques

Currently, the major weapons against the attackers of plants are chemical insecticides. However, chemicals have some impact on the environment. Natural microbial pesticides, such as the species

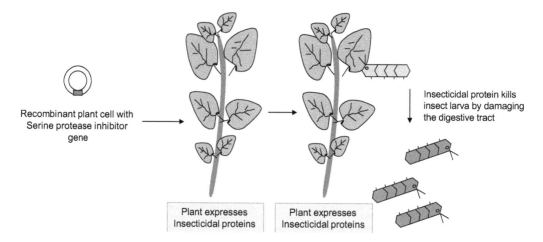

FIGURE 6.10   Transgenic plant with insecticide gene expression.

B. thuringiensis (Bt) have been used in a limited manner for over 38 years. Upon sporulation, these bacteria produce a crystallized protein that is toxic to the larvae of several insects. The toxic protein does not harm non-susceptible insects and has no effect on vertebrates. The crystal protein is normally expressed as a large, inactive protoxin about 1200 amino acids in length and with a molecular weight of 1,200,000 Da. The toxin acts by binding to receptors on the surface of midgut cells and blocking the functioning of these cells.

A second approach to the development of insect-resistant plants has been through the transgenic expression of serine protease inhibitors. These proteins are present in several plants and act to deter insects by inhibiting serine proteases in the insect digestive system (Figure 6.10).

6.4   HERBICIDE-TOLERANT PLANTS

The presence of weeds in a crop field reduces the yield and it is very important to remove these weeds without affecting the crops. Weed killers or herbicides are not very selective and their current use relies on the differential uptake between the weed and the crop plant or on the application of the herbicide before planting a field, altering the food content of plants. With the ability to introduce DNA into plants, researchers are trying to create herbicide-tolerant crops using three strategies: (1) by increasing the level of the target enzyme for a particular herbicide, (2) by expressing a mutant enzyme that is not affected by the compound, and (3) by expressing an enzyme that detoxifies the herbicide. Of the large number of herbicides in use today, only a few of the cellular targets have been characterized. One strategy is to clone the cellular target genes into the plant so that they are produced in large amounts. Another strategy for creating herbicide-tolerant plants is by using mutant forms of bacterial 5-enolpyruvylshikimate-3-phosphate synthase enzymes. Genes encoding these mutant enzymes have been derived from glycophosphate-resistant bacteria and expressed in plants. These enzymes lessen the inhibitory effect of herbicides.

A third strategy is by transgenic expression of enzymes that convert the herbicide to a form that is not toxic to the plant. Some plants have developed their own detoxifying system for certain herbicides. However, these activities in plants are encoded by a complex set of genes that has not yet been fully characterized (Figure 6.11).

6.5   PIGMENTATION IN TRANSGENIC PLANTS

Plants are widely used for ornamental purposes, so it is not surprising that considerable attempts have been made to develop varieties that have flowers of new colors, shapes, and growth properties.

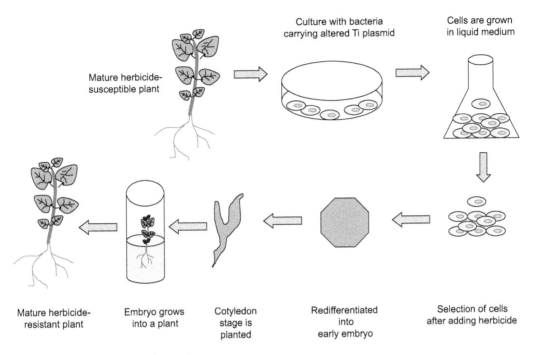

FIGURE 6.11   Herbicide-resistant plant.

Pigmentation in flowers is mainly due to three classes of compounds: the flavonoids, the carotenoids, and the betalains. Of these, the flavonoids are the best characterized with much information now available concerning their chemistry, biochemistry, and molecular genetics. Experiments are underway to expand the spectrum of coloring of certain floral species by introducing genes for the entire pigment biosynthesis pathway. A blue rose was never obtained because rose plants lack the enzymes that synthesize the pigment for blue flower coloration. However, introducing the genes for blue color gave very few successful results. This is because of a phenomenon called *co-suppression*, in which an extra copy of the gene suppresses the expression of the endogenous genes. An experiment was performed in which a second copy of a petunia pigment gene was introduced into a petunia plant with colored flowers. It was expected that increased production of the encoded enzyme might produce flowers with a deeper purple color. However, white-colored flowers were produced because of co-suppression. Co-suppression has now been demonstrated in numerous other systems. It does not appear to be a dosage effect resulting from competition for TFs, nor is it a result of a system that detects specific duplicate plant genes. Rather, it appears to be the result of a homology-dependent interaction between homologous sequences.

6.6   ALTERING THE FOOD CONTENT OF PLANTS

Starch is the major storage carbohydrate in higher plants. A wide range of different starches are used by the food and other industries. These are obtained by sourcing starch from different plant varieties coupled with new enzymatic or chemical approaches to create novel starches with new functional properties. Higher plants produce over 200 kinds of fatty acids, some of which have value as food. However, many are likely to have industrial (not food) uses of higher value than edible fatty acids. These are widely used in detergent synthesis. *Phytate* is the main storage form of phosphorus in many plant seeds, but bound in this form, phosphorus is a poor nutrient for monogastric animals. Plants with the phytase gene will produce seeds with lower phytate content and higher phosphorus content. Supplementation of broiler diets with transgenic seeds resulted in an improved growth rate comparable to diets supplemented with phosphate or fungal phytase.

6.7 GENE TRANSFER METHODS IN PLANTS

Recall that for production of transgenic animals, DNA is usually microinjected into pronuclei of embryonic cells at a very early stage after fertilization, or alternatively, gene targeting of embryo stem (ES) cells is employed. This is possible in animals because of the availability of specialized *in vitro* fertilization technology, which allows manipulation of ovule, zygote, or early embryo. Such techniques are not available in plants. In contrast to this in higher plants, cells or protoplasts can be cultured and used for regeneration of whole plants. Therefore, these protoplasts can be used for gene transfer followed by regeneration, leading to the production of transgenic plants. Besides cultured cells and protoplasts, other meristem cells (immature embryos or organs), pollens, or zygotes can also be used for gene transfer in plants. The enormous diversity of plant species and the availability of diverse genotypes in a species made it necessary to develop a variety of techniques suiting different situations. These different methods of gene transfer in plants will be discussed in this chapter.

6.8 TARGET CELLS FOR GENE TRANSFORMATION

The first step in gene transfer technology is to select cells that can give rise to whole transformed plants. Transformation without regeneration and regeneration without transformation are of limited value. In many species, identification of these cell types is difficult. This is unlike the situation in animals, because plant cells are totipotent and can be stimulated to regenerate into whole plants *in vitro* via *organogenesis* or *embryogenesis*. However, *in vitro* plant regeneration imposes a degree of "genome stress," especially if plants are regenerated via a callus phase. This may lead to chromosomal or genetic abnormalities in regenerated plants, a phenomenon referred to as *somaclonal variation*. In contrast to this, gene transfer into pollen (or possibly egg cells) may give rise to genetically transformed gametes, which if used for fertilization (*in vivo*) may give rise to transformed whole plants. Similarly, insertion of DNA into zygote (*in vivo* or *in vitro*) followed by embryo rescue, may also be used to produce transgenic plants. An alternative approach is the use of individual cells in embryos or meristems, which may be grown *in vitro* or may be allowed to develop normally to produce transgenic plants.

6.9 VECTORS FOR GENE TRANSFER

One common feature of vectors used for transformation is that they carry marker genes, which allow recognition of transformed cells (other cells die because of the action of an antibiotic or herbicide) and are described as selectable markers. Among these marker genes, the most common selectable marker is nptII, which provides kanamycin resistance. Other common features of suitable transformation vector are as follows: (i) multiple unique restriction sites (a synthetic polylinker) and (ii) bacterial origins of replication (e.g., ColE1). Vectors having these properties may not necessarily have features that facilitate their transfer to plant cells or integration into the plant nuclear genome. Therefore, *Agrobacterium* Ti plasmid is preferred over all other vectors because of the wide host range of this bacterial system and the capacity to transfer genes owing to the presence of T-DNA border sequences (Figure 6.12).

6.10 TRANSFORMATION TECHNIQUES USING *AGROBACTERIUM*

Agrobacterium infection (utilizing its plasmids as vectors) has been extensively utilized for transfer of foreign DNA into several dicotyledonous species. The only important species that have not responded well are major seed legumes, even though transgenic soybean (Glycine mar)

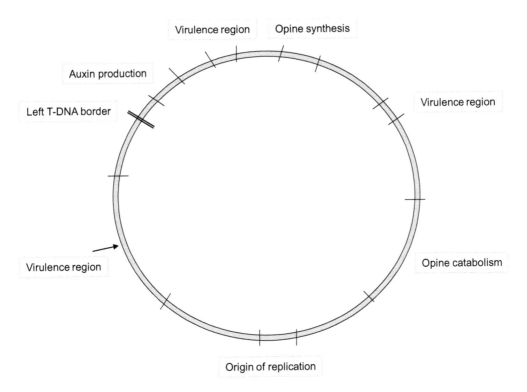

FIGURE 6.12   Gene transfer vectors in plants.

plants have been obtained. The success in this approach for gene transfer has resulted from improvement in tissue culture technology. However, monocotyledons could not be successfully utilized for *Agrobacterium*-mediated gene transfer except a solitary example of asparagus. The reasons for this are not fully understood because T-DNA transfer does occur at the cellular level. It is possible that the failure in monocots lies in the lack of wound response of monocotyledonous cells.

6.10.1   Requirements for Transgenic Plants

The important requirements for *Agrobacterium*-mediated gene transfer in higher plants include the following:

1. The plant explants must produce acetosyringone or other active compounds to induce vir genes for virulence. Alternatively, *Agrobacterium* may be preinduced with synthetic acetosyringone.
2. The induced agrobacteria should have access to cells that are competent for transformation. For gene transfer to occur, cells must be replicating DNA or undergoing mitosis (wounded and dedifferentiated cells, fresh explants, or protoplasts have these properties).
3. Often, transformed tissues or explants do not regenerate and it is difficult to combine transformation competence with totipotency (regeneration ability). Therefore, the transformation competent cells should be able to regenerate in whole plants, a combination that can be easily achieved only in some species such as tobacco. In some cases, undifferentiated cells of embryos may undergo transformation, so that the embryos may develop into chimeric plants.

6.10.2 EXPLANTS USED FOR TRANSFORMATION

The explants used for inoculation or cocultivation with *Agrobacterium* carrying the vector include protoplasts, suspension-cultured cells, callus cell clumps (undifferentiated and proembryogenic), thin cell layers (epidermis), tissue slices, whole organ sections (such as leaf discs, sections of roots, stems or floral tissues), etc. Wounding and inoculation of whole plants may also be used.

6.10.3 MARKER GENES FOR SELECTION AND SCORING OF CELLS

After explants are inoculated with *Agrobacterium* carrying the requisite vector having the gene of interest, transformed cells/tissues then need to be selected. This is facilitated by the presence of selectable marker genes available in the vector. The selectable marker genes enable the transformed cells to survive in media containing toxic levels of the selection agent, which is usually an antibiotic or an herbicide. Tobacco is used as a model transformation system, where explants (such as leaf discs) are placed on regeneration medium containing an antibiotic like kanamycin and transformed shoots can be obtained directly. Any cells that are not transformed die because of the presence of kanamycin. Other antibiotics and herbicides may require more judicious use, as even low concentrations can cause rapid cell death. In some cases, selection is exercised only after the regeneration is achieved because adventitious root formation is sensitive to antibiotics.

6.10.4 NEOMYCIN PHOSPHOTRANSFERASE GENE

This gene is used as both a selectable and a scoreable marker in experiments involving transfer of genes leading to the production of transgenic plants. It imparts kanamycin resistance, so that the transformed tissue can be selected on kanamycin. An assay for NPT II enzyme is also used to detect its presence in transformed tissue or transgenic plants. The gene for NPT II enzyme is often used with nos promoter, which drives its synthesis. In some cases, nptII gene had an adverse effect on the expression of the desirable gene introduced (such as the Bt2 gene for insect resistance), so that alternative approaches for improving its expression had to be used. To assay an NPT II enzyme, the enzyme is first fractionated using non-denaturing polyacrylamide gel electrophoresis (PAGE). Because the enzyme detoxifies kanamycin by phosphorylation, radioactively labeled ATP (^{32}P) is used with kanamycin in an agar layer, which is used to cover the gel containing the enzyme. The whole set is incubated at 35°C and the phosphorylation leading to incorporation of ^{32}P in kanamycin can be detected by autoradiography. The filter with dot blots is incubated with the substrates and is then subjected to autoradiography to detect the presence NPT II enzyme.

6.11 β-GLUCURONIDASE GUS GENE

The enzyme β-glucuronidase, popularly described as GUS, breaks down glucuronides giving a colored reaction, so that its presence can be detected *in situ* (inside the plant tissue), used either as thin section or in any other form. Several glucuronidase, which can be used as substrates, include p-nitro phenyl glucuronide (PNPG), 5-bromo, 4-chloro, 3-indolyl/glucuronide (BCIG), naphthol AS-B1 glucuronide (NAG), and resorufin glucuronide (REG). The enzyme GUS is coded by a gene gus, first isolated from *Escherichia coli*. The major advantage of using this reporter gene (scoreable marker) lies in its assay, which requires no DNA extraction, electrophoresis, or autoradiography.

6.12 AGROINFECTION AND GENE TRANSFER

Agroinfection is a phenomenon in which a virus infects a host as a part of the T-DNA of the Ti plasmid carried by *Agrobacterium*. Viral DNA can be integrated into the T-DNA and can be delivered into plant cells with the normal *Agrobacterium* T-DNA transfer process. After infection, viral DNA is released to form a functional virus that replicates and spreads systemically. Agroinfection may also lead to the integration of viral DNA so that transgenic plants containing integrated viral DNA can be produced. In maize, agroinfection with maize streak virus has been demonstrated. This suggested that the *Agrobacterium*-based vector system can be used for genetic engineering in cereals, although ordinarily *Agrobacterium* does not infect monocotyledons. Thus, agroinfection can lead to the production of transgenic plants, even though it has no better chances of yielding transgenic cereals than does *Agrobacterium* infection alone. However, agroinfection has great potential for studies in virus biology, because it can transfer deletion mutations or even single viral genes.

6.13 DNA-MEDIATED GENE TRANSFER

Agrobacterium-mediated gene transfer has been the most commonly used method of gene transfer in plants, but cereals, comprising the most important food crops, are not amenable to this method of gene transfer. Furthermore, in many crops including cereals and legumes, tissue culture techniques for regeneration are not very successful. These two limitations forced an intense search for alternative methods for gene transfer. Physical delivery of DNA or DNA-mediated gene transfer (DMGT), as it is often described, employs methods that can be grouped according to the type of target cell.

For instance, chemically stimulated endocytosis of plasmids or DNA-loaded liposomes and electroporation are employed for delivery of DNA to protoplasts only. On the other hand, techniques like microinjection, macroinjection, and shooting with microprojectiles can be used with a variety of explants (such as immature embryos, organ meristems, gametes, or zygotes). The latter techniques achieve significance because regeneration of transformed plants from protoplast-derived tissues is still difficult in many species, particularly in cereals (although encouraging results have been recently obtained in the case of maize, rice, wheat, oats, etc.).

6.14 ELECTROPORATION FOR GENE TRANSFER

This method is based on the use of short electrical impulses of high field strength. These impulses increase the permeability of the protoplast membrane and facilitate entry of DNA molecules into the cells if the DNA is in direct contact with the membrane.

In view of this, for delivery of DNA to protoplasts, electroporation is one of several routine techniques for efficient transformation. However, since regeneration from protoplasts is not always possible, cultured cells or tissue explants are often used. Consequently, it is important to test whether electroporation could transfer genes into walled cells. In most of these cases, no proof of transformation was available.

The electroporation pulse is generated by discharging a capacitor across the electrodes in a specially designed electroporation chamber. Either a high-voltage (1.5 kV) rectangular wave pulse of short duration or a low-voltage (350 V) pulse of long duration is used. The latter can be generated by a homemade machine. Protoplasts in an ionic solution containing the vector DNA are suspended between the electrodes, electroporated, and then plated as usual. Transformed colonies are selected as described previously. Using the electroporation method, successful transfer of genes was achieved with the protoplasts of tobacco, petunia, maize, rice, wheat, and sorghum. In most of these cases, the chloramphenicol acetyltransferase gene associated with a suitable promoter sequence was transferred. Transformation frequencies can be further improved by using a field strength of 1.25 kV/cm, adding PEG after adding the DNA, heat shocking protoplasts at 45°C for 5 min before adding the DNA, or by using linear instead of circular DNA.

6.15 LIPOSOME-MEDIATED GENE TRANSFER

Liposomes are small lipid bags in which many plasmids are enclosed. They can be induced to fuse with protoplasts using devices like PEG, and therefore have been used for gene transfer. This technique offers various advantages, which include protection of DNA and RNA from nuclease digestion, low cell toxicity, stability, and storage of nucleic acids because of encapsulation in liposomes, a high degree of reproducibility, and applicability to a wide range of cell types. In this technique, DNA enters the protoplasts because of endocytosis of liposomes, involving adhesion of the liposomes to the protoplast surface and fusion of the liposomes at the site of adhesion and then release of plasmids inside the cell. The technique has been successfully used to deliver DNA into the protoplasts of several plant species (such as tobacco, petunia, and carrot).

6.16   GENE TRANSFORMATION USING POLLEN

There has been a hope that DNA can be taken up by the germinating pollen and can integrate either into sperm nuclei or reach the zygote through the pollen tube pathway. Both these approaches have been tried and interesting phenotypic alterations suggesting gene transfer have been obtained. In no case, however, has unequivocal proof of gene transfer been available. In several experiments, when marker genes were used for transfer, only negative results were obtained. Several problems exist in this method and these include the presence of the cell wall, nucleases, the heterochromatic state of acceptor DNA, callose plugs in the pollen tube, etc. Transgenic plants have never been recovered using this approach and this method, though very attractive, seems to have little potential for gene transfer.

6.17   APPLICATIONS OF TRANSGENIC PLANTS

Transgenic plants have proved to be extremely valuable tools in studies on plant molecular biology, regulation of gene action, and identification of regulatory/promoter sequences. Specific genes have been transferred into plants to improve their agronomic and other features. Genes for resistance to various biotic stresses have been engineered to generate transgenic plants resistant to insects, viruses, etc. Several gene transfers have been aimed at improving quality. Transgenic plants are being used to produce novel biochemicals, such as hirudin, which are not produced by normal plants. Also, transgenic plants are now used to make vaccines for immunization against pathogens. In the following sections we will learn some of the major applications of transgenic plants with suitable examples.

6.17.1   Detoxification or Degradation of Herbicides

Several detoxifying enzymes have been identified in plants as well as in microbes such as the enzyme glutathione-S-transferase (found in maize and other plants), which detoxifies the herbicide atrazine, and nitrilase (encoded by the gene bxn from *Klebsiella pneumoniae*), which detoxifies the herbicide bromoxynil. Similarly, phosphinothricin acetyltransferase encoded by the bar gene from *Streptomyces* spp. detoxifies the herbicide L-phosphinothricin. Transgenic tomato, potato, oilseed rape (*Brassica napus*), and sugar beet plants expressing the bar gene from *Streptomyces* have been obtained. These were found to be resistant to the herbicide phosphinothricin. Similarly, transgenic tomato plants expressing the bxn gene from *Klebsiella* were resistant to the herbicide bromoxynil. Field trials with transgenic herbicide-resistant crops have been very successful, and several such varieties are in commercial cultivation especially in the United States.

6.17.2   Crystal (Cry) Proteins

The cry gene of *B. thuringiensis* produces a protein that forms crystalline inclusions in the bacterial spores. These crystal proteins are responsible for the insecticidal activities of this bacterium. The cry genes (or Cry proteins) have so far been grouped into 16 distinct groups, which either code for a 130 kDa or a 70 kDa protein. These proteins are solubilized in the alkaline environment of the insect midgut and are then prototypically processed to yield a 60 kDa toxic core fragment (except in the case of Cry IVD). The toxin function is localized in the N-terminal half of the 130 kDa proteins; the C-terminal half of these proteins is highly conserved and is most likely involved in crystal formation. The Cry I proteins are insecticidal to Lepidopteran insects. All the proteins, even the Cry IA subfamily, have a distinctive insecticidal spectrum. The Cry IIA proteins are active against both Lepidoptera and Diptera, whereas Cry IIB is specific to Diptera. The Cry III proteins are active against Coleoptera species, whereas Cry IV proteins are specific to Diptera. But the CytA protein does not show any insecticidal activity.

6.17.3   TOXIC ACTION OF CRY PROTEINS

When Cry proteins are ingested by insects, they are dissolved in the alkaline juices present in the midgut lumen. The gut proteases process them hydrolytically to release the core toxic fragments. The toxic fragments are believed to bind to specific high-affinity receptors present in the brush border of midgut epithelial cells. As a result, the brush border membranes develop pores, most likely nonspecific in nature, permitting influx of ions and water into the epithelial cells, which causes their swelling and eventual lysis. The presence of specific receptors in the midgut epithelium is most likely the chief reason for Cry toxin specificity. The specificity seems to be lost upon reduction of the cysteine residues of the protoxin but can be restored by reoxidation of these residues.

6.17.4   EXPRESSION OF CRY GENES IN PLANTS

The Cry genes have been successfully transferred into tobacco, potato, and tomato. Typically, truncated cry genes are used to produce transgenic plants since the level of expression of the complete genes in the transgenic plants is extremely low. In 1987, the first report of the response of transgenic tobacco plants to the insects *Manduca sexta* and *Heliothis virescens* was published. The transgenic tobacco plants expressing Cry protein at about 0.004% of their total leaf protein killed all *M. sexta* larvae within 6 days. Similarly, fruits of transgenic tomatoes grown in the field showed less fruit damage even under heavy infestation by fruitworm and pinworm larvae. Field tests have been quite successful with transgenic cotton as well. This crop requires about 60% of the total insecticides used in agriculture for a successful cultivation.

6.17.5   INSECT RESISTANCE TO CRY PROTEINS

There were some reports on the development of resistance to Cry proteins in some insects. For example, *Plodia interpunctella* and *Cadra cautella* were selected by continuous exposure to high levels of *B. thuringiensis*. There was a >250-fold increase in resistance in *P. interpunctella* after 36 generations, and a sevenfold increase in *C. cautella* after 21 generations. The problem of development of insect resistance to Cry proteins may be managed by combining or alternating two or more kinds of these proteins and by reducing the selection pressure on insects by limiting cry gene expression to only the economically important plant parts. Several important insect pests are not susceptible to the currently available Cry proteins. For such insects, alternative insecticidal proteins will be needed such as inhibitors of digestive enzymes (cowpea trypsin inhibitor or CpTI), serine protease inhibitor (aprotinin), cysteine protease inhibitor, proteinase inhibitor II, and lectins. These genes have been transferred into certain crop plants where they produce resistance to different insects such as members of Lepidoptera and Coleoptera species.

6.17.6   VIRUS RESISTANCE

Several approaches have been used to engineer plants for virus resistance, which include the CP gene, cDNA of satellite RNA, defective viral genome, antisense RNA approach, and ribozyme-mediated protection. Of these strategies, the use of the CP gene has been the most successful. Transgenic plants having virus CP gene linked to a strong promoter have been produced by many crop plants such as tobacco, tomato, alfalfa, sugar beet, and potato. The first transgenic plant of this type was tobacco produced in 1986. It contained the CP gene of tobacco mosaic virus (TMV) strain UI. When these plants were inoculated with TMV UI, symptoms either failed to develop or were considerably delayed. Furthermore, there was much less accumulation of virus in both inoculated and systemically infected leaves than that in the control plants. In addition, these plants showed delayed expression of disease symptoms when inoculated with the related tomato mosaic virus (ToMV) and with tobacco mild green mosaic virus (TMGMV).

It appears to be a common feature that expression of a virus CP gene not only confers resistance to the concerned virus but also gives a measure of resistance to related viruses. The effectiveness of the CP gene in conferring virus resistance can be affected by both the amount of CP produced in transgenic plants and by the concentration of the virus inoculum. Most likely the resistance generated by CP is due to blocking of the process of uncoating of virus particles, which is necessary for viral genome replication as well as expression. However, other effects seem to be involved in producing CP-mediated virus resistance. One such mechanism appears to be the prevention or delay of systemic spread of the viruses. At least in some cases, the resistance mechanism does not involve the CP itself as CP genes even in antisense orientation produce resistance to the virus.

6.17.7 Drought Resistance

Several drought-resistant genes have been identified, isolated, cloned, and expressed in plants. These potential sources of resistance to abiotic stresses include the Rab (responsive to abscisic acid) and SalT (induced in response to salt stress) genes of rice, genes for enzymes involved in proline biosynthesis in bacteria (proBA and proC in *E. coli*) and plants, and spinach genes that are involved in betaine synthesis. In plants, proline is preferentially produced from ornithine under normal conditions. However, under stress, it is made directly from glutamate, the first two reactions of the pathway being catalyzed by a single enzyme Δ^1-pyrroline 5-carboxylate synthetase (P5CS). The gene encoding P5CS has been isolated from soybean and moth bean and cloned. The moth bean P5CS gene has been transferred and overexpressed in tobacco. The transgenic plants produced 10–18-fold more proline than the control plants. The leaves of transgenic plants retained a higher osmotic potential and showed a greater root biomass under water stress than did the control plants.

These findings indicate that overexpression of P5CS in plants enhances their tolerance to osmotic stress. The primary function of accumulation of proline and other solutes such as glycine betaine appears to be the regulation of intracellular water activity. Under water stress, they may induce the formation of strong H-bonded water around proteins, thereby preserving the native state of cell biopolymers. However, it should be kept in mind that accumulation of proline is only one of the factors that enable plants to sustain growth under water stress. Other factors also allow plants to overcome osmotic stress. For example, expression of the *E. coli* gene mtl1D in plants leads to mannitol accumulation and some degree of enhanced growth under stress. An important aspect of such manipulations, however, is that the basal metabolism of the plant should be able to sustain a high rate of accumulation of the concerned osmolytes without too much of a "cost" to the plants.

6.17.8 Modification of Seed Protein Quality

Cereal seed proteins are deficient in lysine, while those of pulses are deficient in sulfur-containing amino acids such as methionine and tryptophan. This limits their nutritional value as these amino acids are essential for man. Therefore, improvement of seed storage protein quality is an important and seemingly feasible objective. The approaches to achieve this objective may be grouped into the following two broad categories, which include introduction of an appropriate transgene and modification of the endogenous protein-encoding gene.

6.18 HOW SAFE ARE TRANSGENIC PLANTS?

The main concerns when taking transgenic plants to the field relate to the possibilities of their becoming persistent weeds, gene transfers from them to other plants making the latter more persistent or invasive, and their being detrimental to the environment. In general, testing of transgenic plants should progress in a stepwise manner from laboratory to growth chamber, to greenhouse, to limited field testing, to large-scale field testing. Many countries have developed their own procedures and policies regulating field tests of such plants. In India, the Department of Biotechnology (DBT), New Delhi, is concerned with the regulation of field testing of transgenic plants.

The antibiotic resistance genes used for the selection of transformed cells are expressed in every cell of the resulting transgenic plants. It has been argued that such genes and their protein products could cause problems for human health and the environment. The protein products of such genes could be toxic to humans/animals. The food from transgenic crops will contain the antibiotic resistance gene. When such food is consumed, the bacteria present in the human intestine could acquire the antibiotic resistance gene present in the food. This would make the bacteria resistant to the antibiotics concerned, and they may become difficult to manage. The antibiotic resistance gene could be passed on from the transgenic crops to some other organisms in the environment, and this could damage the ecosystem.

There are two approaches to resolve these issues, which include excision of the antibiotic resistance gene following transformation and selection and use of nonantibiotic resistance selectable reporter genes. Herbicide resistance markers are similar in action to antibiotic resistance markers in that they save the transformed cells from the killing action of the selection agent.

The strategies of selection based on growth regulator autotrophy or substrate utilization, in contrast, provide the transformed cells with a metabolic advantage over the nontransformed ones, which are not killed by the selection agent. These strategies are based on the consideration that explants of most plant species are in an auxotrophic state, and they are unable to regenerate and grow in the absence of an external supply of several substances.

6.19 BIOENGINEERED PLANTS

After learning about plant-breeding techniques and plant pathology, eradication of various types of plant diseases becomes important (Figures 6.13, 6.14). The advances in modern molecular tools make it possible to create disease-free plants by manipulating their genes. In this section, we will closely look into famous examples of bioengineered agricultural products that have revolutionized modern crop cultivation.

6.19.1 GOLDEN RICE

Dietary micronutrient deficiencies, such as the lack of vitamin A, iodine, iron, or zinc, are a major source of morbidity (increased susceptibility to disease) and mortality worldwide. These deficiencies particularly affect children, impairing their immune systems and normal development and causing disease and ultimately death. The best way to avoid micronutrient deficiencies is by creating GM crops, and golden rice is the best example of such a crop. Golden rice was created by Ingo Potrykus of the Institute of Plant Sciences at the Swiss Federal Institute of Technology together with Peter Beyer of the University of Freiburg. The project started in 1992 and at the time of publication in 2000, golden rice was considered a significant breakthrough in biotechnology as the researchers had engineered an entire biosynthetic pathway. Golden rice was designed to produce beta-carotene, a precursor of vitamin A, in the part of rice that people eat, the endosperm. The rice plant can naturally produce beta-carotene, which is a carotenoid pigment that occurs in the leaves and is involved in photosynthesis. However, the plant does not normally produce this pigment in the endosperm as photosynthesis does not occur in the endosperm (Figure 6.15).

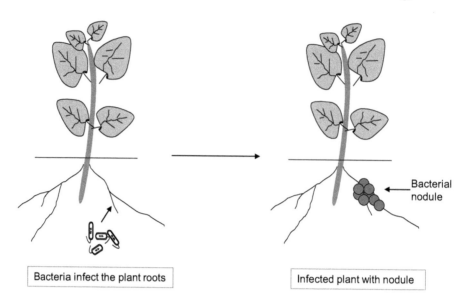

FIGURE 6.13   Plant Disease: Development of a root nodule, a place in the roots of certain plants, most notably legumes (the pea family), where bacteria live symbiotically with the plant.

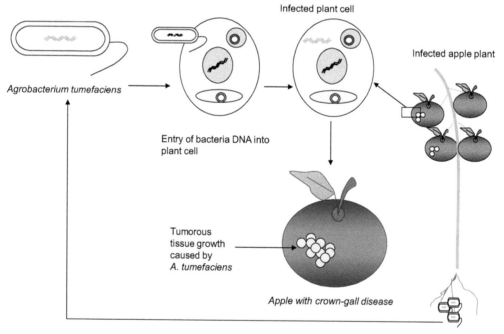

FIGURE 6.14   Plant disease caused by *Agrobacterium tumefaciens*.

Three transgenes providing phytoene synthase, phytoene desaturase, zeta-carotene desaturase, and lycopene cyclase activities were transferred into rice by *Agrobacterium*-mediated transformation. All the transgenes were introduced together in a single co-transformation experiment. The resulting transgenic rice, popularly called "golden rice," contains good quantities of beta-carotene, which gives the grains a golden color. In one transgenic line the beta-carotene content was as high

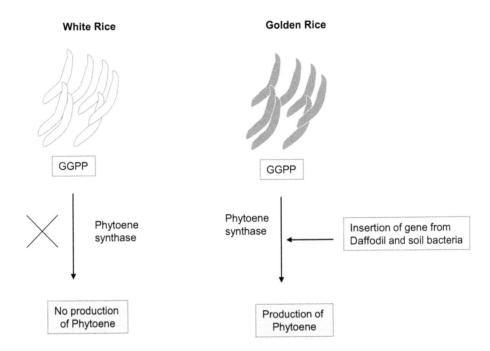

FIGURE 6.15   Bioengineered rice.

as 85% of the total carotenoids present in the grain. The original golden rice was called SGR1, and under greenhouse conditions, it produced 1.6 µg/g of carotenoids. Golden rice has been bred with local rice cultivars in the Philippines, Taiwan, and with the American rice cultivar 'Cocodrie'. The first field trials of these golden rice cultivars were conducted by the Louisiana State University Agricultural Center in 2004. Field testing allowed a more accurate measurement of the nutritional value of golden rice and enabled feeding tests. Preliminary results from the field tests showed that field-grown golden rice produces 4–5 times more beta-carotene than that grown under greenhouse conditions. In 2005, a team of researchers at the biotechnology company Syngenta produced a variety of golden rice called "Golden Rice 2." They combined the phytoene synthase gene from maize with *crt1* from the original golden rice. Golden Rice 2 produces 23 times more carotenoids than golden rice (up to 37 µg/g), and preferentially accumulates beta-carotene (up to 31 µg/g of the 37 µg/g of carotenoids). In June 2005, researcher Peter Beyer received funding from the Bill and Melinda Gates Foundation to further improve golden rice by increasing the levels of or the bioavailability of provitamin A, vitamin E, iron, and zinc and to improve protein quality through genetic modification.

Critics of genetically engineered crops have raised various concerns. One of these is that golden rice originally did not have enough vitamin A, but new strains were developed that solve this problem. However, there are still doubts about the speed at which vitamin A degrades once the plant is harvested, and how much would remain after cooking. Because carotenes are hydrophobic, there needs to be enough fat present in the diet for golden rice (or most other vitamin A supplements) to be able to alleviate vitamin A deficiency. In that respect, it is significant that vitamin A deficiency is rarely an isolated phenomenon but is usually coupled with a general lack of a balanced diet. Hence, assuming a bioavailability on par with other natural sources of provitamin A, Greenpeace estimated that adult humans would need to eat about 9 kg of cooked golden rice of the first breed to receive their recommended daily allowance (RDA) of beta-carotene, whereas a woman who was breastfeeding would need twice that amount; the effects of an unbalanced (fat-deficient) diet were not fully accounted for. In other words, it would probably have been both physically impossible to grow enough as well as to eat enough of the original golden rice to alleviate debilitating vitamin

A deficiency. This claim, however, referred to a prototype cultivar of golden rice. More recent versions have considerably higher quantities of vitamin A in them.

6.19.2 "Flavr Savr" Tomato

The Flavr Savr tomato was the first commercially grown genetically engineered food to be granted a license for human consumption. It was produced by the California-based company Calgene and submitted to the US Food and Drug Administration (FDA) in 1992. It was first sold in 1994 and was only available for a few years before production ceased. Calgene made history but mounting costs prevented it from becoming profitable, and it was eventually acquired by Monsanto. The Flavr Savr tomato was created by using antisense technology. In any gene, the DNA strand oriented as 3′→5′ in relation to its promoter is transcribed; this strand is called the *antisense strand*. The mRNA base sequence, therefore, is complementary to that of the antisense strand. The remaining DNA strand of the gene, called the *sense strand*, is naturally complementary to the antisense strand of the gene. Therefore, the base sequence of the sense strand of a gene is the same as that of the mRNA produced by it. Hence, the mRNA produced by a gene in normal orientation is also known as *sense RNA*. However, an antisense gene is produced by inverting or reversing the orientation of the protein-encoding region of a gene in relation to its promoter and as a result, the natural sense strand of the gene becomes oriented in the 3′→5′ direction regarding its promoter and is transcribed. The normal antisense strand is not transcribed since now its orientation is 5′→3′. The RNA produced by this gene has the same sequence as the antisense strand of the normal gene (except for T in DNA in the place of U in RNA) and is therefore known as *antisense RNA*. When an antisense gene is present in the same nucleus as the normal endogenous gene, transcription of the two genes yields antisense and sense RNA transcripts, respectively (Figure 6.16).

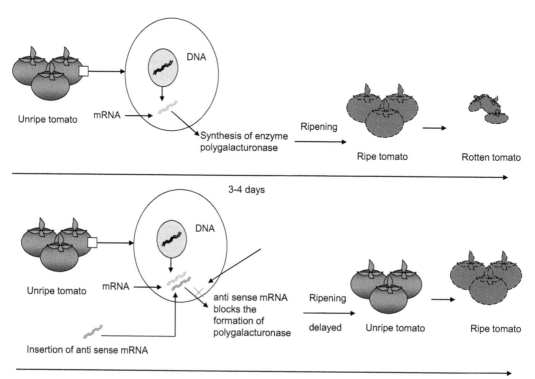

FIGURE 6.16 Making the transgenic tomato.

Because the sense and the antisense RNAs are complementary to each other, they would pair to produce dsRNA molecules. This event makes the mRNA unavailable for translation. At the same time, the RNA double strand is attacked and degraded by dsRNA-specific RNases. Finally, these events may somehow lead to the methylation of the promoter and coding regions of the normal gene, resulting in the silencing of the endogenous gene. The application of antisense RNA technology is explained using the slow ripening of the tomato as an example. In the tomato, the enzyme polygalacturonase (PG) degrades pectin, which is the major component of the fruit cell wall. This leads to the softening of the fruit and deterioration in fruit quality. Transgenic tomatoes have been produced that contain an antisense construct of the gene encoding PG. These transgenics show a drastically reduced expression of PG and markedly slower ripening and fruit softening. This has greatly improved the shelf life and the general quality of tomato fruits. Such tomatoes are being marketed in the United States under the name "Flavr Savr."

6.19.3 GENETICALLY MODIFIED MAIZE

Genetically engineered maize or transgenic maize is created by incorporating desirable traits such as herbicide and pest resistance. Transgenic maize is currently grown commercially in the United States. Maize varieties resistant to glyphosate isopropylamine salt, Liberty herbicides, and Roundup have been produced. There are also maize hybrids with tolerance to imidazoline herbicides marketed by Pioneer Hi-Bred under the trademark Clearfield, but in these the herbicide tolerance trait was bred without the use of genetic engineering. Consequently, the regulatory framework governing the approval, use, trade, and consumption of transgenic crops does not apply for imidazoline-tolerant maize. Herbicide-resistant GM maize is grown in the United States. A variation of herbicide-resistant GM maize was approved for import into the European Union in 2004. Such imports remain highly controversial (The Independent, 2005).

The European maize borer, *Ostrinia nubilalis*, destroys maize crops by burrowing into the stem, causing the plant to fall over. Bt maize is a variant of maize genetically altered to express the bacterial Bt toxin, which is poisonous to insect pests. In the case of maize, the pest is the European maize borer. Expressing the toxin was achieved by inserting a gene from the Lepidoptera pathogen microorganism *B. thuringiensis* into the maize genome. This gene codes for a toxin that causes the formation of pores in the larval digestive tract. These pores allow naturally occurring enteric bacteria such as *E. coli* and *Enterobacter* to enter the hemocoel where they multiply and cause sepsis. This is contrary to the common misconception that Bt toxin kills the larvae by starvation. In 2001, Bt176 varieties were voluntarily withdrawn from the list of approved varieties by the United States Environmental Protection Agency when it was found to have little or no Bt expression in the ears and was not found to be effective against second-generation maize borers.

One of the interesting applications of maize is in making ethanol, which is being used as biofuel especially in the United States and considered to be environmentally friendly with an economical cost. In view of its great potential as a biofuel, farmers have extensively grown maize and earned huge profits. A biomass gasification power plant in Strem near Güssing, Burgenland, Austria was begun in 2005. Research is being done to make diesel out of biogas by the Fischer–Tropsch method. Increasingly, ethanol is being used at low concentrations (10% or less) as an additive in gasoline (gasohol) for motor fuels to increase the octane rating, lower pollutants, and reduce petroleum use. The US federal government announced that production of biofuel may reach a target of 35 billion gallons by 2017.

In the year 2001, the scientific journal *Proceedings of the National Academy of Sciences* (PNAS USA) published six comprehensive studies that showed that Bt maize pollen does not pose a risk to monarch butterfly populations. Monarch populations in the United States during 1999 had increased by 30% despite Bt maize accounting for 30% of all maize grown in the United States that year. The beneficial effects of Bt maize on monarch populations can be attributed to reduced pesticide use. Numerous scientific studies continue to investigate the potential effects of Bt maize on a variety of nontarget invertebrates. A synthesis of data from many such field studies showed that the measured effect depends on the standard of comparison. The overall abundance of nontarget invertebrates in CrylAb variety Bt maize fields is significantly higher compared to non-GM maize fields treated with insecticides, but significantly lower compared to insecticide-free non-GM maize fields. Abundance in fields of another variety, Cry3Bb maize, is not significantly different compared to non-GM maize fields either with or without insecticides. By law, farmers in the United States who plant Bt maize must plant non-Bt maize nearby. These nonmodified fields are to provide a location to harbor pests. The theory behind these refuges is to slow the evolution of pests to the Bt pesticide. Doing so enables an area of the landscape where wild-type pests will not be immediately killed.

6.20   TERMINATOR TECHNOLOGY

Genetic use restriction technology (GURT), also known as *terminator technology*, is the name given to proposed methods for restricting the use of GM plants by causing second-generation seeds to be sterile (Figure 6.17). The technology was developed under a cooperative research and development agreement between the Agricultural Research Service of the USDA and the Delta and Pine Land Company in the 1990s, but it is not yet commercially available. Because some stakeholders expressed concerns that this technology might lead to dependence for poor smallholder farmers, Monsanto Company, an agricultural products company and the world's largest seed supplier, pledged to not commercialize the technology in 1999. Late in 2006, it acquired Delta and Pine Land Company. The technology was discussed during the *8th Conference of the Parties to the UN's Convention on Biological Diversity* in Curitiba, Brazil, March 20–31, 2006.

6.20.1   TYPES OF TERMINATOR TECHNOLOGY

6.20.1.1   V-GURT

This type of GURT produces sterile seeds, which means that a farmer who had purchased seeds containing V-GURT technology could not save the seed from this crop for future planting. This

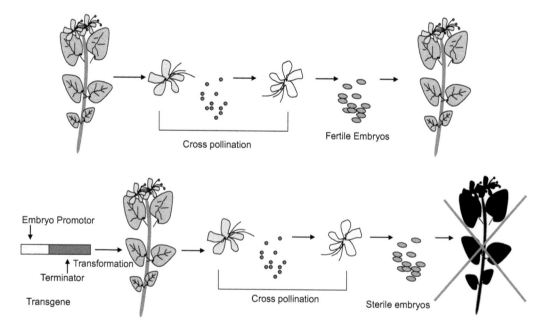

FIGURE 6.17   Terminator technology.

would not have an immediate impact on the large number of primarily western farmers who use hybrid seeds, as they do not produce their own planting seeds and instead buy specialized hybrid seeds from seed production companies. However, currently around 80% of farmers in both Brazil and Pakistan grow crops based on saved seeds from previous harvests. Consequentially, resistance to the introduction of GURT technology in developing countries is strong. The technology is restricted at the plant variety level, hence the term V-GURT. Manufacturers of genetically enhanced crops would use this technology to protect their products from unauthorized use.

6.20.1.2   T-GURT

This type of GURT modifies a crop in such a way that the genetic enhancement engineered into the crop does not function until the crop plant is treated with a chemical that is sold by the biotechnology company. Farmers can save seeds for use each year. However, they do not get to use the enhanced trait in the crop unless they purchase the activator compound. The technology is restricted at the trait level, hence the term T-GURT.

6.20.2   Benefits of Terminal Technology

Where effective intellectual property protection systems do not exist or are not enforced, GURTs could be an alternative to stimulate plant-developing activities by biotech firms.

Nonviable seeds produced on V-GURT plants may reduce the propagation of volunteer plants. Volunteer plants can become an economic problem for larger-scale mechanized farming systems that incorporate crop rotation. Under warm, wet harvest conditions non-V-GURT grain can sprout, which lowers the quality of grain produced. It is likely that this problem would not occur with the use of V-GURT grain varieties. Use of V-GURT technology could prevent the escape of transgenes into wild relatives and prevent any impact on biodiversity. Crops modified to produce nonfood products could be armed with GURT technology to prevent accidental transmission of these traits into crops destined for foods.

6.20.3   Concerns Regarding Terminal Technology

There is a concern that V-GURT plants could cross-pollinate with nongenetically modified plants, either in the wild or in the fields of farmers who do not adopt the technology. Though the V-GURT plants are supposed to produce sterile seeds, there is concern that this trait will not be expressed in the first generation of a small percentage of these plants but will be expressed in later generations. This does not seem to be much of a problem in the wild, as a sterile plant would naturally be selected out of a population within one generation of trait expression. The food safety of GURT technology would need to be assessed if a commercial release of a GURT-containing crop were proposed.

Initially developed by the US Department of Agriculture and multinational seed companies, "suicide seeds" have not been commercialized anywhere in the world because of an avalanche of opposition from farmers, indigenous peoples, and civil society. In 2000, the United Nations Convention on Biological Diversity recommended a *de facto* moratorium on field testing and commercial sale of terminator seeds. The moratorium was reaffirmed in 2006. India and Brazil have already passed national laws to prohibit the technology.

PROBLEMS

Section A: Descriptive Type

Q1. Briefly describe classical breeding.
Q2. How are plants cultivated using somatic hybrid techniques?
Q3. How is plant breeding improved by using the RFLP method?

Q4. Describe various diseases caused by bacteria.
Q5. How is the food content of plants altered?
Q6. How are drought-resistant plants generated?
Q7. Describe biochemical production in plants.
Q8. How are vaccines synthesized from plants?
Q9. How safe are transgenic plants?
Q10. Write an essay on golden rice.

Section B: Multiple Choice

Q1. Domestication of plants is what kind of selection process to get desirable traits of plants?
   a. Natural
   b. Artificial
   c. Classical

Q2. A plant whose origin or selection is due to human activity is known as . . .
   a. Mutagen
   b. Cultigen
   c. Agrogen

Q3. Landrace is a crop that evolved from a wild population because of selective breeding. True/False

Q4. Classical breeding relies largely on recombination between chromosomes to generate genetic diversity.
   a. Heterologous
   b. Homologous
   c. Phytologous
   d. None of the above

Q5. The chromosome numbers can be increased artificially by treating the plant with . . .
   a. Antibiotic
   b. Growth hormone
   c. Colchicine
   d. Oxytocin

Q6. Potato disease in Europe and South America is caused by . . .
   a. Bacteria
   b. Viruses
   c. Nematodes
   d. Oomycetes

Q7. Starch is the major storage of carbohydrates in higher plants. True/False

Q8. The most commonly used vector for gene transfer in higher plants is . . .
   a. *Nocardia*
   b. *Salmonella*
   c. *Agrobacterium tumefaciens*
   d. None of the above

Q9. The T-DNA regions on all Ti and Ri plasmids are flanked by almost 25 bp direct repeat sequences, which are essential for T-DNA transfer. True/False

Q10. The major advantage of using this gene lies in its assay, which requires no DNA extraction, electrophoresis, or autoradiography.
   a. GUS gene
   b. GFAP gene
   c. Bt-2 gene
   d. none of the above

Q11. Golden rice was created by . . .
  a. Ingo Potrykus
  b. Watson
  c. Ian Wilmut
  d. Fleming
Q12. "Flavr Savr" is a genetically modified . . .
  a. Orange
  b. Apple
  c. Potato
  d. Tomato

Section C: Critical Thinking

Q1. Explain how transformation without regeneration and regeneration without transformation are of limited value.
Q2. How would you distinguish a natural plant from a genetically modified plant?
Q3. Do you agree that genetically modified plants that are resistant to pests are harmful to human health? Why or why not?

ASSIGNMENT

Can agriculture biotechnology assist in meeting the food demands of a growing global population? Prepare a report indicating the global status of agricultural products developed from genetically modified plants.

REFERENCES AND FURTHER READING

Briggs, F.N., and Knowles, P.F. *Introduction to Plant Breeding*. Reinhold Publishing Corporation, New York, 1967.
Chavarro, J.E., Toth, T.L., Sadio, S.M., and Hauser, R. Soy food and isoflavone intake in relation to semen quality parameters among men from an infertility clinic. *Hum. Reprod.* 2: 2584–2590, 2008.
Chilcutt, C.F., and Tabashnik, B.E. Contamination of refuges by Bacillus thuringiensis toxin genes from transgenic maize. *Proc. Natl. Acad. Sci. USA* 101: 7526–7529, 2004.
Crawford, G.W. East Asian plant domestication. In: *Archaeology of East Asia*, M. Stark (ed.). Blackwell, Oxford, UK, p. 81, 2006.
Dawe, D., Robertson, R., and Unnevehr, L. Golden rice: What role could it play in alleviation of vitamin A deficiency? *Food Policy* 27: 541–560, 2002.
de Lemos, M.L. Effects of soy phytoestrogens genistein and daidzein on breast cancer growth. *Ann. Pharmacother.* 5: 1118–1121, 2001.
Derbyshire, E. et al. Review: Legumin and vicilin, storage proteins of legume seeds. *Phytochemistry* 15: 3–24, 1976.
Dillingham, B.L., McVeigh, B.L., Lampe, J.W., and Duncan, A.M. Soy protein isolates of varying isoflavone content exert minor effects on serum reproductive hormones in healthy young men. *J. Nutr.* 135: 584–591, 2005.
Fradin, M.S., and Day, J.F. Comparative efficacy of insect repellents against mosquito bites. *N. Engl. J. Med.* 34: 13–18, 2002.
Gepts, P. A comparison between crop domestication, classical plant breeding, and genetic engineering. *Crop Sci.* 42: 1780–1790, 2002.
Giampietro, P.G., Bruno, G., Furcolo, G. et al. Soy protein formulas in children: No hormonal effects in long-term feeding. *J. Pediatr. Endocrinol. Metab.* 17: 191–196, 2004.
Gottstein, N., Ewins, B.A., Eccleston, C. et al. Effect of genistein and daidzein on platelet aggregation and monocyte and endothelial function. *Br. J. Nutr.* 89: 607–616, 2003.
Heald, C.L., Ritchie, M.R., Bolton-Smith, C., Morton, M.S., and Alexander, F.E. Phyto-oestrogens and risk of prostate cancer in Scottish men. *Br. J. Nutr.* 98: 388–396, 2007.

Hirschberg, J. Carotenoid biosynthesis in flowering plants. *Curr. Opin. Plant Biol.* 4: 210–218, 2001.

Hogervorst, E., Sadjimim, T., Yesufu, A., Kreager, P., and Rahardjo, T.B. High tofu intake is associated with worse memory in elderly Indonesian men and women. *Dement. Geriatr. Cogn. Disord.* 26: 50–57, 2008.

Humphrey, J.H., West, K.P. Jr., and Sommer, A. Vitamin A deficiency and attributable mortality in under-5-year-olds. *WHO Bulletin* 70: 225–232, 1992.

International Rice Research Institute. 2005. Program 3. Annual Report of the Director General, 2004–2005. www.irri.org.

King, D., and Gordon, A. Contaminant found in Taco Bell taco shells: Food safety coalition demands recall. Press release, 2001. Friends of the Earth, Washington, DC, 2000. www.foe.org/act/getacobellpr.html.

Martineau, B. First fruit. In: *The Creation of the Flavr Savr Tomato and the Birth of Biotech Food.* McGraw-Hill, New York, 2001.

Maskarinec, G., Morimoto, Y., Hebshi, S., Sharma, S., Franke, A.A., and Stanczyk, F.Z. Serum prostate-specific antigen but not testosterone levels decrease in a randomized soy intervention among men. *Eur. J. Clin. Nutr.* 60: 1423–1429, 2006.

Meagher, R.B. Phytoremediation of toxic elemental and organic pollutants. *Curr. Opin. Plant Biol.* 3: 153–162, 2000.

Merritt, R.J., and Jenks, B.H. Safety of soy-based infant formulas containing isoflavones: The clinical evidence. *J. Nutr.* 134: 1220S–1224S, 2004.

Messina, M., McCaskill-Stevens, W., and Lampe, J.W. Addressing the soy and breast cancer relationship: Review, commentary, and workshop proceedings. *J. Natl. Cancer Inst.* 98: 1275–1284, 2006.

Paine, J.A. et al. Improving the nutritional value of golden rice through increased pro-vitamin A content. *Nat. Biotechnol.* 23(4): 482, 2005.

Potrykus, I. Golden rice and beyond. *Plant Physiol.* 125: 1157–1161, 2001.

Schaub, P. et al. Why is golden rice golden (yellow) instead of red? *Plant Physiol.* 138: 441–450, 2005.

Sears, M.K. et al. Impact of Bt corn pollen on monarch butterfly populations: A risk assessment. *Proc. Natl. Acad. Sci. USA* 98: 11937–11942, October 9, 2001.

Winston, L.R., and Koffler, H. Corn steep liquor in microbiology. *Bacteriol Rev.* 12: 297–311, 1948.

7 Animal Biotechnology

- Discuss the significance of the use of animals in research
- Explain why animal research is important from a historical perspective
- Discuss animal biotechnology and its significance
- Discuss animal testing in pharmaceutical and biotechnology companies
- Discuss animal cloning and transgenic animals
- Discuss the commercial aspects of animal biotechnology

7.1 INTRODUCTION

Animals play an important role in human life as many drugs or medicines that we use today are first tested in animals. Throughout history, scientists have solved many medical problems and developed treatments and cures for diseases by using animals in biomedical research. Some of the most frequently asked questions about animals are generally related to the use of animals in research, and the general impression is that scientists kill animals for no reason. That is why it is necessary to put a balance in the human perspective about the use of animals to understand the ethical as well as the scientific points of view.

Let us first try to understand why researchers or scientists need to kill animals; there are several reasons for that:

- When scientists learned that animals' anatomy and physiology are very similar to humans, it became preferable to use animals rather than humans for their preliminary researches.
- Certain strains or breeds of animals develop the same diseases or conditions as humans. "Animal models" are frequently critical in understanding a disease and in developing appropriate treatments.
- Research means introducing one variable and observing the results of that one item. With animals, we can control their environment (temperature, humidity, etc.), and shield them from diseases or conditions not related to the research (control their health). Although humans and animals develop the disease that may be the subject of a research investigation, their different lifestyles or living conditions make them poor subjects until preliminary research under controlled conditions has been done.
- We can use scientifically valid numbers of animals. Data from one animal or human are not considered reliable data in research. In order to obtain reliable data to scientifically test a hypothesis, an adequate number of animals must be used to statistically test the results of the research

All of the above points clearly suggest the significance of animals in research and experimentation (Figure 7.1). Animals are used as models to test drugs and medical procedures to treat various human diseases. For a new therapy to be developed, it is first tested in animals to check its efficacy. If animal results are positive, then with proper regulatory approvals the drugs are allowed to be tested in humans. Some individuals claim that we should use humans or animals that already have a disease to study that disease. Certainly, epidemiological studies tracking the occurrence of a disease or condition have provided many important insights into the cause of a disease or a

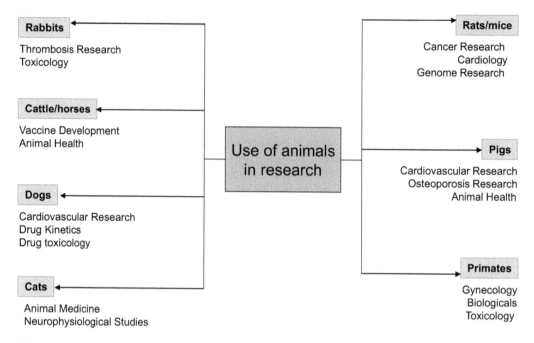

FIGURE 7.1 Application of animals in experimental research.

condition, especially when an environmental aspect is responsible. However, epidemiological studies are successful in only a limited number of situations. As noted previously, the study of a disease is severely hindered or not possible when the research subjects have been/are exposed to a variety of environmental factors. It is important to note that, according to the American Medical Association, humans are the most frequently used subjects in research. However, research studies conducted in humans follow preliminary studies conducted in animals. These animal studies make human studies a reasonable risk. The animal studies are not a guarantee of success, but they do tell us that the human research has a reasonable probability of success. Mice, dogs, rats, rabbits, cats, sheep, pigs, monkeys, and horses are some of the most widely used animals in drug testing. On a different note, we should not forget the sacrifice of thousands of laboratory animals who have been equally contributing to the development of new therapies or new drug molecules for so many years. In the next section, we will learn the historical perspective of animal use in research.

7.2 HISTORY OF THE USE OF ANIMALS IN RESEARCH

It may appear that animal use in research is only happening in modern times, but you may be surprised to know that the information pertaining to animal testing is found in ancient scripts. Aristotle (384–322 BCE) and Erasistratus (304–258 BCE) were among the first to perform experiments on living animals. In the 1880s, Louis Pasteur convincingly demonstrated the germ theory of medicine by inducing anthrax in sheep. In the 1890s, Ivan Pavlov famously used dogs to describe classical conditioning. Insulin was first isolated from dogs in 1922 and revolutionized the treatment of diabetes. In 1957, a Russian dog, Laika, became the first of many animals to orbit the earth. In the 1970s, antibiotic treatments and vaccines for leprosy were developed using armadillos, then given to humans. The ability of humans to change the genetics of animals took a large step forward in 1974 when Rudolf Jaenisch was able to produce the first transgenic mammal by integrating DNA from the SV40 virus into the genome of mice. This genetic research progressed rapidly and, in 1996, Dolly the sheep was born, the first mammal to be cloned from an adult cell.

The controversy surrounding animal testing dates back to the seventeenth century. There were objections on an ethical basis, contending that the benefit to humans did not justify the harm to animals. Early objections to animal testing also came from another angle. Many people believed that animals were inferior to humans and so different that results from animals could not be applied to humans. On the other side of the debate, those in favor of animal testing held that experiments on animals were necessary to advance medical and biological knowledge. Claude Bernard—whose wife, Marie Françoise Martin, founded the first antivivisection society in France in 1883—famously wrote in 1865 that "the science of life is a superb and dazzlingly lighted hall which may be reached only by passing through a long and ghastly kitchen." Arguing that "experiments on animals are entirely conclusive for the toxicology and hygiene of man . . . the effects of these substances are the same on man as on animals, save for differences in degree." In 1883, Claude Bernard, the Father of Physiology, established animal experimentation as part of the standard scientific method.

7.3   DRUG TESTING IN ANIMALS IS MANDATORY

In the nineteenth century, laws regulating drugs were more relaxed in the United States and the federal government could only ban a drug after a company had been prosecuted for selling products that harmed customers. However, in response to a tragedy in 1937 where a drug labeled "Elixir of Sulfanilamide" killed more than 100 people, the US Congress passed laws that required safety testing of drugs in animals before they could be marketed and prescribed for human use. Other countries enacted similar legislation. In the 1960s, in reaction to the thalidomide tragedy, further laws were passed requiring safety testing in pregnant animals before a drug could be sold. Today, the majority of people in our society have agreed with the idea of the humane and responsible use of animals in research. In fact, today's regulatory guidelines pertaining to use of animals as experimental models have been drafted and followed in most of the countries in the world.

7.4   MOST COMMONLY USED ANIMALS IN RESEARCH

One of the major problems in animal testing is the availability of a large number of laboratory animals, especially those that have a long lifespan. Because of this problem, scientists prefer to use laboratory animals with a short lifespan and in a short duration of time.

7.4.1   INVERTEBRATES

Invertebrates are more extensively used in basic research than in applied research. Most of the experiments based on invertebrates are largely conducted without getting animal ethical approvals and are largely unregulated by law. The most used invertebrate species are *Drosophila melanogaster*, a fruit fly, and *Caenorhabditis elegans*, a nematode worm. In the case of *Caenorhabditis elegans*, the worm's body is completely transparent and the precise lineage of all the organism's cells is known, whereas studies in the fly *Drosophila melanogaster* can use an amazing array of genetic tools. These animals offer great advantages over vertebrates, including their short life cycle and the ease with which large numbers may be studied, with thousands of flies or nematodes fitting into a single room. However, the lack of an adaptive immune system and their simple organs prevent worms from being used in medical research such as vaccine development. Similarly, flies are not widely used in applied medical research as their immune system differs greatly from that of humans, and diseases in insects can be very different from diseases in more complex animals.

7.4.2   FISH AND AMPHIBIANS

Besides rodents, fish and amphibians are extensively used in biology education, training, and research. In the United Kingdom alone, nearly 500,000 fish and 9,000 amphibians were used in research experimentations in 2016. The main species used in research are the zebrafish, *Danio rerio*, because it is very easy to study their developmental stages in great detail. Secondly, the body of the zebrafish remains translucent during their embryonic stage. On the other hand, the African clawed frog, *Xenopus laevis*, is used to study regeneration. Another striking difference from vertebrate models is that these animals do not come under strict ethical and animal guidelines and provide an easy access for research experimentation.

7.4.3 Vertebrates

7.4.3.1 Rodents

About 50 to 100 million vertebrate animals are used in experimentations annually. Among vertebrates, rats, mice, and rabbits are the most commonly used animals in biomedical research. In the United States alone, the number of rats and mice used is estimated at 20 million/year. Other commonly used rodents are guinea pigs, hamsters, and gerbils. Of these rodents, mice are the most commonly used because of their size, low cost, ease of handling, and fast reproduction rate. Mice are widely considered to be the best model of inherited human disease and share 99% of their genes with humans. With the advent of genetic engineering technology, genetically modified mice are generated to provide models for a range of human diseases. Rats are also widely used for physiology, toxicology, and cancer research, but genetic manipulation is much harder in rats than in mice, which limits the use of these rodents in basic science. Albino rabbits are used in eye irritancy tests because rabbits have less tear flow than other animals, and the lack of eye pigment in albinos make the effects easier to visualize. Over 20,000 rabbits were used in the UK in 2004. On the other hand, mice are used in monoclonal antibody production, whereas rabbits are frequently used for polyclonal antibody production.

7.4.3.2 Cats and Dogs

Although rats, mice, and other small animals provide basic information on human diseases, they do not have the same neurological characteristics as that of humans. In order to study the pathological conditions associated with neurological diseases, researchers prefer to work on higher vertebrates such as cats and dogs. For neurological pathology associated with humans, the favored animal models are cats. According to the American Anti-Vivisection Society, about 25,500 cats were used in neurological research in the year 2000 in the United States alone and about 18,898 cats were used in 2016. Most of them were used to study the potential cause of pain and/or distress. On the other hand, dogs are widely used in biomedical research, testing, and education, particularly beagles because they are gentle and easy to handle. Dogs are commonly used animal models for human diseases in cardiology, endocrinology, and bone and joint studies, researches that tend to be highly invasive according to the Humane Society of the United States. The US Department of Agriculture's Animal Welfare Report shows that about 66,000 dogs were used in USDA-registered facilities in the year 2005 and about 60,979 dogs were used in 2016. In the United States, some of the dogs are purpose-bred but most are supplied by so-called Class B dealers licensed by the USDA.

7.4.3.3 Primates and Nonprimates

Researchers find it difficult to do experiments on human cognition and behaviors using small animals. In fact, it is very difficult to create human-like behaviors in animals, but there are higher-order animals close to humans that are being considered for such studies. It has been estimated that around 65,000 primates are being used in research projects in each year in the United States and Europe. Nonhuman primates (NHPs) are used in toxicology tests, studies of AIDS and hepatitis, studies of neurology, behavior and cognition, reproduction, genetics, and xenotransplantation. They are caught in the wild or purpose-bred. The European Commission reported that 6,012 monkeys were experimented on in European laboratories in 2011 and according to the US Department of Agriculture, 71,188 monkeys were tested in laboratories in 2016. In the US and China, most primates are domestically purpose-bred, whereas in Europe, the majority are imported purpose-bred animals such as rhesus monkeys, cynomolgus monkeys, squirrel monkeys, and owl monkeys. Around 12,000–15,000 monkeys are imported into the United States annually. In total, around 70,000 NHPs are used each year in the United States and European Union (EU) countries. Most of the NHPs used are macaques, although marmosets, spider monkeys, squirrel monkeys, baboons, and chimpanzees are also used in the United States. In 2006, there were 1133 chimpanzees in US primate centers. The first transgenic primate was produced in 2001, with the development of a method that could introduce new genes into a rhesus macaque. This transgenic technology is now being applied in the search for a treatment for the genetic disorder Huntington's disease. In addition, NHPs were part of the polio vaccine development.

7.5   APPLICATION OF ANIMAL MODELS

In the previous section, we learned about the uses of different types of animals in research and for testing purposes. With the discovery of molecular and genetic tools, it became possible to create specific kinds of animals by manipulating their genes using genetic engineering technology or cloning. These genetically designed animals are regularly being used in laboratory research and for drug testing. In the next section, we will learn various applications of animals in laboratory research. We have broadly classified the application of animals into two subcategories, the use of animals in basic research and the use of animals in applied or industry research.

7.5.1   Use of Animals in Basic Research

Animals are extensively used in researches that have been mostly carried out by university laboratories and research centers funded by the federal government or agencies. The university scientists usually work with animals to determine the cause or progression of diseases. When they come up with any new information, they publish their research in scientific journals. In basic research, scientists use large numbers and a greater variety of animals than in applied research. Fruit flies, nematode worms, mice, and rats together account for the vast majority of animals, though small numbers of other species are used, ranging from sea slugs through to armadillos. We have listed a few examples of types of animals and experiments used in basic research: (1) Animals are used to study *embryogenesis* and developmental aspects of the human body or tissues. Researchers have created genetically designed fruit flies by adding or deleting genes. By studying the changes in development that these changes produce, scientists aim to understand both how organisms normally develop and what can go wrong in this process. These studies are particularly powerful as the basic controls of development, such as the homeobox genes, have similar functions in organisms as diverse as fruit flies and man. (2) Animals are used to understand how organisms detect and interact with each other and their environment in studies in which fruit flies, worms, mice, and rats are all widely used. Studies of brain function, such as memory and social behavior, often use rats and birds. For some species, behavioral research is combined with enrichment strategies for animals in captivity because it allows them to engage in a wider range of activities. (3) Breeding experiments on animals are used to study *evolution* and *genetics*. Laboratory mice, flies, fish, and worms are inbred through many generations to create strains with defined characteristics. These provide animals of a known genetic background, an important tool for genetic analyses. Larger mammals are rarely bred specifically for such studies because of their slow rate of reproduction, though some scientists take advantage of inbred domesticated animals, such as dog or cattle breeds, for comparative purposes. Scientists also use animals to check genetic mutations in the animal population. One example is sticklebacks, which are now being used to study how many and which types of mutations are selected to produce adaptations in animals' morphology during the evolution of new species.

7.5.2 Use of Animals in Applied Research

Interestingly, most of the industry research for new treatments or therapies are in fact based on the outcomes of basic research. In this section, we discuss some of the areas of industrial researches where animal models have been extensively used.

7.5.2.1 Genetic Diseases

Genetic diseases are caused by mutation in a specific type of gene(s). To develop a treatment or therapy for a genetic disease, it is very important to know the cause of the disease. The cause of a genetic mutation can be studied by creating similar mutations in animals. With the advancement of modern genetic engineering tools, it is now possible to add or delete a specific gene and induce mutations in animals. These animals are known as *transgenic animals*. Transgenic animals have specific genes inserted, modified, or removed to mimic specific conditions such as single gene disorders like Huntington's disease. Other models mimic complex, multifactor diseases with genetic components such as diabetes or involve transgenic mice that carry the same mutations that occur during the development of cancer. These models allow investigations on how and why the disease develops. They also provide ways to develop and test new treatments. The vast majority of these transgenic models of human diseases are lines of mice and mammalian species in which genetic modification is most efficient. Smaller numbers of other animals are also used, including rats, pigs, sheep, fish, birds, and amphibians.

7.5.2.2 Virology

Certain domestic and wild animals have a natural propensity or predisposition for certain conditions that are also found in humans. Cats are used as a model to develop immunodeficiency virus vaccines and to study leukemia because of their natural predisposition to feline leukemia virus. Certain breeds of dog suffer from narcolepsy making them the major model used to study the human condition. Armadillos and humans are among only a few animal species that naturally suffer from leprosy, as the bacteria responsible for this disease are yet to be grown in culture. Armadillos are the primary source of bacilli used in leprosy vaccines.

7.5.2.3 Neurological Disorders

Researchers are trying to find a dream drug or therapy for all neurological disorders, which are not currently curable. In order to make this dream drug, researchers have to first create neurological disorder-like conditions in the animals and test the drug or cells in these animal models. In the last few years, researchers have been able to model human diseases in animals. The stroke model in animals is created by restricting blood flow to the brain, and Parkinson-like syndrome can be created by injecting neurotoxins into the substantia nigra region of the midbrain, which are known to be involved in Parkinson's disease. Such studies can be difficult to interpret, and it is argued that they are not always comparable to human diseases. For example, although such models are now widely used to study Parkinson's disease, the British anti-vivisection interest group (British Union for the Abolition of Vivisection or BUAV) argues that these models only superficially resemble the disease symptoms, without the same time course or cellular pathology. In contrast, scientists, as well as the medical research charity, attest to the usefulness of animal models of Parkinson's disease. The Parkinson's appeal states that these models were invaluable and that they led to improved surgical treatments such as pallidotomy, new drug treatments such as levodopa, and later, deep brain stimulation.

7.5.2.4 Organ Transplantation

With the rampant increase in organ failures in humans, there is a tremendous demand for human organs for transplantation such as the kidney and the heart. Getting these organs is extremely difficult and hundreds of patients are dying because of unavailability of these organs. Against

FIGURE 7.2   Testing of an animal for motor function: The mouse is kept on a Rota rod apparatus and the motor function of the animal is evaluated based on the length of time the animal stays on the rotating rod; the longer the animal stays indicates that the animal has good motor coordination.

Source: Picture courtesy of the Department of Biotechnology, Manipal University Dubai, UAE

this background, researchers are working to determine the suitability of using animal organs in place of human organs. The science of transplanting an organ of one species to another, in this case of human to mouse or mouse to human, is called as *xenotransplantation* (Figure 7.2). Xenotransplantation research involves transplanting tissues or organs from one species to another as a way to overcome the shortage of human organs for use in organ transplants. Current research involves the use of primates as the recipients of organs from pigs that have been genetically modified to reduce the primates' immune response against the pig tissue. Although transplant rejection remains a problem, recent clinical trials that involved implanting pig insulin-secreting cells into diabetic patients did reduce these patient's needs for insulin. As the success rate is a bit low at the moment, xenotransplantation is not yet safe in human and extensive clinical trials must be conducted first to achieve complete success. In 1999, the British Home Office released figures which showed that 270 monkeys had been used in xenotransplantation research from 1995–1999. Scientists used wild baboons imported from Africa for xenotransplantation by grafting pigs' hearts and kidneys. Unfortunately, some baboons died after suffering strokes, vomiting, diarrhea, and paralysis.

7.5.2.5   Drug Efficacy Testing

All pharmaceutical companies ensure that drugs must be properly tested in animals before being used in humans (Figure 7.3). *Drug efficacy* is the effect of the drug on the body or body organ(s) at various time intervals. It also checks whether or not the drug has reached the target site and produced desirable effects. A drug efficacy test is commonly a technical examination of urine, hair, blood, sweat, or oral fluid samples to determine the presence or absence of specified drugs or their metabolized traces. In the early twentieth century, laws regulating drugs were not so stringent, but nowadays all new pharmaceuticals undergo rigorous animal testing before being used in humans. The following are some tests done on pharmaceutical products:

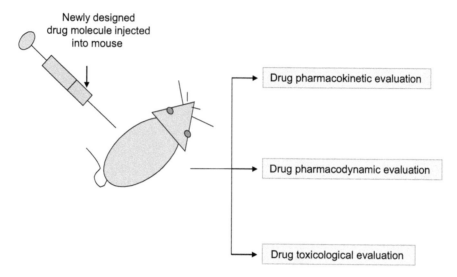

FIGURE 7.3 Drug testing: To evaluate the efficacy and toxicity of drug molecules.

Metabolic tests investigating pharmacokinetics—how drugs are absorbed, metabolized, and excreted by the body when introduced orally, intravenously, intraperitoneally, intramuscularly, or transdermally.

Efficacy studies that test whether experimental drugs work involve inducing the appropriate illness in animals. The drug is then administered in a double-blind controlled trial, which allows researchers to determine the effect of the drug and the dose–response curve.

Specific tests on reproductive function, embryonic toxicity, or carcinogenic potential can all be required by law, depending on the result of other studies and the type of drug being tested.

All of the above tests are very critical to making a successful drug.

7.5.2.6 Toxicological Analysis

As per international drug testing guidelines, it is mandatory to know any toxic effect(s) of a new drug molecule before testing in humans, and a newly synthesized drug must be tested in animals to check for any undesirable effects. *Toxicology* is the study of the adverse effects of drugs or chemicals on living organisms. It is the study of symptoms, mechanisms, treatments, and detection of toxic effects associated with drug or chemical consumption. Pharmaceutical and biotechnology companies normally conduct almost all their toxicological testing in animals. According to 2005 EU figures, around 1 million animals are used every year in Europe in toxicology tests, which are about 10% of all procedures. The toxicological tests are conducted without anesthesia because interactions between drugs may interfere with the results. Toxicology tests are required for products such as pesticides, medications, food additives, packing materials, and air fresheners or their chemical ingredients. Most tests involve testing ingredients rather than finished products. The substances are applied to the skin or dripped into the eyes; injected intravenously, intramuscularly, or subcutaneously; inhaled either by placing a mask over the animals and restraining them or by placing them in an inhalation chamber; or administered orally, through a tube into the stomach or simply in the animal's food.

There are several different types of acute toxicity tests. The lethal dose 50 (LD50) test is used to evaluate the toxicity of a substance by determining the dose required to kill 50% of the test animal population. This test was removed from the Organization for Economic Co-operation and

Development (OECD) international guidelines in 2002 and replaced by methods such as the fixed-dose procedure, which uses fewer animals and causes less suffering. The Humane Society of the United States writes that the procedure can cause redness, ulceration, hemorrhaging, cloudiness, or even blindness in animals. The most stringent tests are reserved for drugs and foodstuffs. For these, a number of tests are performed, lasting less than a month (acute), 1–3 months (subchronic), and more than 3 months (chronic) to test general toxicity (damage to organs), eye and skin irritancy, mutagenicity, carcinogenicity, teratogenicity, and reproductive problems. The cost of the full complement of tests is several million dollars per substance and it may take 3 or 4 years to complete.

7.5.2.7 Cosmetics Testing

There was a surge in cosmetic products and cosmetic-based industries around the world and, like pharmaceutical products, cosmetic products also undergo toxicological tests before use in humans. These cosmetic products are usually made of chemicals or a composition of chemicals derived either from natural sources or synthetic sources. The cosmetic products are first tested in animals to check their efficacy and safety. Using animal testing in the development of cosmetics may involve testing either a finished product or the individual ingredients of a finished product on animals, often rabbits but also mice, rats, and other animals. In some cases, the products or ingredients are applied to the mucous membranes of the animal, including the eyes, nose, and mouth, to determine whether they cause allergic or other reactions (Figure 7.4). Cosmetics testing on animals is controversial, but such tests are still conducted in the United States, and involve general toxicity, eye and skin irritancy, phototoxicity (toxicity triggered by UV light), and mutagenicity. Cosmetics testing is banned in the Netherlands, Belgium, and the United Kingdom. In 2002, after 13 years of discussion, the EU agreed to phase in a near-total ban on the sale of animal-tested cosmetics throughout the EU from 2009, and to ban all cosmetics-related animal testing. France, which is home to the world's largest cosmetics company, L'Oreal, has protested the proposed ban by lodging a case at the European Court of Justice in Luxembourg, asking that the ban be quashed. The ban is also opposed by the European Federation for Cosmetics Ingredients, which represents 70 companies in Switzerland, Belgium, France, Germany, and Italy.

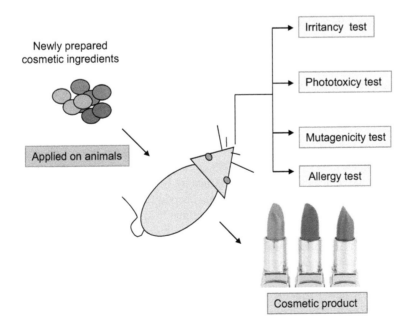

FIGURE 7.4 Cosmetic testing using animals.

7.6 ANIMAL MODELS

Researchers always wanted an ideal situation where they can induce human-like disease conditions and try to reverse it by drug therapy. Studying animal models can be informative, but care must be taken when generalizing from one organism to another. In the next section, we will study some of the most widely used animal models in detail.

7.6.1 *Caenorhabditis elegans*

Caenorhabditis elegans (*C. elegans*) is a free-living, transparent nematode (roundworm), about 1 mm in length, which lives in temperate soil environments. Research into the molecular and developmental biology of *C. elegans* began in 1974 by Sydney Brenner and *C. elegans* has since been used extensively as a model organism. *C. elegans* is studied as a model organism for a variety of reasons. One of the main advantages of *C. elegans* is that these organisms are cheap to breed and can be frozen for further study. *C. elegans* has proven especially useful for studying cellular differentiation, because the complete cell lineage of the species has been determined. From a research perspective, *C. elegans* has the advantage of being a multicellular eukaryotic organism that is simple enough to be studied in detail. In addition, it is transparent, facilitating the study of developmental processes in the intact organism. In addition, *C. elegans* is one of the simplest organisms with a nervous system. In the hermaphrodite, this comprises 302 neurons whose pattern of connectivity has been completely mapped out and shown to be a small-world network. Research has explored the neural mechanisms responsible for several of the more interesting behaviors shown by *C. elegans*, including chemotaxis, thermotaxis, mechanotransduction, and male mating behavior. A useful feature of *C. elegans* is that it is relatively straightforward to disrupt the function of specific genes by RNA interference. Silencing the function of a gene in this way can sometimes allow a researcher to infer what the function of that gene may be. The nematode can either be soaked in (or injected with) a solution of double-stranded RNA, the sequence of which is complementary to the sequence of the gene that the researcher wishes to disable. Alternatively, worms can be fed on genetically transformed bacteria that express the double-stranded RNA of interest.

C. elegans has also been useful in the study of meiosis. As sperm and egg nuclei move down the length of the gonad, they undergo a temporal progression through meiotic events. This progression means that every nucleus at a given position in the gonad will be at roughly the same step in meiosis, eliminating the difficulties of heterogeneous populations of cells. The organism has also been identified as a model for nicotine dependence as it has been found to experience the same symptoms humans experience when they quit smoking. As for most model organisms, there is a dedicated online database for the species that is actively curated by scientists working in this field. The WormBase database attempts to collate all published information on *C. elegans* and other related nematodes.

C. elegans was the first multicellular organism to have its genome completely sequenced. The finished genome sequence was published in 1998, although a number of small gaps were present. The *C. elegans* genome sequence contains approximately 100 million bp and approximately 20,000 genes (Figure 7.5). The vast majority of these genes encode for proteins, but there are likely to be as many as 1000 RNA genes. Scientific curators continue to appraise the set of known genes such that new gene predictions continue to be added and incorrect ones modified or removed. In 2003, the genome sequence of the related nematode *C. briggsae* was also determined, allowing researchers to study the comparative genomics of these two organisms. Work is now ongoing to determine the genome sequences of more nematodes from the same genus such as *C. remanei*, *C. japonica*, and *C. brenneri*. These newer genome sequences are being determined by using the whole genome shotgun technique, which means that the resulting genome sequences are likely to not be as complete or accurate as that of *C. elegans* (which was sequenced using the "hierarchical" or clone-by-clone approach). The official version of the *C. elegans* genome sequence continues to change as and

FIGURE 7.5 Genomes of various species.

when new evidence reveals errors in the original sequencing, and we have to remember that DNA sequencing is not an error-free process.

7.6.2 *DROSOPHILA MELANOGASTER*

Drosophila melanogaster, also known as the fruit fly or vinegar fly, is one of the most studied organisms in biological research, particularly in genetics and developmental biology (Figure 7.5). This is because it is small and easy to grow in the laboratory, and its morphology is easy to identify once they are anesthetized, usually with ether, carbon dioxide gas, or by cooling them. It has a short lifespan of about 10 days at room temperature, so several generations can be studied within a few weeks. It has a high fecundity as females can lay >800 eggs in a lifetime. Males and females are readily distinguished and virgin females are easily isolated, facilitating genetic crossing. The mature larvae show giant chromosomes in the salivary glands called *polytene chromosomes*. It has only four pairs of chromosomes: three autosomes and one sex chromosome. Males do not show meiotic recombination, facilitating genetic studies. Recessive lethal "balancer chromosomes" carrying visible genetic markers can be used to keep stocks of lethal alleles in a heterozygous state without recombination because of multiple inversions in the balancer. Genetic transformation techniques have been available since 1987. *Drosophila* genes are traditionally named after the phenotype they cause when mutated. For example, the absence of a particular gene in *Drosophila* will result in a mutant embryo that does not develop a heart. Scientists have thus called this gene *tinman*, named after the Oz character of the same name.

About 75% of known human disease genes have a recognizable match in the genetic code of fruit flies and 50% of fly protein sequences have mammalian analogues. An online database called Homophila is available to search for human disease gene homologues in flies and vice versa. *Drosophila* is being used as a genetic model for several human diseases including the neurodegenerative disorders Parkinson's, Huntington's, and Alzheimer's disease and spinocerebellar ataxia.

The fly is also being used to study mechanisms underlying aging and oxidative stress, immunity, diabetes, and cancer as well as drug abuse. In 1971, Ron Konopka and Seymour Benzer published "Clock mutants of *Drosophila melanogaster*," a paper describing the first mutations that affected an animal's behavior. Wild-type flies show an activity rhythm with a frequency of about a day. They found mutants with faster and slower rhythms as well as broken rhythms—flies that move and rest in random spurts. Work over the following 30 years has shown that these mutations (and others like them) affect a group of genes and their products that comprise a biochemical or biological clock. This clock is found in a wide range of fly cells, but the clock-bearing cells that control activity are several dozen neurons in the fly's central brain. Since then, Benzer and others have used behavioral screens to isolate genes involved in vision, olfaction, audition, learning/memory, courtship, pain, and other processes such as longevity. The first learning and memory mutants (*dunce, rutabaga,* etc.) were isolated by William "Chip" Quinn while in Benzer's lab, and were eventually shown to encode components of an intracellular signaling pathway involving cyclic AMP, protein kinase A, and a TF known as CREB. These molecules were shown to also be involved in synaptic plasticity in *Aplysia* and mammals. Furthermore, *Drosophila* has been used in neuropharmacological research, including studies of cocaine and alcohol consumption.

7.6.3 Laboratory Mouse

Mice are the most commonly used animal models with hundreds of established inbred, outbred, and transgenic strains. Mice are common experimental animals in biology and psychology primarily because they are mammals and thus share a high degree of homology with humans. The mouse genome has been sequenced, and virtually all mouse genes have human homologs. They can also be manipulated in ways that would be considered unethical in humans. Mice are a primary mammalian model organism, as are rats. There are many additional benefits of mice in laboratory research. Mice are small, inexpensive, easily maintained, and can reproduce quickly. Several generations of mice can be observed in a relatively short period of time. Some mice can become docile if raised from birth and given sufficient human contact. However, certain strains have been known to be quite temperamental.

Most laboratory mice are hybrids of different subspecies, most common of *Mus musculus domesticus* and *Mus musculus musculus*. Laboratory mice come in a variety of coat colors including agouti, black, and albino. Many (but not all) laboratory strains are inbred to make them genetically almost identical. The different strains are identified with specific letter-digit combinations, such as C57BL/6 and BALB/c. The first such inbred strains were produced by Clarence Cook Little in 1909. The sequencing of the mouse genome was completed in late 2002. The haploid genome is about 3 billion bases long (3000 Mb distributed over 20 chromosomes) and therefore equal to the size of the human genome. Estimating the number of genes contained in the mouse genome is difficult, in part because the definition of a gene is still being debated and extended. The current estimated gene count is 23,786. This estimate takes into account knowledge of molecular biology as well as comparative genomic data. For comparison, humans are estimated to have 23,686 genes (Figure 7.5).

7.6.4 Rhesus Monkey

The rhesus macaque (*Macaca mulatta*), often called the rhesus monkey, is one of the best-known species of Old World monkeys. It typically has a lifespan of about 25 years. The species is native to northern India, Bangladesh, Pakistan, Burma, Thailand, Afghanistan, southern China, and some neighboring areas. The rhesus macaque is common in science owing to its relatively easy upkeep in captivity and has been used extensively in medical and biological research. It has given its name to the Rhesus factor, one of the elements of a person's blood group, by the discoverers of the factor, Karl Landsteiner and Alexander Wiener. The rhesus macaque was also used in the well-known experiments on maternal deprivation carried out in the 1950s by the comparative psychologist

Harry Harlow. In January 2000, the rhesus macaque became the first cloned primate with the birth of Tetra. Tetra was cloned using the technique of embryo spitting. January 2001 saw the birth of ANDi, the first transgenic primate. ANDi carries foreign genes originally from a jellyfish. Work on the genome of the rhesus macaque was completed in 2007, making rhesus macaque the second NHP to have its genome sequenced. The study shows that humans and macaques share about 93% of their DNA sequence and shared a common ancestor roughly 25 million years ago (Figure 7.5).

7.6.5   XENOPUS LAEVIS

Xenopus laevis also known as the African clawed frog is a species of South African aquatic frog of the genus *Xenopus*. It can grow up to 12 cm long and has a flattened head and body but no external ear or tongue. The species is found throughout much of Africa and in isolated, introduced populations in North America, South America, and Europe. Although *X. laevis* do not have the short generation time and genetic simplicity generally desired in genetic model organisms, it is an important model organism in developmental biology. *X. laevis* takes 1–2 years to reach sexual maturity and, like most of its genus, it is tetraploid. However, it does have a large and easily manipulatable embryo. The ease of manipulation in amphibian embryos has given them an important place in history and modern developmental biology. A related species, *X. tropicalis*, is now being promoted as a more viable model for genetics. Roger Wolcott Sperry used *X. laevis* for his famous experiments describing the development of the visual system. These experiments led to the formulation of the chemoaffinity hypothesis.

Xenopus oocytes provide an important expression system for molecular biology. By injecting DNA or mRNA into the oocyte or developing embryo, scientists can study the protein products in a controlled system. This allows rapid functional expression of manipulated DNAs (or mRNA). This is particularly useful in electrophysiology, where the ease of recording from the oocyte makes expression of membrane channels attractive. One challenge of oocyte work is to eliminate native proteins that might confound results, such as membrane channels native to the oocyte. Translation of proteins can be blocked or splicing of pre-mRNA can be modified by injection of Morpholino antisense oligos into the oocyte (for distribution throughout the embryo) or early embryo (for distribution only into daughter cells of the injected cell). *X. laevis* is also notable for its use as the first well-documented method of pregnancy testing when it was discovered that the urine from pregnant women induced *X. laevis* oocyte production. Human chorionic gonadotropin (HCG) is a hormone found in substantial quantities in the urine of pregnant women. Today, commercially available HCG is injected into *Xenopus* males and females to induce mating behavior and breed these frogs in captivity at any time of the year.

7.6.6   ZEBRAFISH

The zebrafish, *Danio rerio*, is a tropical freshwater fish belonging to the minnow family (Cyprinidae). It is a popular aquarium fish, frequently sold under the trade name zebra danio, and is an important vertebrate model organism in scientific research. *D. rerio* is a common and useful model organism for studies of vertebrate development and gene function. They may supplement higher vertebrate models, such as rats and mice. The pioneering work of George Streisinger at the University of Oregon established the zebrafish as a model organism; large-scale forward genetic screens consolidated its importance. The Zebrafish embryonic development has advantages over other vertebrate model organisms. Although the overall generation time of zebrafish is comparable to that of mice, zebrafish embryos develop rapidly, progressing from eggs to larvae in less than 3 days. The embryos are large, robust, and transparent and develop external to the mother, all characteristics that facilitate experimental manipulation and observation. Their nearly constant size during early development facilitates simple staining techniques, and drugs may be administered by being directly added to the tank.

Despite the complications of the zebrafish genome (Figure 7.5), a number of commercially available global platforms for analysis of both gene expression by microarrays and promoter regulation using ChIP-on-chip exist. Zebrafish have the ability to regenerate fins, skin, the heart, and the brain. Zebrafish have also been found to regenerate photoreceptors and retinal neurons following injury. The mechanisms of this regeneration are unknown but are currently being studied. Researchers frequently cut the dorsal and ventral tail fins and analyze their regrowth to test for mutations. This research is leading the scientific community in the understanding of healing/repair mechanisms in vertebrates. In December 2005, a study of the *golden* strain identified the gene responsible for the unusual pigmentation of this strain as SLC24A5, a solute carrier that appeared to be required for melanin production, and confirmed its function with a Morpholino knockdown. The orthologous gene was then characterized in humans and a 1 bp difference was found to segregate strongly between fair-skinned Europeans and dark-skinned Africans. This study was featured on the cover of the academic journal *Science* and demonstrates the power of zebrafish as a model organism in the relatively new field of comparative genomics. In January 2007, Chinese researchers at Fudan University raised genetically modified fish that can detect estrogen pollution in lakes and rivers, showing environmental officials what waterways need to be treated for the substance, which is linked to male infertility. Song Houyan and Zhong Tao, professors at Fudan's molecular medicine lab, spent 3 years cloning estrogen-sensitive genes and injecting them into the fertile eggs of zebrafish. The modified fish turn green if they are placed in water that is polluted by estrogen.

Researchers at the University College London grew a type of zebrafish adult stem cell—found in the eyes of fish and mammals—that develops into neurons in the retina, the part of the eye that sends messages to the brain. These cells could be injected in the eye to treat all diseases where the retinal neurons are damaged, nearly all diseases of the eye, including macular degeneration, glaucoma, and diabetes-related blindness. Damage to the retina is responsible for most cases of sight loss. The researchers studied Müller glial cells in the eyes of humans aged from 18 months to 91 years and were able to develop them into all types of neurons found in the retina. They were also able to grow them easily in the lab. The results of their experiments were reported in the journal *Stem Cells*. The cells were tested in rats with diseased retinas, where they successfully migrated into the retina and took on the characteristics of the surrounding neurons. Now, the team is working on the same approach in humans. In February 2008, researchers at Children's Hospital, Boston, reported in the journal *Cell Stem Cell* the development of a new strain of zebrafish, named Casper, with a see-through body. This allows for detailed visualization of individual blood stem cells and metastasizing (spreading) cancer cells within a living adult organism. Because the functions of many genes are shared between fish and humans, this tool is expected to yield insight into human diseases such as leukemia and other cancers.

7.7 ANIMAL BIOTECHNOLOGY

7.7.1 USE OF ANIMALS IN ANTIBODY PRODUCTION

In general, an antigen molecule has antigenic determinants of more than one specificity, this means different determinants will interact with different antibodies. Each distinct antigenic determinant of the antigen will bind to a distinct mature B cell whose surface immunoglobulin (sIg) matches the specificity presented by the concerned determinant. As a result, such a single antigen will activate B cells of more than one sIg specificity. Activated B cells of each sIg specificity will divide and differentiate to give rise to clones of plasma cells producing antibodies of the same specificity. Thus, a single antigen would induce more than one distinct clone of plasma cells, which will produce antibodies of different specificities. Therefore, the serum of an animal immunized by a single antigen will contain antibodies of different specificities but reacting to the same antigen. These are called *polyclonal antibodies*.

In contrast, a hybridoma clone produces antibodies of a single specificity as the clone is derived from the fusion of a single differentiated (antibody-producing) B cell with a myeloma cell, that is, a clone of a single B cell. Therefore, such antibodies are called *monoclonal antibodies* (MABs). Obviously, all the molecules of a MAB will have the same specificity. The chief advantage of MABs is that all the antibody molecules in a single preparation react with a single epitope or antigenic determinant. Therefore, the results obtained by using MABs are clear-cut as there is no background confusion that arises owing to the presence of antibodies of other specificities in the case of conventionally used antisera. The multitude of various applications of MABs may be grouped into the three categories: diagnostic, therapeutic, and purification.

7.7.1.1 Monoclonal Antibodies in Diagnostic Applications

MABs are used to detect the presence of a specific antigen or of antibodies specific to an antigen in a sample or samples—this constitutes a diagnostic application. The presence of antigen is detected by assaying the formation of an antigen–antibody complex (Ag–Ab complex) for which a number of assay techniques have been devised. These assays are highly precise, extremely efficient, rapid, and surprisingly versatile for a large variety of applications. Some examples of diagnostic applications are: (1) MABs are available for unequivocal classification of blood groups such as ABO and Rh and (2) MABs are used for a clear and decisive detection of pathogens involved in diseases (disease diagnosis).

7.7.1.2 Monoclonal Antibodies in Therapeutic Applications

MABs are used for either the treatment of or protection from a disease. Antibodies specific to a cell type, for example, tumor cells, can be linked with a toxin polypeptide to yield a conjugate molecule called an *immunotoxin*. The antibody component of the immunotoxin will ensure that it is bound specifically and only to the target cells, and the attached toxin will kill such cells. Immunotoxins having ricin have been prepared and evaluated for killing of tumor cells with considerable success. *Ricin* is a natural toxin found in the endosperm of castor (*Ricinus communis*). It has two polypeptides called A (toxin peptide) and B (a cell-binding polypeptide, lectin). Ricin A polypeptide enzymatically and irreversibly modifies the larger subunit of ribosomes (in fact, their EF2 binding site) making them incapable of protein synthesis. This toxin is effective against both dividing and nondividing cells as it inhibits protein synthesis.

Antibody–Ricin A conjugate has been shown to reduce protein synthesis in mouse B-cell tumors. The antibody used in the conjugate was specific to the antigen molecules present on the surface of target tumor cells. It is noteworthy that this immunotoxin did not bind to either other tumor cells or the normal cells. The same principle has been used to deliver radioactivity, specifically to target tumor cells. In such cases, radioactivity from ^{131}I (iodine), ^{90}Y (yttrium), ^{67}Cu, ^{212}Pb, etc., is incorporated into the tumor-specific antibody in the place of the toxin. Radiolabeled antibodies have been used in patients having hepatoma, human T-cell leukemia/lymphoma virus-I (HTLV-1), adult T-cell leukemia (ATL), etc. B-lymphocyte proliferation, maturation, and antibody secretion are dependent on interleukin-2 produced by activated T-lymphocytes. Furthermore, T cells mediate graft rejection. Thus, an effective strategy to minimize the rejection of grafts from other individuals would be to eliminate the T cells from their bone marrow/circulatory system (blood stream) by using T-cell specific MABs. T cells exhibit several antigens of which CD3, CD4, CD8, etc. have been the preferred targets for MAB development.

In bone marrow transplantation, the cells of the recipient are inactivated by appropriate irradiation. The donor cells are treated with T-cell specific antibodies to destroy the T cells present in them. The remaining cells are then transplanted into the recipient. Experiments with mice have shown remarkable success. In order to minimize tissue-graft rejection, the T cells present in the circulatory system of the recipient are eliminated prior to the transplant by an administration of T-cell specific MABs. This treatment abolishes, though temporarily, the ability of the recipient to mount an immune response against any foreign antigens, including those present in the graft tissue. MAB OKT3 is the most widely used for treatment of acute cases of rejection of kidney transplants. MABs can be administered to provide passive immunity against diseases. In the case of active immunity, the immunized individual will itself produce the antibodies against the concerned pathogen. However, in the case of passive immunity, antibodies produced elsewhere are introduced into the body of an individual to provide immunity against a pathogen. MABs are very useful in the purification of antigens specific to pathogens. These purified antigens are used as vaccines.

7.7.1.3　Recombinant Antibodies

When antibody molecules are modified or designed using recombinant DNA technology to suit specific applications, such antibodies are called *recombinant* or *hybrid antibodies* and the approach itself is called *antibody engineering.* A recombinant antibody could be constructed so that the constant end of its heavy chain is fused to a polypeptide chain having an enzymatic function. Such an antibody is extremely useful for ELISA as there will be no need for a second antibody. A gene producing such an antibody can be produced by fusing the heavy chain gene having the appropriate L-V-D-J and yl sequences with the sequence coding for the selected enzyme function. The heavy chain gene of an antibody specific to a tumor-specific antigen may be fused with a gene encoding a toxin polypeptide. Such hybrid antibodies will carry the toxin specifically to the tumor cells and, thereby, kill them. Gene segments encoding the variable region or V-region (involved in antigen–antibody interaction) of an antibody have been fused with the constant region or C-region of another antibody to yield a hybrid antibody. A hybrid antibody has the antigen specificity of the first antibody (which contributed the V-region segments), but its other properties are because of the second antibody (contributing the C-region segments).

7.7.2　*In Vitro* Fertilization and Embryo Transfer

The union of an egg cell with sperm occurs outside the body in a culture vessel—this is known as *in vitro* fertilization. This involves collection of healthy ova and sperms from healthy females and males, respectively, and their fusion under appropriate conditions *in vitro.* The resulting zygotes may be cultured *in vitro* for a period of time to obtain young embryos, which ultimately are implanted in the uterus of healthy females to complete their development. The implementation of young embryos developed *in vitro* or obtained from the uterus of different donor females into the womb of selected females is called *embryo transplantation.* The techniques of *in vitro* fertilization and embryo transfer are being applied to animals for a rapid multiplication of desirable genotypes of animals and in cases of infertility of certain types in humans.

7.7.2.1　Embryo Transfer in Cattle

Young embryos of cattle of superior genotype are collected prior to their implantation in the uterus and are implanted in the uterus of other females of inferior genotype, where they complete development; this is called *embryo transfer.* The chief objective of embryo transfer is to obtain several progenies per year from a single female of superior genotype. In a country like India, most cattle are of inferior genotype with rather low productivity and superior genotype females are limited in number and of high price. Therefore, a program of artificial insemination (AI) was widely used in an effort to improve cattle breeds. A limitation of AI is that the superior genes (50%) are contributed by only the male side, while the female side contributes the inferior genes (50%).

In contrast, in the embryo transfer technique, the inferior females used as surrogate or substitute mothers do not contribute any genes to the progeny. They only serve as extremely sophisticated natural incubators for the normal development of young embryos. As a result, the progeny obtained by embryo transfer are of superior genotype.

In embryo transfer, a genetically superior and highly productive female serves as the donor of embryos to be transferred and healthy, young females of inferior genotype are selected to be the recipients of embryos to be transferred. These females are called *surrogate* or *substitute mothers*. The donor females are treated with appropriate doses of the selected *gonadotrophin*, a follicle stimulating hormone (FSH) or luteinizing hormone (LH), to increase the number of ova released at the time of ovulation; this process is called *superovulation*. Under optimum treatment conditions, a single female can provide up to 15 embryos in a single cycle. The chief objective of superovulation is to greatly increase the number of embryos recovered per female in a single cycle. When the donor female is in heat, it is artificially inseminated using semen from a genetically superior bull of top pedigree. The fertilized eggs/young embryos are collected by flushing the uterus of donor females with a special nutrient solution 7 days after insemination. The embryos are examined under a stereoscopic microscope and normal-looking healthy embryos are selected. The selected embryos are incubated in a special nutrient medium at 37°C until they are transferred into the surrogate mothers. Alternatively, they may be frozen and stored in liquid nitrogen for future use. A single embryo is transferred into the uterus of each surrogate mother. It is important that the estrus cycles of the donor and surrogate mothers are synchronized by administering prostaglandins to provide the optimum uterine environment for survival, establishment, and normal development of the young embryos.

This technology achieves a surprisingly rapid rate of multiplication of animals of the selected superior genotype. In the natural course, a single female will produce a single progeny in about a year. However, using superovulation and embryo transfer technology, it is feasible to collect around 36 embryos from one female in 1 year. Assuming an average success rate of 50% in the embryo transfer, an average of 18 progeny can be derived from one superior female in 1 year. Each young embryo can be split into 2–4 parts, each of which would develop into a separate progeny. This process is called *embryo splitting*. By combining embryo splitting with superovulation, the rate of multiplication can be further increased. The young embryos can be frozen and stored in liquid

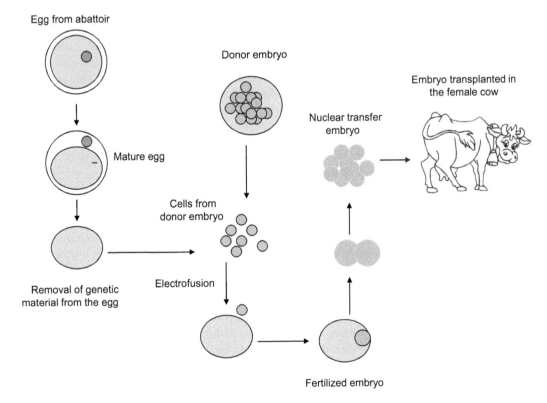

FIGURE 7.6   Animal breeding by nuclear transfer technology.

nitrogen (at −196°C; cryopreservation) for up to 10 years or more and used on a subsequent date. The frozen embryos are far easier to transport and present negligible quarantine problems as compared to the animals themselves. Superior cows that are unfit to carry the fetus to full term can serve as donors of the young embryos. In spite of the various benefits of embryo transfer technology, there are also some limitations. A high degree of expertise is required for an efficient and successful embryo transfer operation. The cost of producing each progeny is several-fold higher than that from the natural process. The donor females are removed from production for the period they are used as donors of young embryos (Figure 7.6).

7.7.3 ANIMAL CELL CULTURE PRODUCTS

Animal cell cultures are used to produce virus vaccines as well as a variety of useful biochemicals, which are mainly high-molecular-weight proteins like enzymes, hormones, cellular biochemical like interferon's, and immunobiological compounds including MABs. Animal cells are also good hosts for the expression of recombinant DNA molecules and a number of commercial products have been/are being developed. Initially, virus vaccines were the dominant commercial products from cell cultures, but at present, monoclonal antibody production is the chief commercial activity. It is expected that recombinant proteins will become the prime product from cell cultures in the near future. Transplantable tissues and organs are another very valuable product from cell cultures. Artificial skins are already in use for grafting in burn and other patients, and efforts are focused on developing transplantable cartilage and other tissues.

7.7.4 ANIMAL CLONING

Animal *cloning* is the process by which an entire organism is reproduced from a single cell taken from the parent organism and in a genetically identical manner. Cloning in biotechnology refers to processes used to create copies of DNA fragments (molecular cloning), cells (cell cloning), or organisms. Scientists have been attempting to clone animals for a very long time. Many of the early attempts came to nothing. The first successful result in animal cloning was seen when tadpoles were cloned from frog embryonic cells in 1952. This was the first vertebrate to be cloned. This cloning was done by the process of nuclear transfer. The tadpoles that were created by this method did not survive to grow into mature frogs, but it was a major breakthrough nevertheless. After this, using the process of nuclear transfer of embryonic cells, scientists managed to produce clones of mammals. Again, the cloned animals did not live very long (Figure 7.7).

The first successful instance of mammal cloning was that of Dolly the sheep. Dolly (05-07-1996-14-02-2003), a Finn Dorsett ewe, was the first mammal to have been successfully cloned from an adult cell. Dolly was cloned at the Roslin Institute in Scotland and lived there until her death when she was six. On April 9, 2003, her stuffed remains were placed at Edinburgh's Royal Museum, part of the National Museums of Scotland. Dolly was publicly significant because the effort showed that the genetic material from a specific adult cell, programmed to express only a distinct subset of its genes, can be reprogrammed to grow an entire new organism. Cloning Dolly had a low success rate per fertilized egg. She was born after 237 eggs were used to create 29 embryos, which only produced three lambs at birth, only one of which lived. Seventy calves have been created from 9000 attempts and one-third of them died young. Prometea took 328 attempts. Notably, although the first clones were frogs, no adult cloned frog have yet been produced from a somatic adult nucleus donor cell. There were early claims that Dolly the sheep had pathologies resembling accelerated aging. Scientists speculated that Dolly's death in 2003 was related to the shortening of telomeres, DNA-protein complexes that protect the end of linear chromosomes. However, other researchers, including Ian Wilmut who led the team that successfully cloned Dolly, argued that Dolly died of respiratory infection.

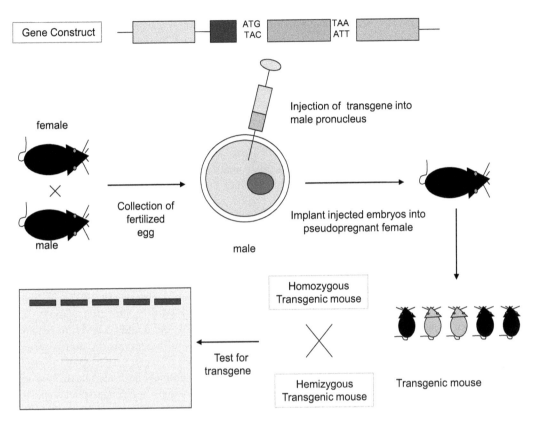

FIGURE 7.7   Generation of transgenic mouse.

7.7.5   TRANSGENIC ANIMALS

Transgenic animals are those animals whose genes have been deliberately modified in order to change their morphological appearance or their physiological functions. A transgenic animal is one whose genome has been changed to carry genes from other species. The nucleus of all cells in every living organism contains genes made up of DNA. These genes store information that regulates how our bodies form and function. Genes can be altered artificially so that some characteristics of an animal are changed. For example, an embryo can have an extra, functioning gene from another source artificially introduced into it or have a gene introduced that can knock out the functioning of another particular gene in the embryo. The majority of transgenic animals produced so far are mice, the animal that pioneered the technology. The first successful transgenic animal was a mouse. A few years later, it was followed by rabbits, pigs, sheep, and cattle.

   Transgenic animals are useful as disease models and producers of substances for human welfare. Some transgenic animals are produced for specific economic traits. For example, transgenic cattle were created to produce milk containing particular human proteins, which may help in the treatment of human emphysema. Other transgenic animals are produced as disease models (animals genetically manipulated to exhibit disease symptoms, so that effective treatment can be studied). For example, Harvard scientists made a major scientific breakthrough when they received a US patent and gave the company DuPont exclusive rights to a genetically engineered mouse, called OncoMouse®, or the Harvard mouse, which carries a gene that promotes the development of various human cancers.

Transgenic mice have also been used in scientific research. Normal mice cannot be infected with the polio virus. They lack the cell-surface molecule that, in humans, serves as the receptor for the virus. Therefore, normal mice cannot serve as an inexpensive, easily manipulated model for studying the disease. However, transgenic mice expressing the human gene for the polio-virus receptor can be infected with the polio virus and even develop paralysis and other pathological changes characteristic of the disease in humans.

Farmers have always used selective breeding to produce animals that exhibit desired traits (such as increased milk production or high growth rate). Traditional breeding is a time-consuming, difficult task. When technology using molecular biology was developed, it became possible to develop traits in animals in a shorter time and with more precision. In addition, it offers the farmer an easy way to increase yields. Transgenic cows exist that produce more milk or milk with less lactose or cholesterol.

In the past, farmers used growth hormones (GHs) to spur the development of animals but this technique was problematic, especially because residue of the hormones remained in the animal product. Products such as insulin, GH, and blood anticlotting factors may soon be or have already been obtained from the milk of transgenic cows, sheep, or goats. Research is also underway to manufacture milk through transgenesis for the treatment of debilitating diseases such as phenylketonuria (PKU), hereditary emphysema, and cystic fibrosis. In 1997, the first transgenic cow, Rosie, produced human protein-enriched milk at 2.4 g/L. This transgenic milk is a more nutritionally balanced product than natural bovine milk and could be given to babies or the elderly with special nutritional or digestive needs. Rosie's milk contains the human gene alpha-lactalbumin.

Besides the agricultural, dairy, and medical fields, transgenic animals have also been used in industrial applications. In 2001, two scientists at Nexia Biotechnologies in Canada spliced spider genes into the cells of lactating goats. The goats began to manufacture silk along with their milk and secreted tiny silk strands from their body by the bucketful. By extracting polymer strands from the milk and weaving them into thread, the scientists can create a light, tough, flexible material that could be used in such applications as military uniforms, medical microsutures, and tennis racket strings. Toxicity-sensitive transgenic animals have been produced for chemical safety testing. Microorganisms have been engineered to produce a wide variety of proteins, which in turn can produce enzymes that can speed up industrial chemical reactions.

To date, there are three basic methods of producing transgenic animals: DNA microinjection, retrovirus-mediated gene transfer, and embryonic stem (ES) cell-mediated gene transfer. Gene transfer by microinjection is the predominant method used to produce transgenic farm animals. Since the insertion of DNA results in a random process, transgenic animals are mated to ensure that their offspring acquire the desired transgene. However, the success rate of producing transgenic animals individually by these methods is very low and it may be more efficient to use cloning techniques to increase their numbers. For example, gene transfer studies revealed that only 0.6% of transgenic pigs were born with a desired gene after 7000 eggs were injected with a specific transgene. Although we benefit a lot from transgenic animals, there are still issues on how they are created, and these issues need to be resolved.

7.8   BIOTECHNOLOGY AND FISH FARMING

In addition to the use of animals in drug testing, they are also used in other applications such as fish farming. Aquaculture is the farming of freshwater and saltwater organisms such as finfish, mollusks, crustaceans, and aquatic plants. Also known as aquafarming, aquaculture involves cultivating aquatic organisms under controlled conditions. One-half of the world's commercial production of fish and shellfish that is directly consumed by humans comes from aquaculture. Aquaculture is a very old fish farming technique and it has been in practice since 2500 BC.

7.8.1 Mariculture

Mariculture is a specialized branch of aquaculture involving the cultivation of marine organisms in the open ocean, an enclosed section of the ocean, or in tanks, ponds, or raceways that are filled with seawater. An example of the latter is the farming of marine fish, prawns, or oysters in saltwater ponds. Nonfood products produced by mariculture include fish meal, nutrient agar, jewelry (such as cultured pearls), and cosmetics. With fishery catches on the decline and aquaculture products on the increase, mariculture holds a great promise for the future, both economically and environmentally. The broad perception by a large majority of fish consumers is that fish is healthy and nutritious, which is an advantage to mariculture. Another advantage is that the naturalness that farmed fish possess is comparable to those that are harvested from the ocean. Mariculture farming helps the species that may be depleting in the wild such as trout, sea bass, and salmon. The consistency of supply all year round and more routine quality control has enabled mariculture supply to be integrated in other food market channels. These benefits have also been able to reach different socioeconomic classes who may not have been able to purchase fish because of high prices.

7.8.2 Polyculture

Polyculture is agriculture using multiple crops in the same space, in imitation of the diversity of natural ecosystems, and avoiding large stands of single crops, or monoculture. It includes crop rotation, multicropping, intercropping, companion planting, beneficial weeds, and alley cropping. Polyculture, though it often requires more labor, has several advantages over monoculture. The diversity of crops avoids the susceptibility of monocultures to disease. For example, a study in China, reported in *Nature*, showed that planting several varieties of rice in the same field increased yields by 89%, largely because of a dramatic (94%) decrease in the incidence of disease, which made pesticides redundant. Also, the greater variety of crops provides habitat for more species, increasing local biodiversity. This is one example of reconciliation ecology, or accommodating biodiversity within human landscapes.

7.8.3 Aquatic Biotechnology

With the advancement of molecular and genetic tools, there is also the development of transgenic aquatic animals to understand their genetic information. The genetic information received from aquatic animals is used to study how the animals live and survive in both extreme cold water and hot water. In this section, we will discuss how different useful genes are isolated from aquatic animals, which are being later used in various applications.

7.8.3.1 Transgenic Fish

Attempts to produce transgenic fish started in 1985, and some encouraging results have been obtained. The genes that have been introduced by microinjection in fish are (1) human or rat gene for GH, (2) chicken gene for delta-crystallin protein, (3) *E. coli* gene for β-galactosidase, (4) *E. coli* gene for neomycin resistance, (5) winter flounder gene for antifreeze protein (AFP), and (6) rainbow trout gene for GH. The technique of microinjection has been successfully used to generate transgenic fish in many species such as common carp, catfish, goldfish, loach, medaka, salmon, tilapia, rainbow trout, and zebrafish.

In other animals (such as mice, cows, pigs, sheep, and rabbits), usually direct microinjection of cloned DNA into male pronuclei of fertilized eggs has proved very successful, but in most fish species studied so far, pronuclei cannot be easily visualized (except in medaka); therefore, the DNA needs to be injected into the cytoplasm. Eggs and sperms from mature individuals are collected and placed into a separate dry container. Fertilization is initiated by adding water and sperm to the eggs, with gentle stirring to facilitate the fertilization process. Egg shells are hardened in water. About 106–108 molecules of linearized DNA in a volume of 20 mL or less is microinjected into each egg (1–4 cells stage) within the first few hours after fertilization. Following microinjection, eggs are incubated in appropriate hatching trays and dead embryos are removed daily. Because in fish fertilization is external, *in vitro* culturing of embryos and their subsequent transfer into foster mothers (required in mammalian systems) is not required. Further, the injection into the cytoplasm is not as harmful as that into the nucleus; therefore, the survival rate in fish is much higher (35%–80%). Human GH genes transferred to transgenic fish allowed growth to twice the size of their corresponding nontransgenic fish (goldfish, rainbow trout, and salmon).

7.8.3.2   Green Fluorescent Protein

Green fluorescent protein (GFP) is a protein isolated from coelenterates such as the Pacific jellyfish, *Aequoria victoria*, or from the sea pansy, *Renilla reniformis*. Although many other marine organisms have similar GFPs, GFP traditionally refers to the protein first isolated from the jellyfish *Aequorea victoria*. The GFP from *A. victoria* has a major excitation peak at a wavelength of 395 nm and a minor one at 475 nm. Its emission peak is at 509 nm, which is in the lower green portion of the visible spectrum. The GFP from the sea pansy (*Renilla reniformis*) has a single major excitation peak at 498 nm. In cell and molecular biology, the GFP gene is frequently used as a reporter of expression. In modified forms, it has been used to make biosensors, and many animals have been created that express GFP as a proof-of-concept that a gene can be expressed throughout a given organism. The GFP gene can be introduced into organisms and maintained in their genome through breeding, injection with a viral vector, or cell transformation. To date, the GFP gene has been introduced and expressed in many bacteria, yeast and other fungi, fish (such as zebrafish), plants, flies, and mammalian cells, including human. Martin Chalfie, Osamu Shimomura, and Roger Y. Tsien were awarded the 2008 Nobel Prize in Chemistry on October 10, 2008, for their discovery and development of the GFP.

7.8.3.3   Antifreeze Proteins

Scientists found out that aquatic animals living in extreme cold conditions (such as whales and sharks) without having any problems produce some kind of proteins that antagonize the effect of cold. They isolated the protein and gene responsible for the antifreezing properties and called these specialized proteins AFPs or ice-structuring proteins (ISPs). These AFPs refer to a class of polypeptides produced by certain vertebrates, plants, fungi, and bacteria that permit their survival in subzero environments. AFPs bind to small ice crystals to inhibit growth and recrystallization of ice that would otherwise be fatal. There is also increasing evidence that AFPs interact with mammalian cell membranes to protect from cold damage. In the 1950s, Canadian scientist Scholander set out to explain how Arctic fish can survive in water colder than the freezing point of their blood. His experiments led him to believe that there was "antifreeze" in the blood of Arctic fish. Then, in the late 1960s, animal biologist Arthur DeVries was able to isolate the AFP through his investigation of Antarctic fish. Antifreeze glycoproteins (AFGPs) were the first AFPs to be discovered. At that time, they were called "glycoproteins as biological antifreeze agents." These proteins were later called AFGPs or antifreeze glycopeptides to distinguish them from newly discovered nonglycoprotein biological AFPs.

7.8.3.4   Transgenic Salmon

Two potential ways in which transgenic technologies can be used to solve the problem of overwintering salmon in sea cages in Atlantic Canada are: (1) Produce freeze-resistant salmon by giving them a set of AFP genes, and (2) enhance growth rates by GH gene transfer so that overwintering may not be necessary. Many commercially important fish (such as salmon and halibut) lack these proteins and their genes and, as a consequence, they will not survive if cultured in icy sea water. In 1982, transgenic studies were initiated by injecting winter flounder antifreeze genes into the fertilized eggs of Atlantic salmon. A full-length gene encoding the major liver secretory AFP was used and the AFP transgene was successfully integrated into the salmon chromosomes, expressed, and found to exhibit Mendelian inheritance over five generations to date. Expressed levels of AFP in the blood of these fish are quite low (100–400 µg/mL) and is insufficient to confer any significant increase in freeze resistance to the salmon. However, the "proof-of-the-concept" that salmon and other fish species can be rendered more freeze resistant by gene transfer has been established. The challenge now is to design an antifreeze gene construct(s) that will result in enhanced expression in appropriate tissues—epithelia and liver.

7.9   REGULATION OF ANIMAL TESTING

Animal testing is one of the most strictly regulated affairs where researchers need to get approval from regulators and ethical organizations first before they are allowed to carry out research work (Figure 7.8). Different agencies regulate a broad range of products from cosmetics to drugs to food additives that routinely undergo animal testing by manufacturers before they can be used by people. Failure to follow the regulations means that test results may be faulty, the data may be rejected, and whatever the submitter was seeking will be denied. The regulations call for humane treatment of the animals with close attention to housing, bedding, food, and water. They stress that animals must be carefully identified and that identity maintained, as losing track of test animals can void a study. Animals from one study must not be mingled with others, and no more animals than are needed are to be used. Regulatory authorities generally ensure that researchers or companies should use the animals with great care and they insist that animals should be properly housed, fed, and kept under

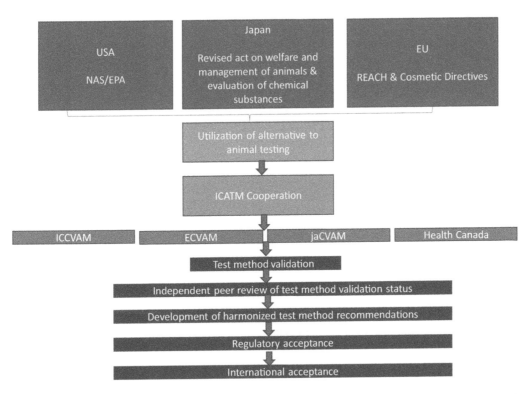

FIGURE 7.8   The process of regulatory approvals nationally and internationally. The Environmental Protection Agency (EPA); National Academy of Sciences (NAS); International Cooperation on Alternative Test Methods (ICATM); Interagency Coordinating Committee on the Validation of Alternative Methods (ICCVAM); European Center for the Validation of Alternative Methods (ECVAM); Japan Center for the Validation of Alternative Methods (JaCVAM).

healthy and hygienic conditions. They also make sure the number of animals used in the study is properly justified and required. The US FDA has a responsibility to assure that products are safe and effective for the public. A heated debate has developed over the use of animal testing to ensure the safety of human beings. The FDA has placed requirements on the use of animal testing and alternative methods. The US Food, Drug, and Cosmetic Act (FD&C Act) does not require the use of animals in testing cosmetic products for safety, nor does it require premarket approval of cosmetics by the FDA. The agency advises cosmetics manufacturers to use whatever form of testing is appropriate and effective for proving the safety of their products.

PROBLEMS

Section A: Descriptive Type

Q1. Why do scientists prefer to work on animal models rather than humans?

Q2. Do you think selling drugs without their being tested in animals is unlawful? Why or why not?

Q3. What are the most commonly used animals in research? Briefly explain their uses.

Q4. How are animals used in basic research?

Q5. Why do animal models of neurological disorders fail to create the same syndromes as those in humans?

Q6. Is it mandatory to know the toxic effect(s) of a new drug molecule before it is tested in humans? Why or why not?

Q7. Explain briefly the use of zebrafish as a research model.

Q8. What are the benefits of animal cloning?

Q9. How are transgenic animals produced?

Q10. What are the ethical concerns related to transgenic animals?

Q11. What is a Limulus Amebocyte Lysate test?

Q12. How are anti-freezing proteins isolated from fish?

Section B: Multiple Choices

Q1. How many reasons will scientists give in favor of the use of animals in research?
   a. 4
   b. 5
   c. 7
   d. 10

Q2. On which year did the US government make animal testing mandatory for all medicines before their use in humans?
   a. 1937
   b. 1940
   c. 1945
   d. 1949

Q3. Which vertebrate animal is the most commonly used in research?
   a. Rodents
   b. Monkeys
   c. Cats
   d. Dogs

Q4. Which is the most commonly used animals for neurological research?
   a. Dog
   b. Cat
   c. Fish
   d. Mouse

Q5. Drug efficacy is the effect of what at various time intervals?
   a. Drug on body
   b. Body on drug
   c. None of these

Q6. What is an LD50?
   a. Dose at which 50% of test animals were killed
   b. Dose at which 50% of test animals survived
   c. Dose at which 50 test animals were killed
   d. None of the above

Q7. Make the right pairing.
   a. Chronic   i. effect lasting for less than 1 month
   b. Acute   ii. effect lasting for 1–3 months
   c. Subacute   iii. effect lasting for more than 3 months

Q8. In 2002, cosmetic testing on animals was banned in these countries except . . .
   a. Belgium
   b. United Kingdom
   c. United States
   d. Netherlands

Q9. What was the first multicellular organism to have its genome completely sequenced?
   a. *Drosophila melanogaster*
   b. *C. elegans*
   c. Zebrafish
   d. *C. brenneri*

Q10. What was the first learning and memory mutant fruit fly?
   a. Dunce
   b. Tetra
   c. Pheta
   d. TTcc

Q11. The first successful instance of animal cloning was the . . .
   a. Goat
   b. Cat
   c. Sheep
   d. Mule

Section C: Critical Thinking

Q1. Is it possible to test a drug without using animals? Explain.

Q2. Do you agree with the ban imposed by some European countries on the use of animals for cosmetics testing? Why or why not?

ASSIGNMENT

Organize a class debate on the pros and cons of using transgenic animals for various applications.

REFERENCES AND FURTHER READING

Butler, M. *Animal Cell Technology: From Biopharmaceuticals to Gene Therapy.* Taylor & Francis Books, Oxford, UK, 2009.

Cohn, M. Alternatives to animal testing gaining ground. *The Baltimore Sun*, August 26, 2010.

Dahm, R. The zebrafish exposed. *Am. Sci.* 94: 446–453, 2006.

GBD 2015 Mortality and Causes of Death, Collaborators. Global, regional, and national life expectancy, all-cause mortality, and cause-specific mortality for 249 causes of death, 1980–2015: A systematic analysis for the global burden of disease study 2015. *Lancet* 388(10053): 1459–1544, October 8, 2016.

Kimmel, C.B., and Law, R.D. Cell lineage of zebrafish blastomeres. I: Cleavage pattern and cytoplasmic bridges between cells. *Dev. Biol.* 10: 78–85, 1985.

Kimmel, C.B., and Law, R.D. Cell lineage of zebrafish blastomeres. III: Clonal analyses of the blastula and gastrula stages. *Dev. Biol.* 108: 94–101, 1985.

Lamason, R.L., Mohideen, M.A., Mest, J.R. et al. SLC24A5, a putative cation exchanger, affects pigmentation in zebrafish and humans. *Science* 310: 1782–1786, 2005.

Mills, D. *Eyewitness Hnbk Aquarium Fish.* Harper Collins, Sydney, Australia, 1993. ISBN: 0-7322-5012-9.

Portner, R. *Animal Cell Biotechnology: Methods and Protocols.* Humana Press, Totowa, NJ, 2007.

Seventh report on the statistics on the number of animals used for experimental and other scientific purposes in the member states of the European Union. Report from the Commission to the Council and the European Parliament, May 12, 2013 (retrieved on July 9, 2015).

Statistics of scientific procedures on living animals, Great Britain (PDF). British Government, 2004 (retrieved on July 13, 2012).

Statistics of scientific procedures on living animals, Great Britain (PDF). UK Home Office, 2017 (retrieved on July 23, 2018).

Twine, R. *Animals as Biotechnology: Ethics, Sustainability and Critical Animal Studies.* Earthscan Publications, Ltd., London, UK, 2010.

USDA Statistics for Animals Used in Research in the US Speaking of Research. https://www.aphis.usda.gov/animal_welfare/downloads/reports/Annual-Report-Animal-Usage-by-FY2016.pdf

Watanabe, M.M., Iwashita, M., Ishii, M. et al. Spot pattern of Leopard Danio is caused by mutation in the zebrafish connexin 41.8 gene. *EMBO Rep.* 7: 893–897, 2006.

White, R.M., Sessa, A., Burke, C. et al. Transparent adult zebrafish as a tool for in vivo transplantation analysis. *Cell Stem Cell* 2: 183–189, 2008.

World Malaria Report 2017 (PDF). World Health Organization, 2017. ISBN: 978-92-4-156552-3.

8 Environmental Biotechnology

LEARNING OBJECTIVES

- Discuss the importance of a healthy environment
- Discuss some factors affecting the environment
- Explain the role of biotechnology in maintaining the planet
- Explain how bioremediation helps clean wastewater
- Discuss some environmentally friendly products

8.1 INTRODUCTION

The environment plays a very important role in all living organisms and it is equally important to know the environment and its components before we discuss the factors adversely affecting it (Figure 8.1). These components can be classified into two major types: living and non-living components. The first component pertains to natural systems without massive human intervention, such as vegetation, animals, microorganisms, soil, rocks, the atmosphere, and the natural phenomena that occur within their boundaries. The second component pertains to universal natural resources and physical phenomena that lack clear-cut boundaries, such as air, water, and the climate, as well as energy, radiation, electric charge, and magnetism that do not originate from human activity.

8.1.1 ECOSYSTEM

An *ecosystem* is a natural unit consisting of all plants, animals, and microorganisms in an area functioning together with all the nonliving physical (abiotic) factors of the environment. The living organisms are continually engaged in a highly interrelated set of relationships with every other element constituting the environment in which they exist. The human ecosystem concept is then grounded in the deconstruction of the human/nature dichotomy and the emergent premise that all species are ecologically integrated with each other, as well as with the abiotic constituents of their biotope. Ecosystems can be grouped and discussed with a tremendous variety of scopes. An ecosystem can also describe any situation where there is a relationship between organisms and their environment. If humans are part of the organism, one can speak of a "human ecosystem." As virtually no surface of the earth today is free of human contact, all ecosystems can be more accurately considered human ecosystems or more neutrally human-influenced ecosystems.

8.1.2 BIOMES

Biomes are terminologically like the concept of ecosystems and are climatically and geographically defined areas of ecological similar climatic conditions, such as communities of plants, animals, and soil organisms. Biomes are defined based on factors such as plant structure (such as trees, shrubs, and grasses), leaf types (such as broadleaf and needle leaf), plant spacing (forest, woodland, savanna), and climate. Unlike *ecozones*, biomes are not defined by genetic, taxonomic, or historical similarities. Biomes are often identified with patterns of ecological succession and climate vegetation.

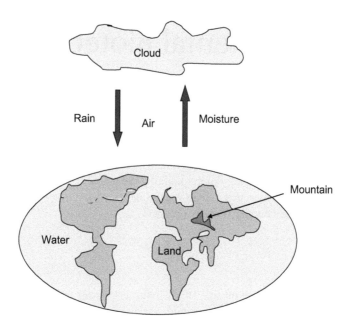

FIGURE 8.1 Our planet and environment.

8.1.3 Wilderness

Wilderness is generally defined as a natural environment on earth that has not been significantly modified by human activity. The WILD Foundation defines wilderness as "the most intact, undisturbed wild natural areas left on our planet—those last truly wild places that humans do not control and have not developed with roads, pipelines, or other industrial infrastructure." Wilderness areas and protected parks are considered important for the survival of certain species, ecological studies, conservation, solitude, and recreation. Wilderness is deeply valued for cultural, spiritual, moral, and aesthetic reasons. Some nature writers believe wilderness areas are vital for the human spirit and creativity. The word "wilderness" was derived from the notion of wildness or, in other words, that which is not controllable by humans. The mere presence or activity of people does not disqualify an area from being a "wilderness." Many ecosystems that are, or have been, inhabited or influenced by activities of people may still be considered "wild." This way of looking at wilderness includes areas within which natural processes operate without very noticeable human interference.

8.1.4 Geological Activity

Another important feature of our planet is its geological activity such as volcanic fissures and lava eruption. The earth's crust (or continental crust) is the outermost solid land surface of the earth. It is chemically and mechanically different from the underlying mantle and has been generated largely by igneous processes in which magma (molten rock) cools and solidifies to form solid land. Plate tectonics, mountain ranges, volcanoes, and earthquakes are geological phenomena that can be explained in terms of energy transformations in the earth's crust and might be thought of as the process by which the earth resurfaces itself. Beneath the earth's crust lies the mantle, which is heated by the radioactive decay of heavy elements. The mantle is not quite solid and consists of magma, which is in a state of semi-perpetual convection. This convection process causes the lithospheric plates to move, although slowly. The resulting process is known as *plate tectonics*. Volcanoes result primarily from the melting of subducted crust material. Crust material that is forced into the asthenosphere melts, and some portion of the melted material becomes light enough to rise to the surface, giving birth to volcanoes.

8.1.5 OCEANIC ACTIVITY

Approximately 71% of the earth's surface (an area of some 361 million square kilometers) is covered by oceans, a continuous body of water that is customarily divided into several principal oceans and smaller seas. More than half of this area is over 3000 meters deep. Average oceanic salinity is around 35 parts per thousand (ppt) (3.5%), and nearly all seawater has a salinity in the range of 30–38 ppt. Though generally recognized as several "separate" oceans, these waters comprise one global interconnected body of salt water, often referred to as the World Ocean or global ocean. This concept of a global ocean as a continuous body of water with relatively free interchange among its parts is of fundamental importance to oceanography. The major oceanic divisions are defined in part by the continents, various archipelagos, and other criteria. These divisions are (in descending order of size) the Pacific Ocean, the Atlantic Ocean, the Indian Ocean, the Southern Ocean (which is sometimes subsumed as the southern portions of the Pacific, Atlantic, and Indian Oceans), and the Arctic Ocean (which is sometimes considered a sea of the Atlantic). The Pacific and the Atlantic oceans may be further subdivided by the equator into northern and southern portions. Smaller regions of the oceans are called seas, gulfs, bays, and other names. There are also salt lakes, which are smaller bodies of landlocked saltwater that are not interconnected with the World Ocean. Two notable examples of salt lakes are the Aral Sea and the Great Salt Lake.

8.2 FACTORS AFFECTING THE ENVIRONMENT

Although the industrial revolution created useful products for humans, it also created problems like global warming; iceberg melting, depletion of the ozone layer, contamination of air, water, and land, and destruction of forests, to name a few. The initiation of large-scale projects such as dams and power plants poses special and growing challenges and risks to the natural environment. In this section, we will discuss some factors that negatively affect our environment, directly or indirectly.

8.2.1 GLOBAL WARMING

Global warming is a serious threat to the earth. The potential dangers of global warming are being increasingly studied by a wide global consortium of scientists that is concerned about the potential long-term effects of global warming on our natural environment and on the earth in general. One concern is how climate change and global warming caused by anthropogenic or human-made releases of GHGs, most notably carbon dioxide (CO_2), can act interactively and have adverse effects upon the planet, its natural environment, and human existence. Efforts have been increasingly focused on the mitigation of GHGs that are causing climatic changes and on developing adaptive strategies to assist humans, animal and plant species, ecosystems, regions, and nations to adjust to the effects of global warming (Figure 8.2).

8.2.1.1 Carbon Footprint

A *carbon footprint* is "the total set of GHG emissions caused directly and indirectly by an individual, organization, event, or product" (U.K. Carbon Trust 2008). The carbon footprint of an individual, a nation, or an organization is measured by undertaking a GHG emissions assessment. Once the size of the carbon footprint is known, a strategy can be devised to reduce it. Carbon offsets, or the mitigation of carbon emissions through the development of alternative projects such as solar or wind energy or reforestation, represent one way of managing a carbon footprint. The term and concept of the carbon footprint originate from the ecological footprint discussion. The carbon footprint is a subset of the ecological footprint.

8.2.1.2 Destruction of Forests

Deforestation can occur in two ways: by natural means or by human intervention. Deforestation happens naturally from time to time via wildfires. Interestingly, trees, plants, and animals all recover

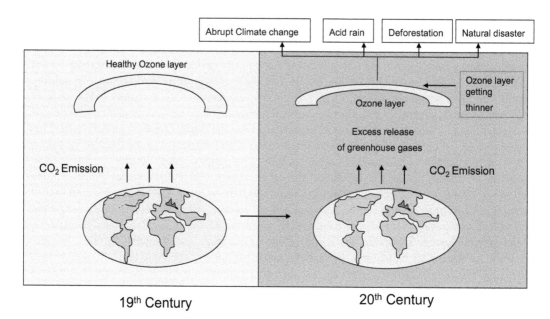

FIGURE 8.2 Global warming.

TABLE 8.1
Deforestation

Year	Loss of Forest in Millions of Hectares
2011	17.5
2012	23.3
2013	20.4
2014	23.5
2015	19.8
2016	29.7
2017	29.4

---

from such events naturally, so there are indeed some benefits from a wildfire. In addition, birds such as the black-backed woodpecker thrive only in freshly burned areas where they eat insects that bore into the burned trees. Some trees, such as the lodgepole pine, produce serotinous cones that only open when heat cooks the cone, thereby spreading the seeds in a freshly burned area with little other competition. Over time, burned areas regrow into forests. This has been observed in the case of Yellowstone National Park, where wildfires burnt all the trees and 20 years later the park was once again filled with medium-height lodgepole pines. Table 8.1 shows more facts about deforestation.

In the second scenario, people cause deforestation for several reasons. The action is not always permanent; some countries are better at replanting forests than others. Trees can be converted into paper or wood, two products that human civilization uses daily. In some parts of the world, wood is still a major source of fuel for cooking and heating. Chances are that wood is a major component of your home's structure, the main use of timber in all countries. Wood is also used to build furniture, cabinetry, and other products. Paper is used in many ways. Trees may also be cut down for forest management reasons. One reason is to limit a wildfire's ability to spread. This may be done in an emergency to combat an active fire or in a methodically planned long-term harvest. Typically, the forest is allowed to recover and it does so in a period of decades. Certain types of wildlife benefit from recovering forests after a harvest or wildfire. Many animals benefit from an edge habitat created by responsible logging or smaller-scale fires. Plants and animals in the natural world typically benefit from one type of habitat or another, or else benefit from living along the boundaries between two habitat types (known as an "edge" habitat). Animals that are specifically adapted to living in the forest cannot usually survive if their habitat is taken away. However, deforestation may benefit certain other animals, particularly grazing animals. It is for this reason that human's clear forests such as the Amazon for cattle grazing. Many birds benefit from having two habitats next to each other; the forest provides security but little food, whereas the open field provides food but relatively little security. Living along the boundary allows these types of animals to benefit from the strengths of both habitats. Deer are another example of an animal that may benefit from an edge habitat.

8.2.1.3 Air Pollution

Air pollution is the introduction of chemicals, particulate matter, or biological materials that cause harm or discomfort to humans or other living organisms or that damage the natural environment into the atmosphere. The atmosphere is a complex, dynamic natural gaseous system that is essential to support life on earth. Stratospheric ozone depletion because of air pollution has long been recognized as a threat to human health as well as to the earth's ecosystems. According to the Environmental Science Engineering Program at the Harvard School of Public Health, approximately 4% of deaths in the United States can be attributed to air pollution. The pollutants can be classified into two major categories: primary and secondary.

Primary pollutants produced by human activity include toxic metals (such as lead, cadmium, and copper), odors (from garbage, sewage, and industrial processes), and radioactive pollutants produced by nuclear explosions, war explosives, and natural processes such as the radioactive decay

TABLE 8.2
Impact of Pollutants on Our Environment

Pollutants	Toxic Concentration (ppm/h)
Sulfur dioxide	0.1–0.5
Oxides of nitrogen	0.21–100
Ozone	0.04–0.70
Fluorides	0.001–0.10
Ammonia	0.001–0.1
Ethylene & Propylene	0.0005–10
Particulates	100–500

TABLE 8.3
Coal Burning and Its Impact

Emission Sectors	Release of SO_2 (kt)
Coal-fired power station	918.11
Coad-fired industrial boilers	311.03
Residential coal combustion	68.48
Iron & steel production	207.48
Ammonia	480.97
Transportation	38.43
Others	348.83

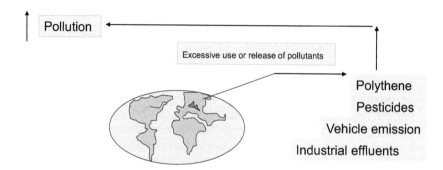

FIGURE 8.3 Global pollution issues.

of radon. Secondary pollutants include particulate matter formed from gaseous primary pollutants and compounds in photochemical smog. Smog is a kind of air pollution; the word "smog" is a portmanteau of smoke and fog. Classic smog results from large amounts of coal burning in an area and is caused by a mixture of smoke and sulfur dioxide (Table 8.3). Modern smog does not usually come from coal but from vehicular and industrial emissions that are acted on in the atmosphere by sunlight to form secondary pollutants that also combine with the primary emissions to form photochemical smog (Figure 8.3).

8.2.1.4 Major Carbon Dioxide Emission Countries

The following is a list of the world's top CO_2 emitting countries. The data itself were collected in 2007 by the Carbon Dioxide Information Analysis Center (CDIAC) for the United Nations (UN). The data considers only CO_2 emissions from the burning of fossil fuels, not emissions from deforestation, fossil fuel exporters, etc. These statistics are rapidly dated owing to the huge recent growth of emissions in Asia. The United States is the tenth largest emitter of CO_2 emissions per capita as of 2004. According to preliminary estimates, China has had a higher total emission since 2006 because of its much larger population and an increase in emissions from power generation. The list of world top 10 CO_2 releasing countries is presented in the Table 8.3.

8.2.1.5 Greenhouse Effect

GHGs are gases in an atmosphere that absorb and emit radiation within the thermal infrared range. This process is the fundamental cause of the greenhouse effect. Common GHGs in the earth's atmosphere include water vapor, CO_2, methane, nitrous oxide, ozone (O_3), and chlorofluorocarbons (CFCs). In our solar system, the atmospheres of Venus, Mars, and Titan also contain gases that cause greenhouse effects. GHGs, mainly water vapor, are essential to help determine the temperature of the earth. Without them, this planet would likely be so cold as to be uninhabitable. Although many factors such as the sun and the water cycle are responsible for the earth's weather and energy balance, if all else were held equal and stable, the planet's average temperature should be considerably lower without GHGs. Human activities have an impact upon the levels of GHGs in the atmosphere, which has other effects upon the system with their own possible repercussions. The most recent assessment report compiled by the Intergovernmental Panel on Climate Change (IPCC) observed that "changes in atmospheric concentrations of GHGs and aerosols, land cover, and solar radiation alter the energy balance of the climate system" and concluded that "increases in anthropogenic GHG concentrations are very likely to have caused most of the increase in global average temperatures since the mid-twentieth century."

8.2.1.6 Acid Rain

The term *acid rain* refers to what scientists call acid deposition. It is caused by airborne acidic pollutants and has highly destructive results. Scientists first discovered acid rain in 1852, when the

English chemist Robert Agnus invented the term. From then until now, acid rain has been an issue of intense debate among scientists and policy makers. Acid rain, one of the most important environmental problems of all, cannot be seen. The invisible gases that cause acid rain usually, come from automobiles or coal-burning power plants. Acid rain moves easily, affecting locations far beyond those that emit pollution. As a result, this global pollution issue causes great debate between countries that fight over polluting each other's environment.

For years, science studied the true causes of acid rain. Some scientists concluded that human production was primarily responsible, but others cited natural causes as well. Recently, more intensive research has been conducted so that countries have the information they need to prevent acid rain and its dangerous effects. The levels of acid rain vary from region to region. In Third World nations without pollution restrictions, acid rain tends to be very high. In Eastern Europe, China, and the Soviet Union, acid rain levels have also risen greatly. However, because acid rain can move about so easily, the problem is a global one.

8.2.1.7 Ocean Acidification

Ocean acidification is the name given to the ongoing decrease in the pH of the earth's oceans caused by their uptake of anthropogenic CO_2 from the atmosphere. One of the first detailed data sets examining temporal variations in pH at a temperate coastal location found that acidification was occurring at a rate much higher than that previously predicted, with consequences for nearshore benthic ecosystems. A December 2009 National Geographic report quoted Thomas Lovejoy, former chief biodiversity advisor to the World Bank, on recent research suggesting "the acidity of the oceans will more than double in the next 40 years. This rate is 100 times faster than any changes in ocean acidity in the last 20 million years, making it unlikely that marine life can somehow adapt to the changes." According to research from the University of Bristol, published in the journal *Nature Geoscience* in February 2010 and comparing current rates of ocean acidification with the greenhouse event at the Paleocene–Eocene boundary, when surface ocean temperatures rose by 5°C–6°C approximately 55 million years ago. During this time, no catastrophe was seen in surface ecosystems, yet bottom-dwelling organisms in the deep ocean experienced a major extinction. The study concluded that the current acidification is headed to reach levels higher than any seen in the last 65 million years. The study also found that the current rate of acidification is "10 times the rate that preceded the mass extinction 55 million years ago."

The main cause of acidification is excess CO_2 levels in the atmosphere. The carbon cycle describes the fluxes of CO_2 between the oceans, terrestrial biosphere, lithosphere, and the atmosphere. Human activities such as land-use changes, the combustion of fossil fuels, and the production of cement have led to a new flux of CO_2 into the atmosphere. Some of this has remained there, some has been taken up by terrestrial plants, and some has been absorbed by the oceans. The carbon cycle comes in two forms: the organic carbon cycle and the inorganic carbon cycle. The inorganic carbon cycle is particularly relevant when discussing ocean acidification because it includes the many forms of dissolved CO_2 present in the earth's oceans. When CO_2 dissolves, it reacts with water to form a balance of ionic and nonionic chemical species: dissolved free CO_2, carbonic acid, bicarbonate, and carbonate. The ratio of these species depends on factors such as seawater temperature and alkalinity.

8.2.1.8 Health Hazards due to Pollution

The World Health Organization states that 2.4 million people die each year from causes directly attributable to air pollution, with 1.5 million of these deaths attributable to indoor air pollution. "Epidemiological studies suggest that more than 500,000 Americans die each year from cardio-pulmonary disease linked to breathing fine particle air pollution." A study by the University of Birmingham has shown a strong correlation between pneumonia-related deaths and air pollution from motor vehicles. The worst short-term civilian pollution crisis in India was the 1984 Bhopal disaster. Leaked industrial vapors from the Union Carbide Factory, belonging to Union Carbide, Inc., United States, killed more than 2000 people outright and injured anywhere from 150,000–600,000 others, some 6000 of whom would later die from their injuries. The United Kingdom suffered its worst air pollution event when the Great Smog of 1952 formed over London. In 6 days more than 4000 died, and 8000 more died within the following months. The worst single incident of air pollution to occur in the United States occurred in Donora, Pennsylvania, in late 1948, when 20 people died and over 7000 were injured. Diesel exhaust (DE) is a major contributor to combustion-derived particulate matter air pollution. In several human experimental studies using a well-validated exposure chamber setup, DE was linked to acute vascular dysfunction and increased thrombus formation. This serves as a plausible mechanistic link between the previously described association between particulate matter air pollution and increased cardiovascular morbidity and mortality. Table 8.2 shows the impact of pollutants on our environment.

8.3 ENVIRONMENTAL PROTECTION BY BIOTECHNOLOGY

In the previous section, we learned about the environment and its various attributes. We also learned how different kinds of pollution are responsible for deteriorating the quality of water, air, and land, thus causing health hazards. Air and water pollution released from industrial, agricultural, and household wastes has contaminated our environment to such an extent that it requires a rescue plan to save our planet. The toxic effluents have not only affected human life but have also affected other living organisms, and quite a few species are now on the verge of extinction. It is true that humans are responsible for deteriorating the quality of the water, air, and land, but they themselves find ways to solve the problems they have created. In the process of solving these problems, biotechnologists play a vital role in providing arrays of solutions to reduce pollution. In this section, we discuss some tools of biotechnology that are used to create environmentally friendly technology and products (Table 8.4).

8.3.1 BIOREMEDIATION

Bioremediation can be defined as any process that uses microorganisms such as fungi, green plants, or their enzymes to return the natural environment altered by contaminants to its original condition. Bioremediation may be employed to attack specific soil contaminants, such as degradation of chlorinated hydrocarbons by bacteria. An example of a more general approach is the cleanup of oil spills by the addition of nitrate and/or sulfate fertilizers to facilitate the decomposition of crude oil by indigenous or exogenous bacteria. Naturally occurring bioremediation and phytoremediation have been used for centuries. For example, desalination of agricultural land by phyto-extraction has a long tradition.

Bioremediation technology using microorganisms was reportedly invented by George M. Robinson. Bioremediation technologies can be generally classified as in situ or ex situ. In situ bioremediation involves treating the contaminated material at the site, whereas ex situ bioremediation involves the removal of the contaminated material to be treated elsewhere. Some examples of bioremediation technologies are bioventing, land farming, bioreactor, composting, bio-augmentation,

TABLE 8.4
Eco-Friendly Products

Eco-Friendly Products	Frequency of Usage in %
Medicine	36.6
Cosmetic and hygiene products	68.4
Cleaning products	49.
Clothes & footwear	27.5
Other	7.0

rhizofiltration, and biostimulation. However, not all contaminants are easily treated by bioremediation using microorganisms. For example, heavy metals such as cadmium and lead are not readily absorbed or captured by organisms. The assimilation of metals such as mercury into the food chain may worsen matters. Phytoremediation is useful in these circumstances, because natural plants or transgenic plants can bioaccumulate these toxins in their above-ground parts, which are then harvested for removal. The heavy metals in the harvested biomass may be further concentrated by incineration or even recycled for industrial use. The elimination of a wide range of pollutants and wastes from the environment requires increasing our understanding of the relative importance of different pathways and regulatory networks to carbon flux environments and for particular compounds, and this will certainly accelerate the development of bioremediation technologies and biotransformation processes.

The use of genetic engineering to create organisms specifically designed for bioremediation has great potential. The bacterium *Deinococcus radiodurans* has been modified to consume and digest toluene and ionic mercury from highly radioactive nuclear waste. There are several cost-efficiency advantages to bioremediation that can be employed in areas that are inaccessible without excavation. For example, hydrocarbon spills (specifically, petrol spills) or certain chlorinated solvents may contaminate groundwater and introducing the right electron acceptor or electron donor amendment, as appropriate, may significantly reduce contaminant concentrations after a lag time allowing for acclimation. This is typically much less expensive than excavation followed by disposal elsewhere, incineration, or other ex situ treatment strategies. It also reduces or eliminates the need for "pump and treat," a common practice at sites where hydrocarbons have contaminated clean groundwater.

8.3.1.1 Mycoremediation

Mycoremediation, a form of bioremediation, is the process of using fungi to change contaminated soil to a less contaminated state. The term *mycoremediation* refers specifically to the use of fungal mycelia in bioremediation. One of the primary roles of fungi in the ecosystem is decomposition, which is performed by the mycelium. The mycelium secretes extracellular enzymes and acids that break down lignin and cellulose, the two main building blocks of plant fiber. These are organic compounds composed of long chains of carbon and hydrogen, structurally similar to many organic pollutants. The key to mycoremediation is determining the right fungal species to target a specific pollutant. Certain strains have been reported to successfully degrade certain nerve gases. With a view to establishing the bioremediation capability of fungi, scientists conducted an experiment where a plot of soil contaminated with diesel oil was inoculated with mycelia of oyster mushrooms. Traditional bioremediation techniques (bacteria) were used on control plots. After 4 weeks, more than 95% of many of the polycyclic aromatic hydrocarbons (PAHs) had been reduced to nontoxic components in the mycelial-inoculated plots. It appears that the natural microbial community participates with the fungi to break down contaminants, eventually into CO_2 and water. Moreover, wood-degrading fungi are particularly effective in breaking down aromatic pollutants such as toxic components of petroleum as well as chlorinated compounds and certain persistent pesticides. The process of bioremediation can be monitored indirectly by measuring the oxidation reduction potential, or redox, in soil and groundwater, together with pH, temperature, oxygen content, electron acceptor/donor concentrations, and concentration of breakdown products such as CO_2 (Figure 8.4).

8.3.2 Wastewater Treatment

Sewage is created by humans, institutions, hospitals, and commercial and industrial establishments. Raw influent includes household waste liquid from toilets, baths, showers, kitchens, sinks, and so forth that is disposed of via sewers. In many areas, sewage also includes liquid waste from industry and commercial entities. The separation and draining of household waste into gray water and black water is becoming more common in the developed world, with gray water being permitted to be used for watering plants or recycled for flushing toilets. Sewage water can be treated close to where it is created, in septic tanks, biofilters, or aerobic treatment systems, or it can be collected

Plot A: Soil contaminated with diesel oil   **Plot B: Soil contaminated with diesel oil**

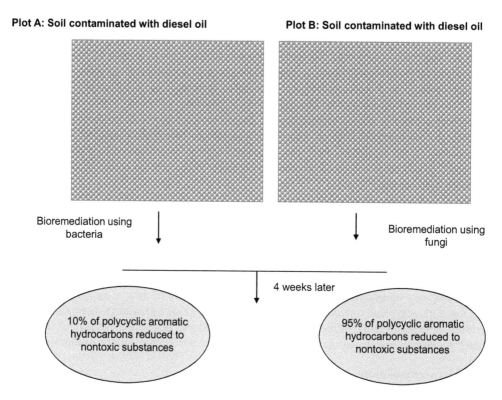

Bioremediation using bacteria

Bioremediation using fungi

4 weeks later

10% of polycyclic aromatic hydrocarbons reduced to nontoxic substances

95% of polycyclic aromatic hydrocarbons reduced to nontoxic substances

FIGURE 8.4   Bioremediation using fungi.

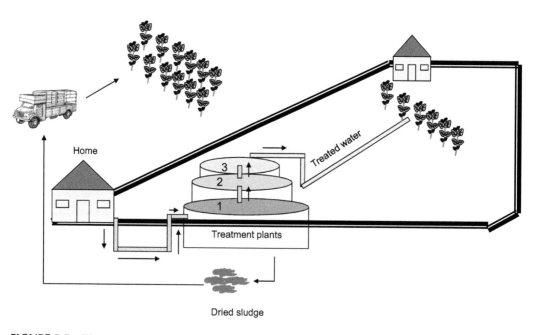

FIGURE 8.5   Wastewater treatment process.

and transported via a network of pipes and pump stations to a municipal treatment plant. Sewage collection and treatment is typically subject to local, state, and federal regulations and standards. Industrial sources of wastewater often require specialized treatment processes. Conventional sewage treatment involves three stages: primary, secondary, and tertiary treatments. First, the solids are separated from the wastewater stream. Then, dissolved biological matter is progressively converted into a solid mass by using indigenous, waterborne microorganisms. Finally, the biological solids are neutralized, then disposed of or reused, and the treated water may be disinfected chemically or physically—for example, by lagoons and microfiltration. The final effluent can be discharged into a stream, river, bay, lagoon, or wetland, or it can be used for the irrigation of golf courses, greenways, or parks. If it is sufficiently clean, it can also be used for groundwater recharge or for agricultural purposes. Wastewater treatment (Figure 8.5) is a critical component of the health and hygienic status of any city or town; we have therefore described different phases of sewage treatment.

8.3.2.1 Pretreatment Phase

The pretreatment step removes the materials that can be easily collected from the raw wastewater and raw solid. The typical materials that are removed during primary treatment include fats, oils, and greases; sand, gravel, and rocks; larger settleable solids; and floating materials such as rags and flushed feminine hygiene products. In developed countries, sophisticated equipment with remote operation and control are employed, whereas developing countries still rely on low-cost equipment for cleaning.

8.3.2.2 Screening Phase

The sewage water is strained to remove all large objects carried in the sewage stream, such as rags, sticks, tampons, cans, fruit, etc. This is done with a manual or automated mechanically raked bar screen. The raking action of a mechanical bar screen is typically paced according to the accumulation on the bar screen. The bar screen is used because large solids can damage or clog the equipment used later in the sewage treatment plant. The large solids can also hinder the biological process. The solids are collected and later disposed of in a landfill or incinerator. During the screening phase, incoming wastewater is carefully controlled to allow sand, grit, and stones to settle, while keeping most of the suspended organic material in the water column. This equipment is called a *degritter* or *sand catcher*. Sand, grit, and stones need to be removed early in the process to avoid damage to pumps and other equipment in the remaining treatment stages. Sometimes there is a sand washer followed by a conveyor that transports the sand to a container for disposal. The contents from the sand catcher may be fed into the incinerator in a sludge processing plant, but in many cases, the sand and grit are sent to a landfill.

8.3.2.3 Sedimentation Phase

In this phase, sewage flows through large tanks, commonly called *primary clarifiers* or *primary sedimentation tanks*. The tanks are large enough for sludge to settle and for floating materials (grease and oils) to be skimmed off. The main purpose of the primary sedimentation stage is to produce both a generally homogeneous liquid capable of being treated biologically and a sludge that can be separately treated or processed. Primary settling tanks are usually equipped with mechanically driven scrapers that continually drive the collected sludge toward a hopper in the base of the tank, from where it can be pumped to where subsequent sludge treatment stages are carried out.

8.3.2.4 Secondary Treatment Phase

This phase of treatment is designed to substantially degrade the biological content of the sewage that is derived from human waste, food waste, soaps, and detergents. Most municipal plants treat the settled sewage liquor using aerobic biological processes. For this to be effective, the biota requires both oxygen and a substrate on which to live. There are several ways in which this is done. In all these methods, the bacteria and protozoa consume biodegradable soluble organic contaminants (such as sugars, fats, and organic short-chain carbon molecules) and bind much of the less-soluble fractions into the flock. Secondary treatment systems are classified as fixed film or suspended growth. The fixed-film treatment process includes a trickling filter and rotating biological contactors where the biomass grows on media and the sewage passes over its surface. In suspended growth systems, such as activated sludge, the biomass is well mixed with the sewage; these can be operated in a smaller space than fixed-film systems that treat the same amount of water. However, fixed-film systems are more able to cope with drastic changes in the amount of biological material and can provide higher removal rates for organic material and suspended solids than suspended growth systems. Roughing filters are intended to treat particularly strong or variable organic loads (typically, industrial loads), to allow them to then be treated by conventional secondary treatment processes. Characteristics include typically tall, circular filters filled with open synthetic filter media to which wastewater is applied at a relatively high rate. They are designed to allow high hydraulic loading and a high flow-through of air. On larger installations, air is forced through the media using blowers.

8.3.2.5 Activated Sludge

Activated sludge plants involve processes that use dissolved oxygen to promote the growth of the biological flock that substantially removes organic material. Most biological oxidation processes for treating industrial wastewaters use oxygen (or air) and microbes. Surface-aerated basins achieve 80%–90% removal of biochemical oxygen demand with retention times of 1–10 days.

8.3.2.6 Filter Beds

In older plants and in plants receiving more variable loads, trickling filter beds are used in which settled sewage liquor is spread onto the surface of a deep bed made up of coke or carbonized coal, limestone chips, or specially fabricated plastic media. The surface beds also provide a source of air, which percolates up through the bed, keeping it aerobic. Biological films of bacteria, protozoa, and fungi form on the media's surfaces and eat or otherwise reduce the organic content. This biofilm is grazed by insect larvae and worms, which help maintain an optimal thickness. Overloading of beds increases the thickness of the film, leading to clogging of the filter media.

8.3.2.7 Biological Aerated Filters

A biological aerated filter (BAF) is generally used to provide access to oxygen and induce carbon reduction and nitrification. A BAF usually includes a reactor filled with a filter media; the media is either in suspension or supported by a gravel layer at the foot of the filter. The dual purpose of this media is to support the highly active biomass that is attached to it and to filter suspended solids. Carbon reduction and ammonia conversion occurs in aerobic mode and is sometimes achieved in a single reactor, while nitrate conversion occurs in anoxic mode. A BAF is operated either in upflow or downflow configuration, depending on the design specified by the manufacturer.

8.3.2.8 Nutrient Nitrogen and Phosphorus Removal

One of the major problems with wastewater treatment is handling high levels of the nutrients nitrogen and phosphorus. Their excessive release into the environment can lead to a buildup of nutrients, called *eutrophication*, which can in turn encourage the overgrowth of weeds, algae, and cyanobacteria. This may cause an algal bloom, a rapid growth in the population of algae. The algae numbers are unsustainable and eventually most of them die. The decomposition of the algae by bacteria uses up so much of the oxygen in the water that most or all the animals die, which creates more organic matter for the bacteria to decompose. In addition to causing deoxygenating, some algal species produce toxins that contaminate drinking water supplies. Different treatment processes are required to remove nitrogen and phosphorus.

8.3.2.9 Disinfection of Wastewater

Before releasing the treated wastewater back into the environment, the wastewater needs to be disinfected to substantially reduce the number of microorganisms in the water. Common methods of disinfection include ozone (O_3), chlorine, or UV light. Chloramine, which is used for drinking water, is not used in wastewater treatment because of its persistence. Chlorination remains the most common form of wastewater disinfection in North America owing to its low cost and long history of effectiveness. One disadvantage is that chlorination of residual organic material can generate chlorinated organic compounds that may be carcinogenic or may be harmful to the environment. Residual chlorine or chloramines may also be capable of chlorinating organic material in the natural aquatic environment. Further, because residual chlorine is toxic to aquatic species, the treated effluent must also be chemically dechlorinated, adding to the complexity and cost of treatment. UV light can be used instead of chlorine, iodine, or other chemicals. Because no chemicals are used, the treated water has no adverse effect on organisms that later consume it as may be the case with other methods. The cities of Edmonton and Calgary in Alberta, Canada, use UV light for their water treatment. O_3 is generated by passing oxygen through a high voltage potential, resulting in a third oxygen atom becoming attached and forming O_3. O_3 is very unstable and reactive and oxidizes most organic material it meets, thereby destroying many pathogenic microorganisms. O_3 is safer than chlorine because, unlike chlorine, which must be stored on site, O_3 is generated on site as needed. O_3 also produces fewer disinfection by-products than chlorination. A disadvantage of O_3 disinfection is the high cost of the O_3 generation equipment and the requirement for special operators.

8.3.2.10   Sludge Disposal

The liquid sludge produced after various treatments is further processed to make it suitable for final disposal. Typically, sludge is thickened to reduce the volumes transported off-site for disposal. There is no process that eliminates the need to dispose of biosolids. There is, however, an additional step some cities are taking to superheat the wastewater sludge and convert it into small pellets that are high in nitrogen and other organic materials. In New York City, for example, several sewage treatment plants have dewatering facilities that use large centrifuges along with the addition of chemicals such as polymer to further remove liquid from the sludge. The removed fluid is typically reintroduced into the wastewater process. The product that is left is called "cake," which is picked up by companies that turn it into fertilizer pellets. This product is then sold to local farmers and turf farms as a soil fertilizer. The process of industrial or chemical wastewater treatment is shown in Figure 8.6 where the pretreatment process starts with flow equalization treatment followed by biological treatment then pretreated water is released into the sea.

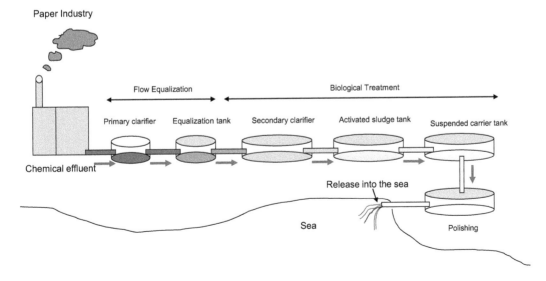

FIGURE 8.6   Industrial wastewater treatment plant.

8.3.3 BIOFUELS

The world's fossil fuel reservoirs are quickly diminishing and may be exhausted within the next 100 years. In recent years, researchers have succeeded in finding an alternate source of energy that is called *biofuel*. A biofuel is a liquid or gaseous fuel obtained from lifeless biological materials such as corn and soybean. In addition, various plants and plant-derived materials are used for biofuel manufacturing. Globally, biofuels are most commonly used to power vehicles, to heat homes, and for cooking.

Perhaps we must say that a biofuel is one that does not add to the stock of total CO_2 in the atmosphere. These are plant forms that typically remove CO_2 from the atmosphere and give up the same amount when burnt. The time duration required to produce biofuel is 1 or 2 years, whereas the fossil fuels can reach millions of years. Biofuel industries are expanding in Europe, Asia, and the United States. Recent technology developed at Los Alamos National Lab even allows for the conversion of pollution into renewable biofuel. *Agrofuels* are biofuels that are produced from specific crops rather than from waste processes such as landfill off-gassing or recycled vegetable oil. The process of biofuel production from sugarcane is shown in Figure 8.7.

Biomass or biofuel is material derived from recently living organisms. This includes plants, animals, and their by-products; for example, manure, garden waste, and crop residues are all sources of biomass. It is a renewable energy source based on the carbon cycle, unlike other natural resources such as petroleum, coal, and nuclear fuels. Biomass is made from many types of waste organic matter such as crop stalks, tree thinning, wooden pallets, construction waste, chicken and pig waste, and agricultural waste and lawn trimmings. It is used to produce power, heat and steam, and fuel through several different processes. Although renewable, biomass often involves a burning process that produces emissions such as sulfur dioxide, nitrogen oxide, and CO_2, but fortunately in quantities far less than those emitted by coal plants. However, proponents of coal plants feel that their way of doing it is a lot cheaper and there is a lot of dispute over this.

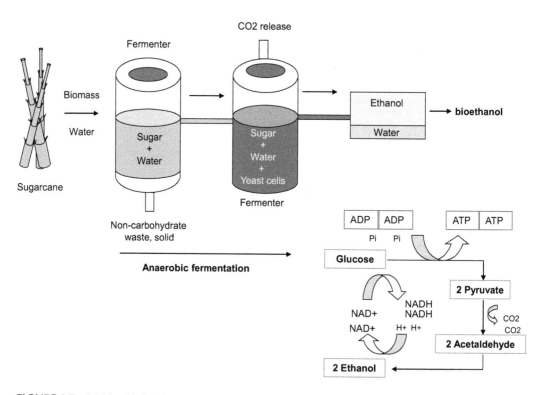

FIGURE 8.7 Making biofuel from plants.

Animal waste is a persistent and unavoidable pollutant produced primarily by the animals housed in an industrial-sized farm. Researchers from Washington University have figured out a way to turn manure into biomass. In April 2008, with the help of imaging technology, they noticed that vigorous mixing helps microorganisms turn farm waste into alternative energy, providing farmers with a simple way to treat their waste and convert it into energy. There are also agricultural products specifically grown for biofuel production, including corn, switch grass, and soybeans, primarily in the United States; rapeseed, wheat, and sugar beet, primarily in Europe; sugar cane in Brazil; palm oil and miscanthus in Southeast Asia; sorghum and cassava in China; jatropha and pongamia pinnata in India; and pongamia pinnata in Australia and the tropics. Hemp has also been proven to work as a biofuel. Biodegradable outputs from industry, agriculture, forestry, and households can be used for biofuel production. Biomass can come from waste plant material. The use of biomass fuels can therefore contribute to waste management as well as fuel security and help to prevent global warming, though alone, they are not a comprehensive solution to these problems.

8.3.3.1 Biodiesel

Biodiesel was probably the first of the alternative fuels to really become known to the public. The great advantage of biodiesel is that it can be used in existing vehicles with little or no adaptation necessary. Cars running on bioethanol, which is produced from agricultural crops, sugar cane, or biomass are governed by the same laws of physics as those using gasoline. This means that although both emit CO_2 as an inevitable consequence of the combustion process, there is a crucial difference: burning ethanol, in effect, recycles the CO_2 because it has already been removed from the atmosphere by photosynthesis during the natural growth process. In contrast, the use of gasoline or diesel injects into the atmosphere additional new quantities of CO_2 that have lain fixed underground in oil deposits for millions of years.

Biodiesel is the most common biofuel in Europe. It is produced from oils or fats using transesterification and is a liquid similar in composition to fossil/mineral diesel. Its chemical name is fatty acid methyl (or ethyl) ester (FAME). Oils are mixed with sodium hydroxide and methanol and the chemical reaction produces biodiesel (FAME) and glycerol. One-part glycerol is produced for every 10 parts biodiesel. Feedstock for biodiesel includes animal fats, vegetable oils, soy, rapeseed, jatropha, mahua, mustard, flax, sunflower, palm oil, hemp, field pennycress, pongamia pinnata, and algae. Because biodiesel is an effective solvent and cleans residues deposited by mineral diesel, engine filters may need to be replaced more often as the biofuel dissolves old deposits in the fuel tank and pipes. It also effectively cleans the engine combustion chamber of carbon deposits, helping to maintain efficiency. In many European countries, a 5% biodiesel blend is widely used and is available at thousands of gas stations. Biodiesel is also an *oxygenated fuel*, meaning that it contains a reduced amount of carbon and higher hydrogen and oxygen content than fossil diesel. This improves the combustion of fossil diesel and reduces the particulate emissions from unburnt carbon.

8.3.3.2 Biogas

Biogas is an alternative to natural gas. It is especially useful that the composition of biogas and natural gas is practically identical, so the same burners can be used for both fuels. Biogas can be produced from plant or animal waste or a combination of both. A mixture of both has proven to be the best method. There are many different methods used, depending on the starting material and quantity involved. Animal waste produces the nitrogen needed for growth of the bacteria, and vegetable waste supplies most of the carbon and hydrogen necessary.

8.3.3.3 Bioethanol

Bioethanol is produced by the action of microorganisms and enzymes through the fermentation of sugars or starches (which is easier), or cellulose (which is more difficult). Biobutanol (also called *biogasoline*) is often claimed to provide a direct replacement for gasoline because it can be used directly in a gasoline engine (in a similar way to biodiesel in diesel engines). Ethanol fuel is the most common biofuel worldwide, particularly in Brazil. Alcohol fuels are produced by fermentation of sugars derived from wheat, corn, sugar beets, sugar cane, molasses, and any sugar or starch that alcoholic beverages can be made from, like potato and fruit waste. The ethanol production methods used are enzyme digestion (to release sugars from stored starches), fermentation of the sugars, distillation, and drying. The distillation process requires significant energy input for heat (often unsustainable natural gas fossil fuel, but cellulosic biomass such as bagasse, the waste left after sugar cane is pressed to extract its juice, can also be used and is more sustainable). Ethanol can be used in petrol engines as a replacement for gasoline. It can be mixed with gasoline to any percentage. Most existing automobile petrol engines can run on blends of up to 15% bioethanol with petroleum/gasoline. Gasoline with ethanol added has higher octane, which means that an engine can typically burn hotter and more efficiently. In high-altitude (thin air) locations, some states mandate a mix of gasoline and ethanol as a winter oxidizer to reduce atmospheric pollution emissions.

Many car manufacturers are now producing flexible-fuel vehicles (FFVs), which can safely run on any combination of bioethanol and petrol up to 100% bioethanol. They dynamically sense exhaust oxygen content and adjust the engine's computer systems, spark, and fuel injection accordingly. Alcohol mixes with both petroleum and with water, so ethanol fuels are often diluted after the drying process by absorbing environmental moisture from the atmosphere. Water in alcohol-mix fuels reduces efficiency, makes engines harder to start, causes intermittent operation (sputtering), oxidizes aluminum (in carburetors), and rusts steel components. Methanol is currently produced from natural gas, a nonrenewable fossil fuel. It can also be produced from biomass as biomethanol. The methanol economy is an interesting alternative to the hydrogen economy compared to today's hydrogen produced from natural gas, but not hydrogen production directly from water and state-of-the-art clean solar thermal energy processes.

8.3.3.4 Bioethers

Bioethers, also referred to as fuel ethers, are cost-effective compounds that act as octane enhancers. They also enhance engine performance whilst significantly reducing engine wear and toxic exhaust emissions. Greatly reducing the amount of ground-level O_3, they contribute to the quality of the air we breathe.

8.3.3.5 Syngas

Syngas, a mixture of carbon monoxide and hydrogen, is produced by partial combustion of biomass—that is, combustion with an amount of oxygen that is not enough to convert the biomass completely to CO_2 and water. Before partial combustion, the biomass is dried and sometimes pyrolyzed. The resulting gas mixture, syngas, is itself a fuel. Using syngas is more efficient than direct combustion of the original biofuel; more of the energy contained in the fuel is extracted. Syngas may be burned directly in internal combustion engines or turbines. The wood gas generator is a wood-fueled gasification reactor mounted on an internal combustion engine. Syngas can be used to produce methanol and hydrogen or converted via the Fischer–Tropsch process to produce a synthetic diesel substitute or a mixture of alcohols that can be blended into gasoline. Gasification normally relies on temperatures >700°C. Lower temperature gasification is desirable when co-producing biochar but results in a syngas polluted with tar.

8.3.3.6 Second-Generation Biofuels

Supporters of biofuels claim that a more viable solution is to increase industrial support to produce second-generation biofuel from nonfood crops, including cellulosic biofuels. Second-generation biofuel production processes can use a variety of nonfood crops. These include waste biomass, the stalks of wheat, corn, wood, and special energy or biomass crops (e.g., Miscanthus). Second-generation biofuels use biomass-to-liquid technology, including cellulosic biofuels from nonfood crops. Many second-generation biofuels are under development, such as biohydrogen, biomethanol,

2,5-dimethylfuran (DMF), dimethylether (Bio-DME), Fischer–Tropsch diesel, biohydrogen diesel, mixed alcohols, and wood diesel.

8.3.3.7 Third-Generation Biofuels

Algae fuel, also called *oilgae* or *third-generation biofuel*, is a biofuel from algae. Algae are low-input, high-yield feedstock that produce biofuels. They produce 30 times more energy per acre than land crops such as soybeans. With the higher prices of fossil fuels (petroleum), there is much interest in algaculture (farming algae). One advantage of many biofuels over most other fuel types is that they are biodegradable and so relatively harmless to the environment if spilled. The US Department of Energy estimated that if algae fuel replaced all the petroleum fuel in the United States, it would require 15,000 square miles, which is roughly the size of Maryland. Algae, such as *Botryococcus braunii* and *Chlorella vulgaris*, are relatively easy to grow, but the algal oil is hard to extract. There are several approaches, some of which work better than others.

8.3.4 Biodegradable Plastic

Plastics are durable and degrade very slowly, and the molecular bonds that make plastic so durable make it equally resistant to natural processes of degradation. Since the 1950s, one billion tons of plastic have been discarded and may persist for hundreds or even thousands of years. In some cases, burning plastic can release toxic fumes. In addition, the manufacture of plastics often creates large quantities of chemical pollutants as well.

Prior to the ban on the use of CFCs in extrusion of polystyrene (and general use, except in life-critical fire suppression systems), the production of polystyrene contributed to the depletion of the O_3 layer. However, non-CFCs are currently used in the extrusion process. By 1995, plastic recycling programs were common in the United States and elsewhere.

On one hand, researchers have been working to find a method to degrade the existing plastic; on the other hand, they are also working on the development of biodegradable plastic. Recently, there have been reports of making biodegradable plastics using polyhydroxy alkanoates (PHAs), such as polyhydroxybutyrate (PHB), which are synthesized from acetyl-CoA used as a precursor and are used for the synthesis of biodegradable plastics with thermoplastic properties. At present, PHAs are produced by bacterial fermentation and the cost of biodegradable plastic is substantially higher than that of synthetic plastics. To reduce the cost, attempts are being made to produce PHAs in transgenic plants. Genes encoding the two enzymes acetoacetyl-CoA reductase (PhbB) and PHB synthase (phbC)—which are involved in PHB synthesis from the precursor acetyl-CoA—have been transferred from the bacterium *Alcaligenes eutrophus* and expressed in *Arabidopsis thaliana*. When the two enzymes were targeted into the plastids, PHB accumulated in leaves. PHB production of transgenic plants provides an example of a novel compound synthesized in plants. Transgenic trees, such as poplars, expressing phbB and phbC accumulate PHB in their leaves. The leaves are then collected and used for PHB extraction (Figure 8.8).

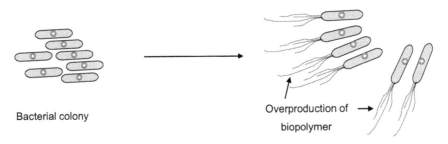

Bacterial colony

Overproduction of → biopolymer

FIGURE 8.8 Bioengineered bacteria overproducing biopolymer.

8.3.5 Biodegradation by Bacteria

Pseudomonas putida F1 is a Gram-negative rod-shaped saprotrophic soil bacterium. This bacterium was isolated from a polluted creek in Urbana, Illinois, by enrichment culture, with ethylbenzene as the sole source of carbon and energy. *Pseudomonas putida* F1 is one of the most well-studied aromatic hydrocarbon-degrading bacterial strains. Over 200 articles have been written about various aspects of *Pseudomonas putida* F1 physiology, enzymology, and genetics by microbiologists and biochemists, in addition to more applied studies by chemists and environmental engineers utilizing *Pseudomonas putida* F1 and its enzymes for green chemistry application and bioremediation. Strain F1 grows well with benzene, toluene, ethylbenzene, and *p*-cymene. Mutants of strain F1 that are capable of growth with *n*-propylbenzene, *n*-butylbenzene, isopropylbenzene, and biphenyl are easily obtained. In addition to aromatic hydrocarbons, the broad substrate toluene dioxygenase in strain F1 can oxidize trichloroethylene (TCE), indole, nitrotoluenes, chlorobenzenes, chlorophenols, and many other aromatic substrates. Although *Pseudomonas putida* F1 cannot use TCE as a source of carbon and energy, it is capable of degrading and detoxifying TCE in the presence of an additional carbon source. The ability of *Pseudomonas putida* F1 to degrade benzene, toluene, and ethylbenzene has a direct bearing on the development of strategies for dealing with environmental pollution. For example, many underground gasoline storage tanks leak and contaminate groundwater supplies with benzene, toluene, ethylbenzene, and xylenes (BTEX). *Pseudomonas putida* F1 can oxidize all these compounds, and thus a detailed knowledge of the physiology and biodegradation capabilities of this organism will be essential in providing the scientific foundations for the emerging bioremediation industry (Figure 8.9).

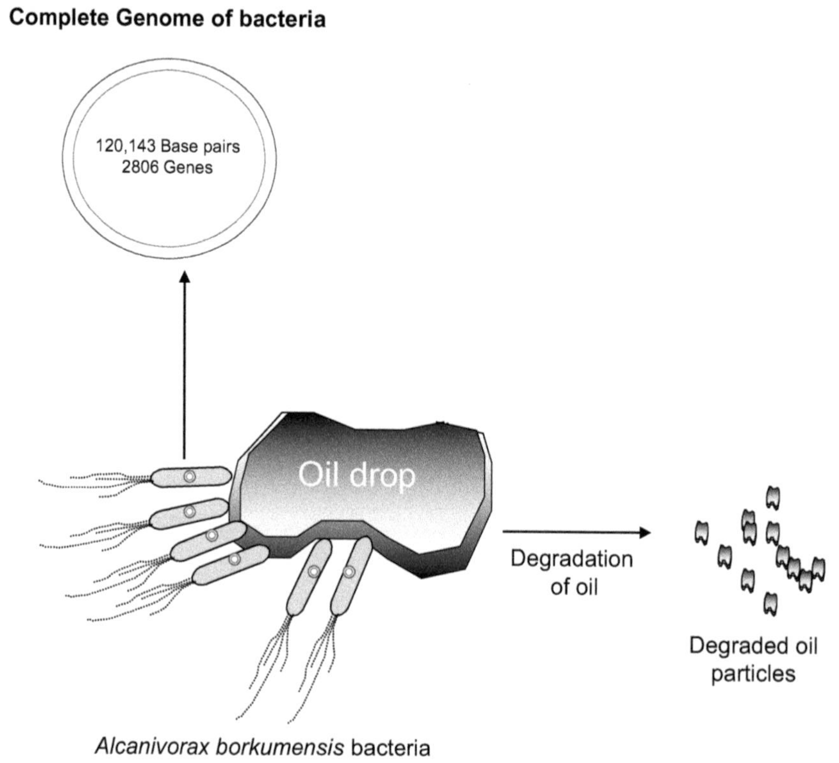

Complete Genome of bacteria

120,143 Base pairs
2806 Genes

Oil drop

Degradation
of oil

Degraded oil
particles

Alcanivorax borkumensis bacteria

FIGURE 8.9   Oil biodegrading by bacteria.

8.3.6 OIL-EATING BACTERIA

The transportation of oil has also caused incidences of oil spillage around the world. An *oil spill* is a release of a liquid petroleum hydrocarbon into the environment in the form of pollution. The term often refers to marine oil spills where oil is released into the ocean or coastal waters. Oil spills include releases of crude oil from tankers, offshore platforms, drilling rigs and wells, as well as spills of refined petroleum products (such as gasoline, diesel) and their by-products, and heavier fuels used by large ships such as bunker fuel, or the spillage of any oily refuse or waste oil. The oil penetrates the structure of the plumage of birds, reducing its insulating ability, thus making the birds more vulnerable to temperature fluctuations and much less buoyant in the water. It also impairs the bird's flight abilities to forage and escape from predators. Ingestion of the oil causes dehydration and impaired digestion. Because oil floats on top of water, less sunlight penetrates the water, limiting the photosynthesis of marine plants and phytoplankton. This, as well as decreasing fauna populations, affects the food chain in the ecosystem. Besides natural calamities, companies that drill oil (Table 8.5) from the seas also get into trouble because of oil spillage, and one such recent example was the oil spillage that killed 11 people and that contaminated a large area of the US marine environment along the Gulf of Mexico. BP's bill for containing and cleaning up the oil spill has reached nearly $10 billion (£6.4 billion), as the US government declared that the blown-out well has finally been plugged, 5 months after the explosion on the Deepwater Horizon rig. The beleaguered oil company revealed that its total cost for the spill had climbed to $9.5 billion. BP also said payouts to people affected by the spill, such as fishermen, hoteliers, and retailers, had dramatically increased since it handed over authority for dispensing funds to a White House appointee. BP has set up a $20 billion compensation fund, which has so far paid out 19,000 claims totaling more than $240 million.

Cleanup and recovery from an oil spill is difficult and depends upon many factors, including the type of oil spilled, the temperature of the water (affecting evaporation and biodegradation), and the types of shorelines and beaches involved. Among various cleanup methods, bioremediation has attracted a lot of attention and has been found to be the best method to clean up oil spills. Bioremediation may be employed to attack specific soil contaminants, such as degradation of chlorinated hydrocarbons by bacteria. An example of a more general approach is the cleanup of oil spills by the addition of nitrate and/or sulfate fertilizers to facilitate the decomposition of crude oil by indigenous or exogenous bacteria. The use of genetic engineering to create organisms specifically designed for bioremediation has great potential. The bacterium *Deinococcus radiodurans* (the most radiation-resistant organism known) has been modified to consume and digest toluene and ionic mercury from highly radioactive nuclear waste.

8.3.7 BIOLEACHING

Microbial leaching involves the process of dissolution of metals from ore-bearing rocks using microorganisms. Recently, bacterial activity has been implicated in the weathering, leaching, and

TABLE 8.5
Oil Drilling Companies

Companies	Revenues
Schlumberger	$ 30.44 Billion
Halliburton	$ 20.62 Billion
Fluor Corporation	$ 19.52 Billion
Baker Hughes	$ 17.25 Billion
Transocean Ltd	$ 7.38 Billion

deposition of mineral ores. The new discipline created by the marriage between biotechnology and metallurgy is known as *biohydrometallurgy, bioleaching*, or *biomining*. Conventional metallurgy involves smelting of ores at high temperatures. This involves high energy costs and leads to pollution. That copper could be leached from its ores by the activity of a bacterium, *Acidithiobacillus ferroxidans*, was discovered in 1947. This discovery opened the way to biohydrometallurgy for extraction of uranium in Canada and gold in South Africa. This technology is now commercially exploited in several countries for the extraction of copper, arsenic, nickel, zinc, etc. Microbes useful for metal recovery depend upon the temperature of the recovery process. Microbial technology is helpful in recovery of ores that cannot be economically processed with chemical methods as they contain only low-grade ores. During the separation of high-grade ores, large quantities of low-grade ore are produced and are discarded into waste heaps, from where they reach the environment.

Microbial leaching plays its role here, to retrieve lead, nickel, and zinc ores, which are there in significant amounts. Large-scale chemical processing causes environmental problems when the dump is not managed properly. Leach fluid containing a large quantity of metals at very low pH seeps into nearby natural water supplies and ground water. Biomining is thus an economically sound hydrometallurgical process that creates fewer environmental problems than conventional chemical processing. The vast unexploited mineral treasure of India is the treasure arena of action for biomining technology.

8.3.8 Single-Cell Protein and Biomass from Waste

In highly populated countries like India, the high quantity of waste that is generated can be effectively utilized as the growth medium for single-cell protein (SCP). *SCP* typically refers to sources of mixed protein extracted from pure or mixed cultures of algae, yeasts, fungi, or bacteria (grown on agricultural wastes) used as a substitute for protein-rich foods in human and animal feeds. In the 1960s, researchers at British Petroleum developed what they called "proteins-from-oil process," a technology for producing SCP by yeast fed waxy n-paraffins, a product produced by oil refineries. Initial research work was done by Alfred Champagnat at BP's Lavera Oil Refinery in France. A small pilot plant there started operation in March 1963, and the construction of the second pilot plant was authorized at Grangemouth Oil Refinery in Britain.

SCPs develop when microbes ferment waste materials (including wood, straw, cannery, and food processing wastes, residues from alcohol production, hydrocarbons, or human and animal excreta). The problem with extracting SCPs from the wastes is the dilution and cost. They are found in very low concentrations, usually less than 5%. Engineers have developed ways to increase the concentrations, including centrifugation, flotation, precipitation, coagulation and filtration or the use of semipermeable membranes. SCPs need to be dehydrated to approximately 10% moisture content and/or acidified to aid in storage and prevent spoilage. The methods to increase the concentrations to adequate levels and the dewatering process require equipment that is expensive and not always suitable for small-scale operations. It is economically prudent to feed the product locally and shortly after it is produced. Microbes employed include yeasts (*Saccharomyces cerevisiae, Candida utilis (Torulopsis), Geotrichum candidum*, and *Oidium lactis*), other fungi (*Aspergillus oryzae, Sclerotium rolfsii, Polyporus*, and *Trichoderma*), bacteria (*Rhodopseudomonas capsulata*), and algae (*Chlorella and Spirulina*).

8.3.9 Vermitechnology

Vermitechnology or *vermicomposting* is an effective way to recycle agricultural waste, city garbage, kitchen waste, etc. by converting these organic waste materials into nutritious compost by using earthworms. Worms transform the waste into high-quality fertilizer. Vermicast is a valuable soil amendment and can replace chemical fertilizers to some extent. Vermicompost is potential organic manure, rich in plant nutrients. It also contains micronutrients, certain hormones and enzymes, etc., which have stimulatory effects on plant growth. It also lodges many beneficial bacteria and actinomycetes.

Vermicompost containing a large number of earthworm eggs that hatch out within a month is equivalent to a mini-fertilizer factory in the soil. The earthworms eat biomass and excrete it in a digested form, generally called *vermicompost*. Vermicompost increases the water-holding capacity and reduces the irrigation water requirement of crops. The worm species most commonly used are *Eisenia foetida, Lubricus rubellus, Eudrilus eugeniae, Perionyx excavatus, P. arbricola,* and *P. sansibaricus.* Vermicompost influences the physicochemical as well as the biological properties of the soil, which in turn improves its fertility. It is rich in micronutrients like magnesium, iron, molybdenum, boron, copper, zinc, and also some growth regulators. It enhances the water-holding capacity of light-textured sandy soil. Vermicompost is rich in several microflora, several enzymes, auxins, and complex growth regulators like gibberellins, which are present in the earthworm castings. It also has a buffering action and so neutralizes soil pH and helps in making minerals and trace elements more easily available to crops. It also enhances soil fertility and reduces toxicity. The quality, shelf life, and nutritive value of horticultural crops are enhanced, enabling value addition to the produce. Vermitechnology is an essential tool of organic farming that increases crop productivity in a sustainable manner, thus resulting in quality produce, and reduces the cost of agricultural inputs in addition to improving the inherent capacity of the soil without deleterious effects on the environment.

8.3.10 BIOSORPTION

Pollution interacts naturally with biological systems. It is currently uncontrolled, seeping into any biological entity within the range of exposure. The most problematic contaminants include heavy metals, pesticides, and other organic compounds that can be toxic to wildlife and humans in small concentrations. There are existing methods for remediation, but they are expensive or ineffective. However, an extensive body of research has found that a wide variety of commonly discarded waste, including eggshells, bones, peat, fungi, seaweed, yeast, and carrot peels can efficiently remove toxic heavy metal ions from contaminated water. Ions from metals like mercury can react in the environment to form harmful compounds such as methyl mercury, a compound known to be toxic

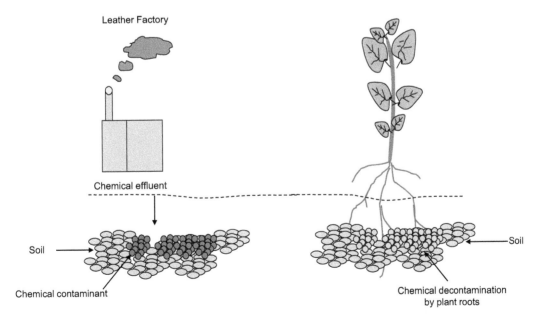

FIGURE 8.10   Pollution removal by plants.

to humans. In addition, adsorbing biomass, or biosorbents, can also remove other harmful metals such as arsenic, lead, cadmium, cobalt, chromium, and uranium. Biosorption is a physiochemical process that occurs naturally in certain biomass, which allows it to passively concentrate and bind contaminants onto its cellular structure. Biosorption may be used as an environmentally friendly filtering technique. There is no doubt that the world would benefit from more rigorous filtering of harmful pollutants created by industrial processes and pervasive human activity. Bacteria, yeast, algae, fungi and plants (Figure 8.10) are found to be most effective in decontamination of pollution.

PROBLEMS

Section A: Descriptive Type

Q1. What are the effects of global warming?

Q2. Describe the various health hazards caused by man-made pollution.

Q3. What is the greenhouse effect?

Q4. What is bioremediation? Describe its significance.

Q5. Describe the various steps involved in the wastewater treatment process.

Q6. Can biofuel replace fossil fuel in the future? Describe the advantages and disadvantages of biofuel.

Q7. How is energy converted from biowaste?

Q8. Do you think it is possible to make biodegradable plastics?

Section B: Multiple Choice

Q1. A carbon footprint is a . . .
 a. Greenhouse gas emission
 b. Hydrogen gas emission
 c. Methane gas emission

Q2. Greenhouse gases are gases in the atmosphere that absorb and emit radiation within the thermal infrared range. True/False

Q3. Acid rain is caused by a reaction between these in the atmosphere.
 a. Hydrogen and nitrogen
 b. Carbon and hydrogen
 c. Sulfur and nitrogen
 d. Hydrogen and sulfur

Q4. This bacterium has been modified to consume and digest toluene and ionic mercury from highly radioactive nuclear waste.
 a. *Deinococus radiodurans*
 b. *Salmonella*
 c. *Nocardia*
 d. *E. coli*

Q5. The term mycoremediation refers to . . .
 a. Bacterial bioremediation
 b. Yeast bioremediation
 c. Fungal bioremediation
 d. None of the above

Q6. The secondary treatment phase of wastewater treatment deals with the . . .
 a. Degradation of biological waste
 b. Degradation of chemical waste
 c. Treatment of microbes
 d. None of the above

Q7. Which among the following is not a biofuel?
   a. Ethanol
   b. Methanol
   c. Butanol
   d. Butane
Q8. FFV is a . . .
   a. Fast-fuel vehicle
   b. Flexible-fuel vehicle
   c. Front-fuel vehicle
   d. None of the above
Q9. Syngas is a mixture of carbon monoxide and hydrogen. True/False
Q10. Third-generation biofuel refers to . . .
   a. Ethanol fuel
   b. Algal fuel
   c. Fungal fuel
   d. Bacterial fuel
Q11. *P. putida* F1 is one of the most well-studied aromatic hydrocarbon-degrading bacterial strains. True/False

Section C: Critical Thinking

Q1. What efforts should we make to prevent the s release of greenhouse gases?
Q2. Do you think biofuel will ever replace fossil fuel? If yes, explain how.

Debate

Organize a debate on the topic "Global warming: myth or reality?" The students may speak about global warming based on scientific data and research.

Field Visit

With the help of your course instructor, arrange a field visit to your city wastewater treatment plant and learn how wastewater is being treated and cleaned in the plant and how the bio-waste-water or sludge is being used for irrigation and to generate energy. Also, write a brief report based on your visual experience of the sewage treatment plant and submit your report to your course instructor.

REFERENCES AND FURTHER READING

Adams, S., and Lambert, D. *Earth Science: An Illustrated Guide to Science*. Chelsea House, New York, p. 20, 2006.
ADM Biodiesel. Hamburg, Leer, Mainz, 2010. http://wwwbiodisel.de.
Bent, F., Bruzelius, N., and Rothengatter, W. *Megaprojects and Risk: An Anatomy of Ambition*. Cambridge University Press, Cambridge, UK, 2003.
Beychok, M.R. Performance of surface-aerated basins. *Chem. Eng. Progr. Symp. Ser.* 67: 322–339, 1971.
Bounds, A. OECD warns against biofuels subsidies. *Financial Times*, 2007. www.ft.com/cms/s/0/e780d216-5fd5-11dc-b0fe-0000779fd2ac.html (retrieved on March 7, 2008).
Caldeira, K., and Wickett, M.E. Anthropogenic carbon and ocean pH. *Nature* 425: 365–365, 2003.
Choi, E.N., Cho, M.C., Kim, Y., Kim, C-K., and Lee, K. Expansion of growth substrate range in *Pseudomonas putida* F1 by mutations in both *cymR* and *todS*, which recruit a ring-fission hydrolase CmtE and induce the *tod* catabolic pathway, respectively. *Microbiology* 149: 795–805, 2003.

Earth's Energy Budget. Oklahoma climatological survey: 1996–2004, 2007. http://okfirst.mesonet.org/train/meteorology/EnergyBudget.html (retrieved on November 17, 2007).

Eaton, R.W. P-Cymene catabolic pathway in *Pseudomonas putida* F1: Cloning and characterization of DNA encoding conversion of p-cymene to p-cumate. *J. Bacteriol.* 179: 3171–3180, 1997.

EurekAlert. Fifteen new highly stable fungal enzyme catalysts that efficiently break down cellulose into sugars at high temperatures, 2009. www.eurekalert.org.

Evans, J. Biofuels aim higher: Biofuels, bioproducts and biorefining (BioFPR), 2008. www.biofpr.com/details/feature/102347/Biofuels_aim_higher.html (retrieved on December 3, 2008).

Farrell, A.E. et al. Ethanol can contribute to energy and environmental goals. *Science* 311: 506–508, 2006.

Gibson, D.T., Koch, J.R., and Kallio, R.E. Oxidative degradation of aromatic hydrocarbons by microorganisms I: Enzymatic formation of catechol from benzene. *Biochemistry* 7: 2653–2661, 1968.

Globeco biodegradable bio-diesel, 2008. www.globeco.co.uk.

Greenfuelonline.com, 2011. www.greenfuelonline.com.

Hammerschlag, R. Ethanol's energy return on investment: A survey of the literature 1999–present. *Environ. Sci. Technol.* 40: 1744–1750, 2006.

Hartman, E. A promising oil alternative: Algae energy. *Washington Post*, 2008. www.washingtonpost.com/wpdyn/content/article/2008/01/03/AR2008010303907.html (retrieved on January 15, 2008).

Hudlicky, T., Gonzalez, D., and Gibson, D.T. Enzymatic dihydroxylation of aromatics in enantioselective synthesis: Expanding asymmetric methodology. *Aldrichim. Acta* 32: 35–62, 1999.

IEA Bioenergy. 2010. www.ieabioenergy.com.

John, F.K., Hamilton, T.G., Bra, M., and Röckmann, T. Methane emissions from terrestrial plants under aerobic conditions. *Nature* 439: 187–191, 2006.

Karl, T.R., and Trenberth, K.E. Modern global climate change. *Science* 302: 1719–1723, 2003.

Key, R.M., Kozyr, A., Sabine, C.L. et al. A global ocean carbon climatology: Results from GLODAP. *Global Biogeochem. Cycles* 18: GB4031, 23, 2004.

Kiehl, J.T., and Trenberth, K.E. Earth's annual global mean energy budget. *Bull. Amer. Meteor. Soc.* 78(2): 197–208, 1997.

Kyoto protocol from United Nations framework convention on climate change, 2008. www.unfccc.int (retrieved on August 2008).

Lau, P.C.K., Wang, Y., Patel, A., Labbé, D. et al. A bacterial basic region leucine zipper histidine kinase regulating toluene degradation. *Proc. Natl. Acad. Sci. USA* 94: 1453–1458, 1997.

Le Treut, H., Somerville, R., Cubasch, U. et al. Historical overview of climate change science. In: *Climate Change 2007: The Physical Science Basis. Contribution of Working Group I to the Fourth Assessment Report of the Intergovernmental Panel on Climate Change*, S. Solomon, D. Qin, M. Manning et al. (eds.). Cambridge University Press, Cambridge, UK, 2007.

Low cost algae production system introduced, 2011. www.energy-arizona.org.

Non-CO_2 gases economic analysis and inventory 2007: Global warming potentials and atmospheric lifetimes. *U.S. Environmental Protection Agency* (retrieved on August 31, 2007).

Ocean. *The Columbia Encyclopedia*. Columbia University Press, New York, 2002.

Odum, E.P. *Fundamentals of Ecology*, 3rd edn. Saunders, New York, 1971.

Oldroyd, D. *Earth Cycles: A Historical Perspective*. Greenwood Press, Westport, CT, 2006.

Orr, J.C., Fabry, V.J., Aumont, O. et al. Anthropogenic ocean acidification over the twenty-first century and its impact on calcifying organisms. *Nature* 437: 681–686, 2005.

Parales, R.E., Ditty, J.L., and Harwood, C.S. Toluene-degrading bacteria are chemotactic to the environmental pollutants benzene, toluene, and trichloroethylene. *Appl. Environ. Microbiol.* 66: 4098–4104, 2000.

Pelletier, J.D. Natural variability of atmospheric temperatures and geomagnetic intensity over a wide range of time scales. *Proc. Natl. Acad. Sci.* 99: 2546–2553, 2002.

Robert, W.C. *Geosystems: An Introduction to Physical Geography*, Prentice Hall, Inc., Upper Saddle River, NJ, 1996.

Simison, W.B. The mechanism behind plate tectonics, 2007. www.ucmp.berkeley.edu/geology/tecmech.html (retrieved on November 17, 2007).

Smith, G.A., and Pun, A. *How Does the Earth Work?* Pearson Prentice Hall, Upper Saddle River, NJ, 2006.

Somerville, C. Development of cellulosic biofuels. *U.S. Department of Agriculture*, 2008. www.usda.gov/oce/forum/2007%20Speeches/PDF%20PPT/CSomerville.pdf (retrieved on January 15, 2008).

Spilhaus, A.F. Distribution of land and water on the planet. *UN Atlas of the Oceans, Maps of the Whole World Ocean, Geogr. Rev.* 32(3): 431–435, 1942.

UN Biofuels Report, 2007. http://esa.in.org.

Wackett, L.P., and Gibson, D.T. Degradation of trichloroethylene by toluene dioxygenase in whole cell studies with *Pseudomonas putida* F1. *Appl. Environ. Microbiol.* 54: 1703–1708, 1988.

Welcome to biodiesel filling stations, 2009. http://biodieselrefillingstations.co.uk.

Western Climate Initiative. 2008. www.westernclimateinitiative.org.

Yunqiao, P., Dongcheng, Z., Singh, P.M., and Ragauskas, A.J. The new forestry biofuels sector. *Biofuel Bioprod. Biorefin.* 2(1): 58–73, 2007.

Zylstra, G.J., and Gibson, D.T. Toluene degradation by *Pseudomonas putida* F1: Nucleotide sequence of the todC1C2BADE genes and their expression. *E. Coli. J. Biol. Chem.* 264: 14940–14946, 1989.

Zylstra, G.J., and Gibson, D.T. Aromatic hydrocarbon degradation: A molecular approach. *Genet. Eng.* 13: 183–203, 1991.

9 Medical Biotechnology

LEARNING OBJECTIVES

- Explain the role of biotechnology in medicine
- Discuss how vaccines are manufactured and explain the related issues
- Discuss the process of making antibodies and hybridoma technology
- Explain the significance of therapeutic protein and its production
- Discuss the success and failure of gene therapy and organ transplantation
- Discuss tools and techniques used in medical biotechnology

9.1 INTRODUCTION

Biotechnology has many applications in medicine. While dealing with diseases, biotechnology tools and techniques have been used in the prevention, diagnosis, and cure of diseases. Through human genetics, biotechnology has found use in genetic counseling, antenatal diagnosis, and gene therapy. In forensic medicine, it has already been used for identification of individuals who could be criminals. The major applications include animal and human healthcare, genetic counseling, and forensic medicine. Biotechnology is useful in providing immunity against a disease through the development of a vaccine and in the diagnosis of a disease at an early stage of its onset. Table 9.1 shows a list of the various applications of medical biotechnology.

9.2 VACCINES

A *vaccine* is a biological preparation that improves immunity to a specific disease. It typically contains an agent that resembles a disease-causing microorganism and is often made from weakened or killed forms of the microbe. The agent stimulates the body's immune system to recognize the agent as foreign, destroy it, and "remember" it so that the immune system can more easily recognize and destroy any of these microorganisms that it later encounters. Sometime during the 1770s, Edward Jenner heard a milkmaid boast that she would never have the often-fatal or disfiguring disease smallpox because she had already had cowpox, which has a very mild effect in humans. In 1796, Jenner took pus from the hand of a milkmaid with cowpox, inoculated an 8-year-old boy with it, and 6 weeks later, variolated the boy's arm with smallpox, afterward observing that the boy did not catch smallpox. Vaccination with cowpox was much safer than smallpox inoculation.

Louis Pasteur generalized Jenner's idea by developing what he called a rabies vaccine (now termed an antitoxin), and in the nineteenth century, vaccines were considered a matter of national prestige and compulsory vaccination laws were passed. The twentieth century saw the introduction of several successful vaccines, including those against diphtheria, measles, mumps, and rubella. Major achievements included the development of the polio vaccine in the 1950s and the eradication of smallpox during the 1960s and 1970s. As vaccines became more common, many people began taking them for granted. However, vaccines remained elusive for many important diseases, including malaria and HIV. There are several types of vaccines currently in use. These represent different strategies used to reduce the risk of illness while retaining the ability to induce a beneficial immune response (Figure 9.1).

TABLE 9.1

Applications of Medical Biotechnology

Class	Products
Hormones	Follicle-stimulating hormones, growth hormones, insulin,
Growth factors	Platelet-derived growth factors, nerve growth factors, insulin growth factor
Cytokines	Interferons, interleukins, erythropoietin
Vaccines	Conventional and recombinant vaccines from virus and bacteria
Nucleic acid-based products	Gene therapy, DNA vaccines
Cell therapy	Autologous transplantation and tissue engineering
Others	Clotting factors, enzymes

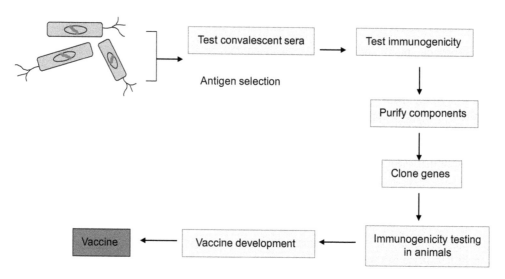

FIGURE 9.1   Conventional vaccine development.

9.2.1   KILLED VACCINES

Some vaccines are being produced by using killed microorganisms. The virulent microorganisms need to first be killed with chemicals or heat to be able to make the vaccine. Some examples are vaccines against influenza, cholera, the bubonic plague, polio, and hepatitis A.

9.2.2   ATTENUATED VACCINES

Some vaccines contain live, attenuated virus microorganisms. These are live microorganisms that have been cultivated under conditions that disable their virulent properties or which use closely-related but less dangerous organisms to produce a broad immune response. They typically provoke more durable immunological responses and are the preferred type for healthy adults. Examples include vaccines against yellow fever, measles, rubella, and mumps. The live *Mycobacterium tuberculosis* vaccine, developed by Calmette and Guérin, is not made of a contagious strain but contains a virulently modified strain called "BCG," used to elicit immunogenicity to the vaccine.

9.2.3 Toxoid Vaccines

Toxoids are inactivated toxic compounds in cases where these (rather than the microorganism itself) cause illness. Examples of toxoid-based vaccines include tetanus and diphtheria. Not all toxoids are for microorganisms; for example, *Crotalus atrox* toxoid is used to vaccinate dogs against rattlesnake bites.

9.2.4 Subunit Vaccines

Rather than introducing an inactivated or attenuated microorganism to an immune system (which would constitute a "whole-agent" vaccine), a fragment of it can create an immune response. These are called *subunit vaccines*. Characteristic examples include the subunit vaccine against the hepatitis B virus, which is composed of only the surface proteins of the virus (produced in yeast) and the virus-like particle (VLP) vaccine against human papillomavirus (HPV), which is composed of the viral major capsid protein.

9.2.5 Conjugate Vaccines

Certain bacteria have polysaccharide outer coats that are poorly immunogenic. By linking these outer coats to proteins (such as toxins), the immune system can be led to recognize the polysaccharide as if it were a protein antigen. This approach is used in the *Haemophilus influenzae* type B vaccine.

9.2.6 Experimental Vaccines

There are several innovative vaccines currently in the developmental stage or which are already in use, such as recombinant vector vaccines, DNA vaccines, and T-cell receptor peptide vaccines. Recall that in recombinant vector vaccines, combining the physiology of one microorganism and the DNA of another can create immunity against diseases that have complex infection processes. On the other hand, a new type of vaccine, called a *DNA vaccine*, has been recently created from an infectious agent's DNA. It works by insertion (and expression, triggering immune system recognition) of viral or bacterial DNA into human or animal cells. Some cells of the immune system that recognize the proteins expressed will mount an attack against these proteins and the cells expressing them. Because these cells live for a very long time, if the pathogen that normally expresses these proteins is encountered at a later time, they will be attacked instantly by the immune system. One advantage of DNA vaccines is that they are very easy to produce and store. As of 2006, DNA vaccination is still experimental. Likewise, T-cell receptor peptide vaccines are under development for several diseases, using models of valley fever, stomatitis, and atopic dermatitis. These peptides have been shown to modulate cytokine production and improve cell-mediated immunity.

Targeting of identified bacterial proteins that are involved in complement inhibition would neutralize the key bacterial virulence mechanism. Although most vaccines are created using inactivated or attenuated compounds from microorganisms, synthetic vaccines are composed mainly or wholly of synthetic peptides, carbohydrates, or antigens.

9.2.7 Valence Vaccines

Valence vaccines are monovalent (also called *univalent*) or multivalent (also called *polyvalent*) in nature. A monovalent vaccine is designed to immunize against a single antigen or a single microorganism. A multivalent vaccine is designed to immunize against two or more strains of the same microorganism or against two or more microorganisms. In certain cases, a monovalent vaccine may be preferable for rapidly developing a strong immune response.

9.3   VACCINE PRODUCTION

Pharmaceutical firms and biotechnology companies have little incentive to develop vaccines for diseases because there are little commercial benefits. Even in more affluent countries, financial returns are usually minimal and the financial and other risks are great. Most vaccine development to date has relied on "push" funding by governments, universities, and nonprofit organizations. Many vaccines have been highly cost-effective and beneficial for public health.

Vaccine production has several stages. First, the antigen itself is generated. Viruses are grown either on primary cells such as chicken eggs (e.g., for influenza), or on continuous cell lines such as cultured human cells (e.g., for hepatitis A). Bacteria such as *Haemophilus influenzae* type b are grown in bioreactors. Alternatively, a recombinant protein derived from the viruses or bacteria can be generated in yeast, bacteria, or cell cultures. After the antigen is generated, it is isolated from the cells used to generate it. A virus may need to be inactivated, possibly with no further purification required. Recombinant proteins (Figure 9.2) need many operations involving ultrafiltration and column chromatography. Finally, the vaccine is formulated by adding adjuvant, stabilizers, and preservatives, as needed. The adjuvant enhances the immune response of the antigen, the stabilizers increase the storage life, and the preservatives allow the use of multidose vials. Combination vaccines are harder to develop and produce because of potential incompatibilities and interactions among the antigens and other ingredients involved.

Vaccine production techniques are evolving. Cultured mammalian cells are expected to become increasingly important compared to conventional options such as chicken eggs because of greater productivity and few problems with contamination. Recombination technology that produces genetically detoxified vaccines is expected to grow in popularity to produce bacterial vaccines that use

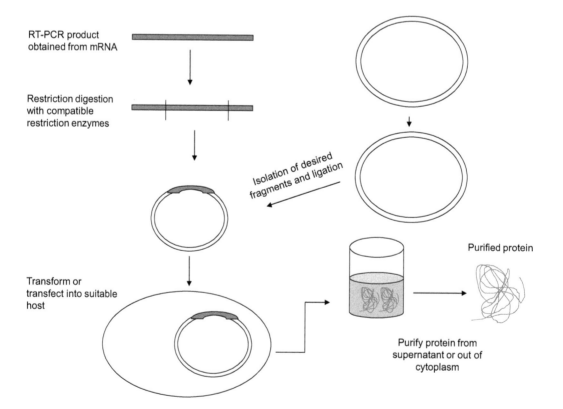

FIGURE 9.2   The production of therapeutic protein.

toxoids. Combination vaccines are expected to reduce the quantities of antigens they contain and thereby decrease undesirable interactions by using pathogen-associated molecular patterns. Many vaccines need preservatives to prevent serious adverse effects such as the *Staphylococcus* infection that, in one 1928 incident, killed 12 of 21 children inoculated with a diphtheria vaccine that lacked a preservative. Several preservatives are available, including thiomersal, phenoxyethanol, and formaldehyde. Thiomersal is more effective against bacteria, has better shelf life, and improves vaccine stability, potency, and safety. However, in the United States, the European Union, and a few other affluent countries, it is no longer used as a preservative in childhood vaccines as a precautionary measure because of its mercury content. Although controversial claims have been made that thiomersal contributes to autism, no convincing scientific evidence supports these claims.

9.3.1 MAKING INFLUENZA VACCINES

Influenza, commonly referred to as the flu, is an infectious disease caused by RNA viruses of the family Orthomyxoviridae (the influenza viruses), which affects birds and mammals. Typically, influenza is transmitted through the air by coughs or sneezes, creating aerosols containing the virus. Influenza can also be transmitted by direct contact with bird droppings or nasal secretions or through contact with contaminated surfaces. Airborne aerosols have been thought to cause most infections, although which means of transmission is most important is not entirely clear. Influenza viruses can be inactivated by sunlight, disinfectants, and detergents. Three influenza pandemics occurred in the twentieth century and killed tens of millions of people, with each of these pandemics being caused by the appearance of a new strain of the virus in humans. Often, these new strains appear when an existing flu virus spreads to humans from other animal species or when an existing human strain picks up new genes from a virus that usually infects birds or pigs. An avian strain named H5N1 raised the concern of a new influenza pandemic after it emerged in Asia in the 1990s, but it has not evolved to a form that spreads easily between people. In April 2009, a novel flu strain evolved that combined genes from human, pig, and bird flu, initially dubbed "swine flu" and known as influenza A/H1N1; it emerged in Mexico, the United States, and several other nations. The World Health Organization officially declared the outbreak to be a pandemic on June 11, 2009.

The following are the steps in producing influenza vaccine (Figure 9.3).

Step 1: A survey must first be done to find out which kinds of influenza strains are dominantly present in geographical locations throughout the year.

Step 2: The surveillance data, based on the presence of dominant influenza strains, are analyzed by researchers and the selected influenza strains are submitted to the Food and Drug Administration (FDA) to recommend which to try to include. The FDA distributes seed viruses to manufacturers to begin the vaccine production.

Step 3: The seed viruses received from the FDA are processed for production in specially designed labs. Each virus strain is produced separately and later combined to make one vaccine. Millions of specially prepared chicken eggs are used to produce the vaccine. For 7 months, fertilized eggs are delivered to the manufacturer. Each egg is cleaned with a disinfectant spray and injected with one strain. The eggs are incubated for several days to allow the virus to multiply. After incubation, the virus-loaded fluid is harvested.

Step 4: The virus fluid undergoes multiple purification steps and a special chemical treatment to ensure that the virus is inactivated or killed. The virus is split by chemically disrupting the whole virus. Viral fragments from all the virus strains are collected from different batches and combined after completion of quality control tests. Manufacturers and the FDA test the vaccine concentrate to determine the amount and yield of the virus to ensure the concentrate is adequate for immunization.

Step 5: Upon receiving FDA approval and licensing, the vaccine is released for distribution in time for immunization. Manufacturers begin filling the doses into vials and syringes,

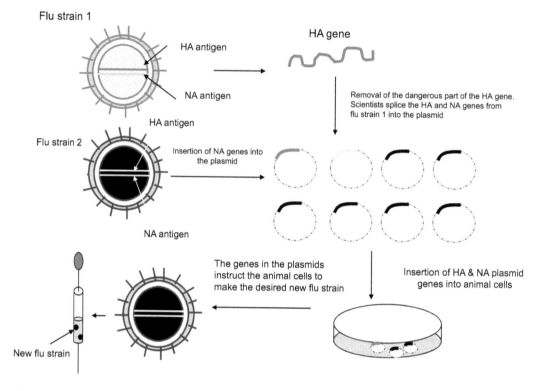

FIGURE 9.3   Making the influenza vaccine.

which are then sealed and carefully inspected before labels are applied to show the vaccine batch, lot numbers, and expiration dates. Each lot must be specially released by the FDA before the manufacturer can ship it.

Step 6: Vaccine shipments typically begin in August/September and continue till November each year. Immunization generally begins in the month of October or as soon as the vaccine becomes available, continues throughout the influenza season, and typically ends in March. Immunity develops approximately 2 weeks following vaccination.

9.4   LARGE-SCALE PRODUCTION OF VACCINES

The first step in making a vaccine is to store the virus of interest under frozen conditions that prevent the virus from becoming either stronger or weaker than desired. After defrosting and warming the seed virus under carefully specified conditions (i.e., at room temperature or in a water bath), a small amount of virus cells is placed into a "cell factory," a small machine that, with the addition of an appropriate medium, allows the virus cells to multiply. Each type of virus grows best in a medium specific to it, established in pre-manufacturing laboratory procedures, but all contain proteins from mammals in one form or another, such as purified protein from cow blood. Mixed with the appropriate medium, at the appropriate temperature, and with a predetermined amount of time, viruses will multiply.

The virus from the cell factory is then separated from the medium and placed in a second medium for additional growth. Early methods dating back 40 or 50 years ago used a bottle to hold the mixture, and the resulting growth was a single layer of viruses floating on the medium. It was soon discovered that if the bottle was turned while the viruses were growing, even more viruses could be produced because a layer of viruses grew on all the inside surfaces of the bottle. An important discovery in the 1940s was that cell growth is greatly stimulated by the addition of enzymes to a medium, of which trypsin is the most commonly used. An enzyme is a protein that also functions as a catalyst in the feeding and growth of cells. In current practice, bottles are not used at all. The growing virus is kept in a container larger than but like the cell factory and mixed with "beads" (near-microscopic particles to which the viruses can attach themselves). The use of the beads provides the virus with a much greater area to attach to and consequently allows a much greater growth of virus. As in the cell factory, temperature and pH are strictly controlled. Time spent in growing viruses varies according to the type of virus being produced and is, in each case, a closely guarded secret of the manufacturer. When there are enough viruses, they are separated from the beads in one or more ways. The broth might be forced through a filter with openings large enough to allow the viruses to pass through, but small enough to prevent the beads from passing. The mixture may be centrifuged several times to separate the virus from the beads in a container from which they can then be drawn off. The eventual vaccine will be made of either attenuated (weakened) virus or a killed virus. The choice of one or the other depends on several factors, including the efficacy of the resulting vaccine and its secondary effects. The vaccine based on rubella is developed almost every year in response to new variants of the causative virus and is always an attenuated vaccine. The virulence of a virus can dictate the choice. The rabies vaccine, for example, is always a killed vaccine.

If the vaccine is attenuated, the virus is usually attenuated before it goes through the production process. Carefully selected strains are cultured (grown) repeatedly in various media. There are strains of viruses that actually become stronger as they grow. These strains are clearly unusable for an attenuated vaccine. Other strains become too weak as they are cultured repeatedly, and these too are unacceptable for vaccine use. Like the porridge, chair, and bed that Goldilocks liked, only some viruses are "just right," reaching a level of attenuation that makes them acceptable for vaccine use and unchanging in strength. Recent molecular technology has made possible the attenuation of live viruses by molecular manipulation, but this method is still rare. The virus is then separated from the medium in which it has grown. Vaccines that are of several types (as most are) are combined before packaging. The actual amount of vaccine given to a patient will be relatively small compared to the medium in which it is given. Decisions about whether to use water, alcohol, or some other solution for an injectable vaccine, for example, are made after repeated tests for safety, sterility, and stability.

9.5   TRENDS IN VACCINE RESEARCH

Research on influenza includes how the virus produces disease–host immune responses, viral genomics, and how the virus spreads. These studies help in developing influenza countermeasures. For example, a better understanding of the body's immune system response helps vaccine development, and a detailed picture of how influenza invades cells aids the development of antiviral drugs. One important basic research program is the Influenza Genome Sequencing Project, which is creating a library of influenza sequences. This library should help clarify which factors make one strain more lethal than another, which genes most affect immunogenicity, and how the virus evolves over time. Research into new vaccines is particularly important, as current vaccines are very slow and expensive to produce and must be reformulated every year. The sequencing of the influenza genome and rDNA technology may accelerate the generation of new vaccine strains by allowing scientists to substitute new antigens into a previously developed vaccine strain. New technologies are also being developed to grow viruses in cell culture, which promises higher yields, less cost, better quality, and surge capacity. Research on a universal influenza A vaccine, targeted against the external domain of the transmembrane viral M2 protein (M2e), is being done at the University of Ghent by Walter Fiers, Xavier Saelens, and their team and has now successfully concluded Phase I clinical trials. Several biologics, therapeutic vaccines, and immunobiologics are also being investigated for treatment of infections caused by viruses. Therapeutic biologics are designed to activate the immune response to viruses or antigens. Typically, biologics do not target metabolic pathways as do antiviral drugs, but instead stimulate immune cells such as lymphocytes, macrophages, and/or antigen-presenting cells to drive an immune response toward a cytotoxic effect against the virus.

Influenza models such as murine influenza are convenient models to test the effects of prophylactic and therapeutic biologics. For example, lymphocyte T-cell immune modulator inhibits viral growth in the murine model of influenza.

When it comes to making an influenza vaccine, there seems to be a crack in the system. Although dependable since the 1970s, the current practice of injecting the flu virus into fertilized hens' eggs requires at least 6 months and hundreds of millions of eggs to produce enough supply of vaccine for the US population. The egg method is not sufficient to ensure rapid vaccine supply, and vaccine manufacturers need to arrange for egg supplies months in advance. ID Biomedical was recently awarded a $9.5 million contract by the National Institute of Allergy and Infectious Diseases (NIAID) to study an alternative method (called cell culture) to rapidly produce large quantities of the flu vaccine.

The cell culture technique is a very interesting alternative, as it is an efficient and flexible strategy for manufacturing influenza vaccines. Instead of injecting the flu virus into eggs, the virus is overlaid onto special dog kidney cells, which, unlike eggs, grow and multiply rapidly in the lab. The inoculated cells are then incubated inside a growth chamber (called a bioreactor), adhering to small round beads (called microcarriers). As with the egg method, once the virus-containing fluid is removed from the bioreactor, the virus is purified, killed, split, and then blended with the two other viruses to make doses of flu vaccine. The advantages of the cell-culture technique are that if the need should arise for increased amounts of vaccine, a manufacturer could simply thaw out more cells and perhaps add more bioreactors. In addition, unlike eggs, the cells are naturally sterile, which helps to control product quality. Finally, the nutrients in which the cells grow contain no animal serum, which reduces the risk of microbial contamination. Currently, the research team is conducting studies to determine the optimal conditions under which the cell-culture vaccine can be produced.

9.6   ISSUES RELATED TO VACCINES

Producing a usable, safe antiviral vaccine involves many steps; unfortunately, these steps cannot always be carried out for each and every virus. There is still much to be done and learned. The new methods of molecular manipulation have caused more than one scientist to believe that vaccine technology is only now entering a "golden age." Refinements of existing vaccines are possible in the future. The rabies vaccine, for example, produces side effects that make the vaccine unsatisfactory for mass immunization. In the United States, the rabies vaccine is now used only on patients who have contracted the virus from an infected animal and are likely to develop the fatal disease without immunization. The HIV virus, which biologists believe causes AIDS, is not currently amenable to traditional vaccine production methods. The AIDS virus rapidly mutates from one strain to another, and any given strain does not seem to confer immunity against other strains. Additionally, a limited immunizing effect of either attenuated or killed virus cannot be demonstrated either in the laboratory or in test animals. No HIV vaccine has yet been developed.

9.7   SYNTHETIC PEPTIDES AS VACCINES

Vaccines can also be prepared through short synthetic peptide chains, which have therefore become a subject of considerable research activity. To synthesize peptides to be used as vaccines, the structure and function of proteins involved should be studied. As it is the three-dimensional structure (TDS) and not the amino acid sequence that is responsible for the immunogenic response, it may be necessary to identify the protein region involved in the immunogenic response. For instance, in foot and mouth disease virus (FMDV), it is the amino acids 114–160 of the virus polypeptide that can produce antibodies neutralizing FMDV and thus provide protection.

Neutralization of FMDV was also possible through the region of 201–213 amino acids of the same protein. It has been shown that small synthetic peptides representing these regions of proteins can show immunogenic response and can therefore be used for development of a vaccine. An alternative approach to identify the immunogenic region of the protein is through the study of the gene encoding the protein.

Recently, it has been shown that a cloned gene of an immunogenic protein of a pathogen (feline leukemia virus or FeLV) can be cut into fragments by DNase, and these fragments can be cloned in lambda phages, where they may express. Phage colonies (plaques) having different cloned fragments are screened with a specific MAB that neutralizes the pathogen. The fragments that react with the antibody must be synthesizing the immunogenic peptide fragment. This cloned DNA fragment can then be sequenced. In this manner, it was possible to identify a 14 amino acid immunogen of the envelope protein of FeLV. The corresponding synthetic peptide was also found to compete with the virus for antibodies. When injected in guinea pigs, such synthetic peptides also elicited a partial immunogenic response. Therefore, there is great promise for the use of such synthetic peptides as vaccines. In fact, a vaccine for malaria in the form of a synthetic peptide has already been prepared and is being tested for its suitability. This is the first example of a vaccine developed in the form of a synthetic peptide. Recently, it has been shown that the immunogenic region of the protein of a pathogen can also be identified by eluting it from purified major histocompatibility complex (MHC) molecules. Different MHC allelic variants are available in cells for binding of different proteins, and they can be purified using specific T-cells. Peptides can be eluted from these purified MHC molecules and the sequence of such peptides can be determined and used for manufacturing synthetic peptides to be used as vaccines. Using the above three approaches, vaccines against several pathogens have either been produced in recent years or are expected to be produced soon.

9.8 ANTIBODY PRODUCTION

The body produces antibodies against antigens to fight back bacterial or viral diseases. These antibodies make our immune system very strong, but in some cases, the body fails to produce such antibodies. In that case, the body can be provided with antibodies produced in animals and bioreactors. In this section, we will discuss various MABs and their production and applications (Figure 9.4).

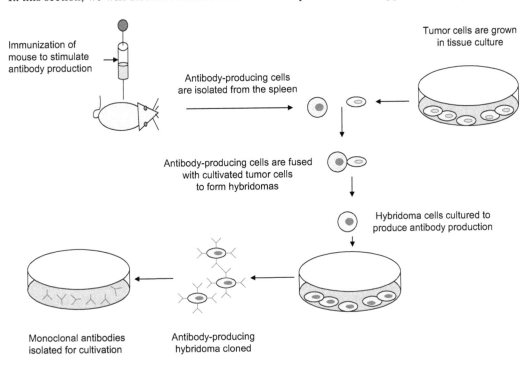

FIGURE 9.4 Monoclonal antibody production.

9.8.1 Monoclonal Antibodies

Although MABs have been long-established as essential research tools, their therapeutic promise has taken considerably longer to realize, requiring further advances such as the humanization of mouse antibodies and recombinant production protocols. The diagnostic and therapeutic implications of MABs were immediately recognized. We will next discuss the various applications of MABs.

9.8.1.1 Diagnostic Test

MABs have been extensively used for diagnostic purposes to identify a disease and its progression. Several antibody tests and kits have been developed for rapid detection of disease using the ELISA technique. The western blot test or immuno-dot blot test detects the protein on a membrane. They are also very useful in immunohistochemistry, which detects antigens in fixed tissue sections, and the immunofluorescence test, which detects the substance in a frozen tissue section or in live cells.

9.8.1.2 Cancer Treatment

One possible treatment for cancer involves MABs that bind only to cancer-cell-specific antigens and induce an immunological response against the target cancer cells. Such MABs could also be modified for delivery of a toxin, radioisotope, cytokine, or other active conjugate. It is also possible to design bispecific antibodies that can bind with their fragment antigen-binding (Fab) regions, both to target antigens and to a conjugate or effector cell. In fact, every intact antibody can bind to cell receptors or other proteins with its fragment crystallizable (Fc) region. Antibody-directed enzyme pro-drug therapy (ADEPT) is a new type of cancer treatment that uses MABs. Now, ADEPT is being used only in clinical trials. The trials aim to find out whether ADEPT may be useful as a new type of treatment for bowel cancer. ADEPT is a type of targeted therapy. It uses a MAB to carry an enzyme directly to the cancer cells. Enzymes are proteins that control chemical reactions in the body. First, the MAB is given (with the enzyme attached). A few hours later, a second drug (the pro-drug) is given. When the pro-drug encounters the enzyme, a reaction takes place. This reaction activates the pro-drug and it is then able to destroy the cancer cells. As the enzyme does not attach to normal cells, this treatment does not affect them.

9.8.2 Hybridoma Technology

We will now discuss the process of generating MABs using hybridoma technology. In hybridoma technology, MABs are typically made by fusing myeloma cells with the spleen cells from a mouse that has been immunized with the desired antigen. However, recent advances have allowed the use of rabbit B-cells. Polyethylene glycol is used to fuse adjacent plasma membranes, but the success rate is low, so a selective medium is used in which only fused cells can grow. This is because myeloma cells have lost the ability to synthesize *hypoxanthine-guanine-phosphoribosyl transferase* (HGPRT), an enzyme necessary for the salvage synthesis of nucleic acids. The absence of HGPRT is not a problem for these cells unless the de novo purine synthesis pathway is also disrupted. By exposing cells to aminopterin, a folic acid analogue that inhibits dihydrofolate reductase (DHFR), they are unable to use the de novo pathway and become fully auxotrophic for nucleic acids, requiring supplementation to survive.

The selective culture medium is called *HAT medium*, because it contains hypoxanthine, aminopterin, and thymidine. This medium is selective for fused (hybridoma) cells. Unfused myeloma cells cannot grow because they lack HGPRT and thus cannot replicate their DNA. Unfused spleen cells cannot grow indefinitely because of their limited life span. Only fused hybrid cells, referred to as *hybridomas*, can grow indefinitely in the media, because the spleen cell partner supplies HGPRT and the myeloma partner has traits that make it immortal (as it is a cancer cell). This mixture of cells is then diluted, and clones are grown from single parent cells on microtiter wells. The antibodies secreted by the different clones are then assayed for their ability to bind to the antigen (with a test such as ELISA or antigen microarray assay or immuno-dot blot). The most productive and stable clone is then selected for future use. The hybridomas can be grown indefinitely in a suitable cell culture media. When the hybridoma cells are injected into mice (in the peritoneal cavity, the gut), they produce tumors containing an antibody-rich fluid called *ascites fluid*. The medium must be enriched during selection to further favor hybridoma growth. This can be achieved using a layer of feeder fibrocyte cells or supplement medium such as briclone.

9.8.2.1 Purification of Monoclonal Antibodies

After obtaining either a media sample of cultured hybridomas or a sample of ascites fluid, the desired antibodies must be extracted. The contaminants in the cell culture sample would consist of media components such as growth factors, hormones, and transferrins. In contrast, the *in vivo* sample is likely to have host antibodies, proteases, nucleases, nucleic acids, and viruses. In both cases, other secretions by the hybridomas, such as cytokines, may be present. There may also be bacterial contamination, resulting in endotoxins being secreted by the bacteria. Depending on the complexity of the media required in cell culture, and thus the contaminants in question, one method (*in vivo* or *in vitro*) may be preferable to the other. The sample is first conditioned or prepared for purification. Cells, cell debris, lipids, and clotted material are first removed, typically by filtration with a 0.45 μm filter. These large particles can cause a phenomenon called *membrane fouling* in subsequent purification steps. Additionally, the concentration of the product in the sample may not be enough, especially in cases where the desired antibody is one produced by a low-secreting cell line. The sample is therefore condensed by ultrafiltration or dialysis.

To achieve maximum purity in a single step, affinity purification can be performed by using the antigen to provide exquisite specificity for the antibody. In this method, the antigen used to generate the antibody is covalently attached to an agarose support. If the antigen is a peptide, it is commonly synthesized with a terminal cysteine that allows selective attachment to a carrier protein, such as KLH, during development and to the support for purification. The antibody-containing media are then incubated with the immobilized antigen, either in batches or as the antibody is passed through a column, where it is selectively retained while impurities are removed. An elution with a low pH buffer, high salt elution buffer is then used to recover the purified antibody from the support. To further select for antibodies, the antibodies can be precipitated out using sodium sulfate or ammonium sulfate. Antibodies precipitate at low concentrations of the salt, while most other proteins precipitate at higher concentrations. The appropriate level of salt is added to achieve the best separation. Excess salt must then be removed by a desalting method such as dialysis. The final purity can be analyzed using a chromatogram. Any impurities will produce peaks, and the volume under the peak indicates the amount of the impurity. Alternatively, gel electrophoresis and capillary electrophoresis (CE) can be carried out. Impurities will produce bands of varying intensity, depending on how much of the impurity is present.

9.8.2.2 Recombinant Monoclonal Antibodies

The need to overcome the immunogenicity problem of rodent antibodies in clinical practice has resulted in a plethora of strategies to isolate human antibodies. If human antibodies are to be used, then it is necessary to understand the basis by which different isotypes interact with host effector systems and, if possible, improve on nature by engineering in desirable modifications. Recombinant antibody technology is a rapidly evolving field that enables the study and improvement of antibody properties by means of genetic engineering. Moreover, the functional expression of antibody fragments in *Escherichia coli* has formed the basis for antibody library generation and selection, a powerful method to produce human antibodies for therapy. Because *in vitro*-generated antibodies offer various advantages over traditionally produced MABs, such molecules are now increasingly used for standard immunological assays. The original humanization strategy described by Dr. Winter and his group exploited knowledge of the solved crystal structure to graft rodent complementarity-determining regions (CDRs) into the defined human frameworks and judicious mutations in critical framework residues.

9.9 GENE THERAPY

A potential approach to the treatment of genetic disorders in humans is gene therapy. This is a technique whereby the absent or faulty gene is replaced by a working gene so that the body can make the correct enzyme or protein and consequently eliminate the root cause of the disease. The most likely candidates for gene therapy trials are rare diseases such as Lesch–Nyhan syndrome, a distressing disease in which patients are unable to manufacture an enzyme. If gene therapy does become practicable, the biggest impact would be the treatment of diseases where the normal gene needs to be introduced into one organ. One such disease is PK), which affects approximately one in 12,000 white children and, if not treated early, can result in severe mental retardation. The disease is caused by a defect in a gene producing a liver enzyme. If detected early enough, the child can be placed on a special diet for its first few years, but this is very unpleasant and can lead to many problems within the family.

The types of gene therapy described so far all have one factor in common, which is that the tissues being treated are somatic. In contrast to this is the replacement of defective genes in the germline cells that contribute to the genetic heritage of the offspring. Gene therapy in germline cells has the potential to affect not only the individual being treated, but also his or her children. Germline therapy would change the genetic pool of the entire human species, and future generations would have to live with that change. In addition to these ethical problems, several technical difficulties make it unlikely that germline therapy would be tried on humans in the near future.

Before treatment for a genetic disease can begin, an accurate diagnosis of the genetic defect needs to be made. It is here that biotechnology is also likely to have a great impact soon. Genetic engineering research has produced a powerful tool to uncover specific diseases rapidly and accurately. Short pieces of DNA called *DNA probes* can be designed to stick very specifically to certain other pieces of DNA. The technique relies upon the fact that complementary pieces of DNA stick together. DNA probes are more specific and have the potential to be more sensitive than conventional diagnostic methods, and it should be possible in the near future to distinguish between defective genes and their normal counterparts, which is an important development.

Somatic cells are nonreproductive. Somatic cell therapy (Figure 9.5) is viewed as a more conservative and safer approach because it affects only the targeted cells in the patient and is not passed on to future generations. In other words, the therapeutic effect ends with the individual who receives the therapy. However, this type of therapy presents unique problems of its own. Often the effects of somatic cell therapy are short-lived. Because the cells of most tissues ultimately die and are replaced by new cells, repeated treatments over the course of the individual's life span are required to maintain the therapeutic effect. Transporting the gene to the target cells or tissue is also problematic. Regardless of these difficulties, however, somatic cell gene therapy is appropriate and acceptable for many disorders, including cystic fibrosis, muscular dystrophy, cancer, and certain infectious diseases. To date, all gene therapy on humans has been directed at somatic cells, while germline engineering in humans remains controversial and prohibited (for instance, in the European Union). There are broad categories of somatic gene therapy, *ex vivo* gene therapy, and *in vivo* gene therapy.

9.9.1 *Ex Vivo* Gene Therapy

In *ex vivo* gene therapy, cells are modified outside the body and then transplanted back in again. In this process, cells from the patient's blood or bone marrow are removed and grown in the laboratory. The cells are then exposed to the virus that is carrying the desired gene. The virus enters the cells and inserts the desired gene into the cells' DNA. The cells grow in the laboratory and are transplanted back into the patient. *Ex vivo* gene therapy can be classified into two types: *ex vivo* gene therapy using viral vectors and *ex vivo* gene therapy using nonviral vectors (Figure 9.6).

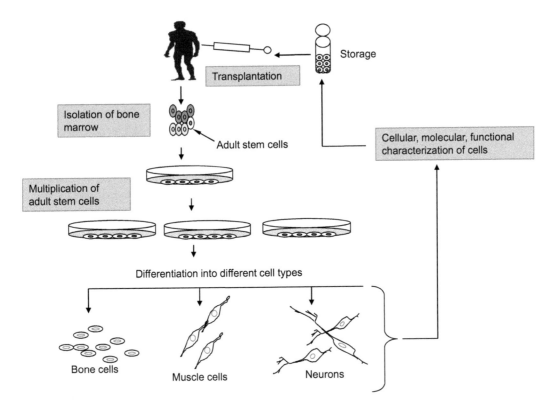

FIGURE 9.5 Adult stem cell transplantation: Bone marrow stem cells are isolated from an adult person and are normally cultured to get an enriched population of stem cells. These isolated stem cells can be differentiated into neurons, muscle, or bone cells and can be transplanted into patients suffering from degenerative diseases.

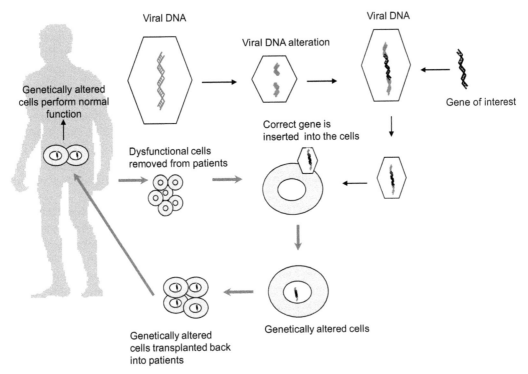

FIGURE 9.6 *Ex vivo* gene therapy.

9.9.2 Gene Therapy Using Viral Vectors

Efforts have also been made to repair genes by using viral vectors. Viruses are obligate intracellular parasites, designed through the course of evolution to infect cells, often with great specificity to a cell. They tend to be very efficient at transfecting their own DNA into the host cell, which is expressed to produce new viral particles. By replacing genes that are needed for the replication phase of their life cycle (the nonessential genes) with foreign genes of interest, the recombinant viral vector can transduce the cell type it would normally infect. To produce such recombinant viral vectors, the nonessential genes are provided in *trans*, either integrated into the genome of the packaging cell line or on a plasmid. As viruses have evolved as parasites, they all elicit a host immune system response to some extent. Though several viruses have been developed, interest has centered on four types: retroviruses (including lentiviruses), adenoviruses, adeno-associated viruses (AAVs), and herpes simplex virus type 1 (HSV-1).

9.9.3 Nonviral Methods of DNA Transfer

Viral vectors all induce an immunological response to some degree and may have safety risks such as insertional mutagenesis and toxicity problems. Furthermore, their capacity is limited, and large-scale production may be difficult to achieve. Nonviral methods of DNA transfer require only a small number of proteins, have a virtually infinite capacity, have no infectious or mutagenic capability, and large-scale production is possible using pharmaceutical techniques.

9.9.4 *In Vivo* Gene Therapy

In vivo gene therapy means to repair or correct genes inside the body. In this section, we will study how *in vivo* gene therapies are being used to treat various genetic diseases. In 2003, a University of California, Los Angeles research team inserted genes into the brain using liposomes coated in a polymer called PEG. The transfer of genes into the brain is a significant achievement, because viral vectors are too big to get across the "blood–brain barrier." This method has potential for treating Parkinson's disease.

RNA interference or gene silencing may be a new way to treat Huntington's disease. Short pieces of double-stranded RNA (short interfering RNAs or siRNAs) are used by cells to degrade RNA of a sequence. If a siRNA is designed to match the RNA copied from a faulty gene, then the abnormal protein product of that gene will not be produced.

In 2006, scientists at the National Institutes of Health (Bethesda, Maryland) successfully treated metastatic melanoma in two patients using killer T-cells genetically retargeted to attack the cancer cells. This study constitutes the first demonstration that gene therapy can be effective in treating cancer. In March 2006, an international group of scientists announced the successful use of gene therapy to treat two adult patients for a disease affecting myeloid cells. The study, published in *Nature Medicine*, is believed to be the first to show that gene therapy can cure diseases of the myeloid system.

In May 2006, a team of scientists led by Dr. Luigi Naldini and Dr. Brian Brown from the San Raffaele Telethon Institute for Gene Therapy (HSR-TIGET) in Milan, Italy, reported a breakthrough for gene therapy in which they developed a way to prevent the immune system from rejecting a newly delivered gene. Like organ transplantation, gene therapy has been plagued by the problem of immune rejection. So far, delivery of the "normal" gene has been difficult, because the immune system recognizes the new gene as foreign and rejects the cells carrying it. To overcome this problem, the HSR-TIGET group utilized a newly uncovered network of genes regulated by molecules known as microRNAs. Dr. Naldini's group reasoned that they could use this natural function of microRNAs to selectively turn off the identity of their therapeutic gene in cells of the immune system and prevent the gene from being found and destroyed. The researchers injected mice with the gene containing an immune-cell microRNA target sequence and, spectacularly, the mice did not reject the gene, as had occurred previously when vectors without the microRNA target sequence were used. This work will have important implications for the treatment of hemophilia and other genetic diseases by gene therapy.

9.9.5 PROBLEMS WITH GENE THERAPY

There are several issues associated with gene therapy that need to be resolved before using it in humans.

9.9.5.1 Short-Lived Nature of Gene Therapy

Before gene therapy can become a permanent cure for any condition, the therapeutic DNA introduced into target cells must remain functional, and the cells containing the therapeutic DNA must be long-lived and stable. Problems with integrating therapeutic DNA into the genome and the rapidly dividing nature of many cells prevent gene therapy from achieving any long-term benefits. Patients will have to undergo multiple rounds of gene therapy.

9.9.5.2 Immune Response

Anytime a foreign object is introduced into human tissues, the immune system evolves to attack the invader. The risk of stimulating the immune system in a way that reduces gene therapy effectiveness is always a possibility. Furthermore, the immune system's enhanced response to invaders it has seen before makes it difficult for gene therapy to be repeated in patients.

9.9.5.3 Virus-Induced Toxicity

The viruses that act as carrier vehicles in gene delivery may cause potential problems in patients, which include toxicity, immune and inflammatory responses, and gene control and targeting issues. In addition, there is always the fear that the viral vector, once inside the patient, may recover its ability to cause disease.

9.9.5.4 Not for Multi-Gene Disorders

Single gene mutations may cause genetic disorders, which are, in fact, considered the best candidates for gene therapy. Unfortunately, some of the most commonly occurring disorders, such as heart disease, high blood pressure, Alzheimer's disease, arthritis, and diabetes, are caused by the combined effects of variations in many genes. Multigene or multifactorial disorders such as these would be especially difficult to treat effectively using gene therapy.

9.9.5.5 Induced Mutagenesis

During gene delivery, there is a high possibility that DNA is delivered to a nontargeted site in the genome, thus causing severe health problems in patients; for example, in a tumor suppressor gene, it could induce a tumor. This has occurred in clinical trials for X-linked severe combined immunodeficiency (X-SCID) patients, in which HSCs were transduced with a corrective transgene using a retrovirus, and this led to the development of T-cell leukemia in 3 of 20 patients.

9.10 ORGAN TRANSPLANT

In general terms, *organ transplant* is the moving of an organ from one body to another to replace the recipient's damaged or failing organ with a working one from the donor. Organ donors can be living or deceased. In the human body, both organs (such as the heart, kidneys, liver, lungs, pancreas, and intestines) and tissues (which include bones, tendons, cornea, heart valves, veins, arms, and skin) can be transplanted. Transplantation medicine is one of the most challenging and complex areas of modern medicine.

9.10.1 History of Organ Transplant

The first successful corneal allograft transplant was performed in 1837 in a gazelle model. The first successful human corneal transplant, a keratoplastic operation, was performed by Eduard Zirm in Olomouc, Czech Republic, in 1905. Pioneering work in the surgical technique of transplantation was done in the early 1900s with the transplantation of arteries or veins by the French surgeon Alexis Carrel with Charles Guthrie. Their skillful anastomosis operations and new suturing techniques laid the groundwork for subsequent transplant surgery and won Carrel the 1912 Nobel Prize in Physiology or Medicine. From 1902, Carrel performed transplant experiments on dogs. Surgically successful in moving kidneys, hearts, and spleens, he was one of the first to identify the problem of rejection, which remained insurmountable for decades. Major steps in skin transplantation occurred during World War I, notably in the work of Harold Gillies at Aldershot. Among his advances was the tubed pedicle graft, maintaining a flesh connection from the donor site until the graft established its own blood flow. Gillies' assistant, Archibald McIndoe, carried on the work into World War II as reconstructive surgery. In 1962, the first successful replantation surgery was performed, reattaching a severed limb and restoring (limited) function and feeling.

Transplant of a single gonad (testis) from a living donor was carried out in early July 1926 by a Russian surgeon, Dr. Peter Vasil'evič Kolesnikov. The donor was a convicted murderer, Ilija Krajan, whose death sentence was commuted to 20 years imprisonment; he was led to believe that it was done because he had donated his testis to an elderly medical doctor. Both the donor and the receiver survived, but charges were brought in a court of law by the public prosecutor against Dr. Kolesnikov—not for performing the operation, but for lying to the donor. The first attempted human deceased-donor transplant was performed by the Ukrainian surgeon Yu Voronoy in the 1930s, where rejection resulted in failure. Joseph Murray performed the first successful transplant, a kidney transplant between identical twins, in 1954, successful because no immunosuppression was necessary in genetically identical twins.

In 1951, Peter Medawar improved the understanding of organ rejection by identifying the immune reactions. He suggested that immunosuppressive drugs could be used to minimize organ rejection. The discovery of cyclosporine in 1970 revolutionized organ transplantation techniques with greater success. Dr. Murray's success with the kidney led to attempts to transplant other organs. There was a successful deceased-donor lung transplant into a lung cancer sufferer in June 1963 by James Hardy in Jackson, Mississippi. The patient survived for 18 days before dying of kidney failure. Thomas Starzl of Denver attempted a liver transplant in the same year but was not successful until 1967. The heart was a major prize for transplant surgeons. However, as well as rejection issues, the heart deteriorates within minutes of death, so any operation would have to be performed at great speed. The development of the heart–lung machine was also needed. Lung pioneer James Hardy attempted a human heart transplant in 1964, but a premature failure of the recipient's heart caught Hardy with no human donor, so he used a chimpanzee heart, which failed very quickly. The first success was achieved on December 3, 1967, by Christiaan Barnard in Cape Town, South Africa. Louis Washkansky, the recipient, survived for 18 days amid what many saw as a distasteful publicity circus. The media interest prompted a spate of heart transplants. Over a hundred were performed in 1968–1969, but almost all the patients died within 60 days. Barnard's second patient, Philip Blaiberg, lived for 19 months.

It was the advent of cyclosporine that altered transplants from research surgery to life-saving treatment. In 1968, surgical pioneer Denton Cooley performed 17 transplants, including the first heart–lung transplant. Fourteen of his patients were dead within 6 months. By 1984, two-thirds of all heart transplant patients survived for 5 years or more. With organ transplants becoming commonplace, limited only by donors, surgeons moved onto more risky fields—multiple organ transplants on humans and whole-body transplant research on animals. On March 9, 1981, the first successful heart–lung transplant took place at Stanford University Hospital. The head surgeon, Bruce Reitz, credited the patient's recovery to cyclosporine A. In most countries, there is a shortage of suitable organs for transplantation. Countries often have formal systems in place to manage the allocation and reduce the risk of rejection. Some countries are associated with international organizations like Eurotransplant to increase the supply of appropriate donor organs and allocate organs to recipients. Transplantation also raises several bioethical issues, including the definition of death, when and how consent should be given for an organ to be transplanted, and payment for organs for transplantation.

9.10.2 TYPES OF TRANSPLANTS

9.10.2.1 Autograft

Tissue transplanted from one part of the body to another in the same individual is called an *autograft* or *autotransplant*. A common example is when a piece of bone (usually from the hip) is removed and ground into a paste when reconstructing another portion of bone. This is sometimes done with surplus tissue, or tissue that can regenerate, or tissues more desperately needed elsewhere (such as skin grafts). In orthopedic medicine, a bone graft can be sourced from a patient's own bone in order to fill space and produce an osteogenic response in a bone defect. However, because of the donor-site morbidity associated with autografts, other methods such as bone allograft and bone morphogenetic proteins and synthetic graft materials are often used as alternatives. Autografts have long been considered the "gold standard" in oral surgery and implant dentistry, because they offered the best regeneration results. Lately, the introduction of morphogen-enhanced bone graft substitutes has shown similar success rates and quality of regeneration. However, their price is still very high. Sometimes, an autograft is done to remove the tissue and then treat it or the person before returning it (examples include stem cell (Figure 9.7) autograft and storing blood in advance of surgery).

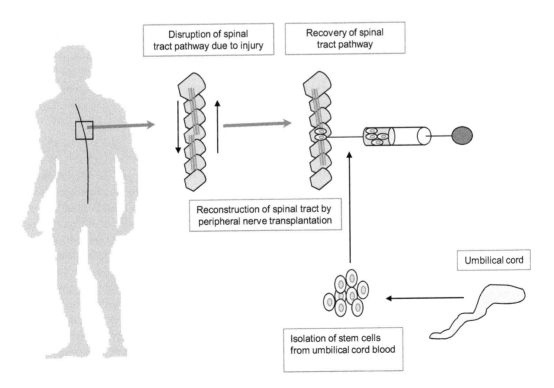

FIGURE 9.7   Repair of spinal cord injury by stem cell therapy.

9.10.2.2 Allograft

An *allograft* or *allotransplantation* is the transplantation of cells, tissues, or organs sourced from a genetically nonidentical member of the same species as the recipient. The transplant is called an allograft or allogeneic transplant, or a homograft. Most human tissue and organ transplants are allografts. In bone marrow transplantation, the term for a genetically identical graft is *syngeneic*, whereas the equivalent of an autograft is termed *autologous transplantation*. When a host mounts an immune response against an allograft or xenograft, the process is termed *rejection*. For any number of reasons, before death, a person may decide to donate tissue from his or her body for transplant to another who is in need. Additionally, consent for donation may also be given by the donor's family if the donor did not specify his or her wishes before death. After consent is obtained, potential donors are thoroughly screened for risk factors and medical conditions that would rule out donation. This screening includes interviews with family members, evaluation of medical and hospital records, and a physical assessment of the donor. Recovery of the tissue is performed with respect for the donor, using surgical techniques.

Personnel from tissue recovery agencies remove the tissue from the donor. These agencies are under the regulation of the FDA and must abide by the Current Good Tissue Practices (cGTP) rule. Once the tissue is removed, it is sent to tissue banks for processing and distribution. Each year, tissue banks accredited by the American Association of Tissue Banks (AATB) distribute 1.5 million bone and tissue allografts. These banks are also regulated by the FDA to ensure the quality of the tissue being distributed. An allogeneic bone marrow transplant can result in an immune attack against the recipient, called *graft-versus-host disease*. Because of the genetic differences between the organ and the recipient, the recipient's immune system will identify the organ as foreign and attempt to destroy it, causing transplant rejection. To prevent this, the organ recipient must take immunosuppressants. This dramatically affects the entire immune system, making the body vulnerable to pathogens.

9.10.2.3 Isograft

An *isograft* is a graft of tissue between two individuals who are genetically identical (monozygotic twins). Transplant rejection between two such individuals virtually never occurs. As monozygotic twins have the same MHC, there is very rarely any rejection of transplanted tissue by the adaptive immune system. Furthermore, there is virtually no incidence of graft-versus-host disease. This forms the basis for why a monozygotic twin will be the preferred choice of a physician considering an organ donor.

9.10.2.4 Xenograft

A *xenograft* is a surgical graft of cells, tissues, or organs from one species to an unlike species, such as from pigs to humans. Human xenotransplantation offers a potential treatment for end-stage organ failure, a significant health problem in parts of the industrialized world. It also raises many novel medical, legal, and ethical issues. A continuing concern is that pigs have a shorter lifespan than humans and their tissues age at a different rate. Disease transmission (xenozoonosis) and permanent alteration to the genetic code of animals are also causes for concern. There are few published cases of successful xenotransplantation. Because there is a worldwide shortage of organs for clinical implantation, approximately 60% of patients awaiting replacement organs die while on the waiting list. Recent advances in understanding the mechanisms of transplant rejection have brought science to a stage where it is reasonable to consider that organs from other species, probably pigs, may soon be engineered to minimize the risk of serious rejection and be used as an alternative to human tissues, possibly ending organ shortages. Other procedures, some of which are being investigated in early clinical trials, aim to use cells or tissues from other species to treat life-threatening and debilitating illnesses such as cancer, diabetes, liver failure, and Parkinson's disease. If vitrification can be perfected, it could allow for long-term storage of xenogeneic cells, tissues, and organs so that they would be more readily available for transplant. Xenotransplants could save thousands of patients waiting for donated organs. The animal organ, probably from a pig or baboon, could be genetically altered with human genes to trick a patient's immune system into accepting it as a part of his own body. Xenotransplants have re-emerged because of the lack of organs available and the constant battle to keep immune systems from rejecting allotransplants.

9.11 CLONING

The word "cloning," which literally means the exact copy of an individual, became a household name ever since researcher Ian Wilmut created a cloned sheep named Dolly. In biology, *cloning* is the process of producing populations of genetically identical individuals and occurs in nature when organisms such as bacteria, insects, or plants reproduce asexually. In this section, we will discuss cloning and its significance, and how cloning has been used in medical biotechnology to produce an array of products. One of the setbacks to animal cloning is that most cloned animals died due to unknown health complications.

9.11.1 HUMAN CLONING

After cloning so many species, the next big step would be cloning humans. *Human cloning* is the creation of a genetically identical copy of an existing or previously existing human. There are two commonly discussed types of human cloning: *therapeutic cloning* and *reproductive cloning*. Therapeutic cloning involves cloning cells from an adult for use in medicine and is an active area of research, while reproductive cloning involves making cloned human beings. Such reproductive cloning has not been performed and is illegal in many countries. A third type of cloning called *replacement cloning* is a theoretical possibility and would be a combination of therapeutic and reproductive cloning. Replacement cloning would entail the replacement of an extensively damaged or a failed or failing body through cloning, followed by whole or partial brain transplant. The various forms of human cloning are controversial. There have been numerous demands for all progress in the human cloning field to be halted. Some people and groups oppose therapeutic cloning, but most scientific, governmental, and religious organizations oppose reproductive cloning. The American Association for the Advancement of Science (AAAS) and other scientific organizations have made public statements suggesting that human reproductive cloning be banned until safety issues are resolved. Serious ethical concerns have been raised by the idea that it might be possible in the future to harvest organs from clones. Some people have considered the idea of growing organs separately from a human organism. In doing this, a new organ supply could be established without the moral implications of harvesting them from humans.

The first hybrid human clone was created in November 1998 by American Cell Technologies (ACT). It was created from a man's leg cell and a cow's egg with the DNA removed. However, it was destroyed after 12 days. While making an embryo, which may have resulted in a complete human had it been allowed to come to term, ACT claimed that their aim was "therapeutic cloning," not "reproductive cloning." In January 2008, Stemagen Corporation Lab, in La Jolla in the United States, announced that they had successfully created the first five mature human embryos using DNA from adult skin cells, aiming to provide a source of viable ES cells; however, the cloned embryos were destroyed for ethical reasons.

9.12 HUMAN EMBRYONIC STEM CELLS

Human embryonic stem cells (hESCs) are stem cells that are derived from the undifferentiated inner mass cells (day 5 of fertilization) of a human embryo. Embryonic stem cells are pluripotent, meaning they can grow and differentiate into all three primary germ layers such as ectoderm, endoderm, and mesoderm (Figure 9.8).

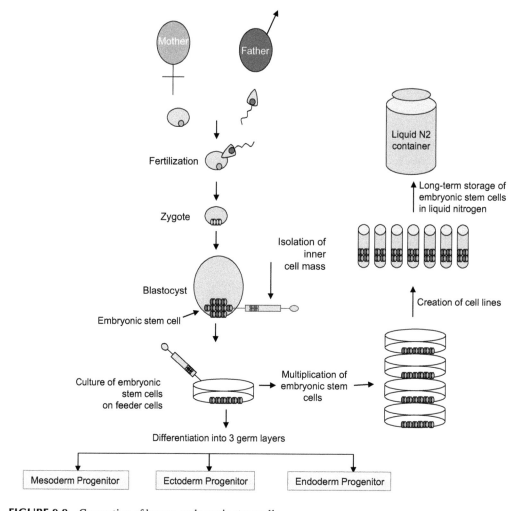

FIGURE 9.8 Generation of human embryonic stem cells.

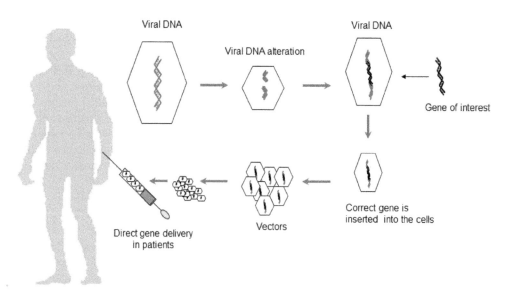

FIGURE 9.9 *In vivo* gene therapy.

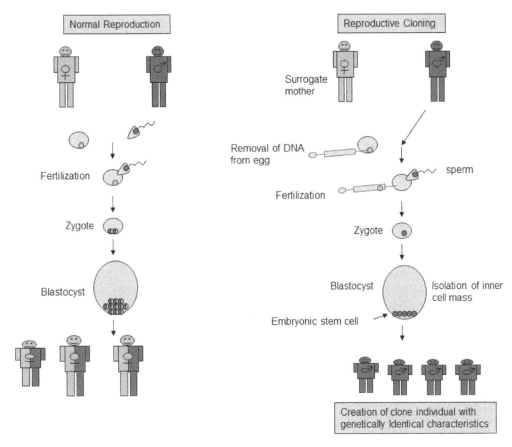

FIGURE 9.10   Reproductive cloning.

QUESTIONS

Section A: Descriptive Type

Q1. What are the different types of vaccines?
Q2. Describe various applications of MABs.
Q3. Explain the significance of adult stem cells.
Q4. Is it possible to isolate stem cells from the human body? Why or why not?
Q5. Explain the issue related to contamination of human ES cells.
Q6. What is reproductive cloning?
Q7. How does gene therapy work?
Q8. Explain various problems associated with gene therapy.
Q9. How is gene sequencing done?
Q10. What is DNA fingerprinting?
Q11. Describe RFLP with suitable example.
Q12. Describe allograft transplantation.
Q13. Discuss various benefits of bioengineered skin.

Section B: Multiple Choice

Q1. Vaccines are often made from weakened or killed forms of . . .
    a. Human cells
    b. Fungi

    c. Microbes

    d. Chicken

Q2. Which of the following has no available vaccine to date?

    a. Diphtheria

    b. Measles

    c. Malaria

    d. Rubella

Q3. Who first proposed the idea of the "magic bullet"?

    a. Alexander Fleming

    b. Paul Ehrlich

    c. Kary Mullis

    d. Georges Kohler

Q4. Hybridoma technology deals with . . .

    a. Vaccine production

    b. Antibiotic production

    c. Polyclonal antibodies production

    d. MABs production

Q5. Recombinant antibody engineering involves the use of viruses or yeast, rather than mice, to create antibodies. True/False

Q6. Therapeutic proteins are also known as . . .

    a. Biologicals

    b. Biomaterials

    c. Biopharmaceuticals

    d. Pharmaceuticals

Q7. Mesenchymal stem cells are . . .

    a. Unipotent stem cells

    b. Multipotent stem cells

    c. Pluripotent stem cells

    d. Totipotent stem cells

Q8. From which embryonic stage are embryonic stem cells derived?

    a. Blastocyst

    b. Gastrula

    c. Morula

    d. Germ layer

Q9. The US FDA approved the first clinical trial using embryonic stem cells in 2009. True/False

Q10. *Ex vivo* gene therapy means . . .

    a. Externally modified cells transplanted back into the same patient

    b. Externally modified cells transplanted into another patient

    c. Internally modified cells

Q11. Xenograft involves a transplant from . . .

    a. Mouse to mouse

    b. Mouse to human

    c. Human to human

    d. None of the above

Section C: Critical Thinking

Q1. Do you think it would be better if recombinant therapeutic proteins are produced using human cells rather than using microbes or animals? If no, explain. If yes, what would be your strategy?

Q2. Will embryonic stem cells ever be a successful cell-based therapy? Why or why not?

Q3. Why does gene therapy fail in many aspects?

ASSIGNMENT

Read more about the human genome project and submit a report on genetic variations and mutations found in the human genome.

REFERENCES AND FURTHER READING

Alexeyev, M.F., LeDoux, S.P., and Wilson, G.L. Mitochondrial DNA and aging. *Clin. Sci.* 107: 355–364, 2004.

Andrews, P., Matin, M., Bahrami, A., Damjanov, I., Gokhale, P., and Draper, J. Embryonic stem (ES) cells and embryonal carcinoma (EC) cells: Opposite sides of the same coin. *Biochem. Soc. Trans.* 33: 1526–1530, 2005.

Bammler, T., Beyer, R.P., Bhattacharya, S., Boorman, G.A., Boyles, A., Bradford, B.U., Bumgarner, R.E., Bushel, P.R., Chaturvedi, K., Choi, D., Cunningham, M.L. et al. Standardizing global gene expression analysis between laboratories and across platforms. *Nat. Methods* 2: 351–356, 2005.

Barrilleaux, B., Phinney, D.G., Prockop, D.G., and O'Connor, K.C. Review: Ex vivo engineering of living tissues with adult stem cells. *Tissue Eng.* 12: 3007–3019, 2006.

Beachy, P.A., Karhadkar, S.S., and Berman, D.M. Tissue repair and stem cell renewal in carcinogenesis. *Nature* 432: 324–331, 2004.

Bhattacharya, S. Killer convicted thanks to relative's DNA. *New Sci.* 2004. https://www.newscientist.com/article/dn4908-killer-convicted-thanks-to-relatives-dna/

Black, L.L., Gaynor, J., Adams, C., Dhupa, S., Sams, A.E., Taylor, R., Harman, S., Gingerich, D.A., and Harman, R. et al. Effect of intraarticular injection of autologous adipose-derived mesenchymal stem and regenerative cells on clinical signs of chronic osteoarthritis of the elbow joint in dogs. *Vet. Therap.* 9: 192–200, 2008.

Burt, R.K., Loh, Y., Pearce, W. et al. Clinical applications of blood-derived and marrow-derived stem cells for nonmalignant diseases. *JAMA* 299: 925–936, 2008.

Carson, H.L. Chromosomal tracers of evolution. *Science* 168: 1414–1418, 1970.

Carson, H.L. Chromosomal sequences and interisland colonizations in Hawaiian Drosophila. *Genetics* 103: 465–482, 1983.

Chaudhary, P.M., and Roninson, I.B. Expression and activity of P-glycoprotein, a multidrug efflux pump, in human hematopoietic stem cells. *Cell* 66: 85–94, 1991.

Churchill, G.A. Fundamentals of experimental design for cDNA microarrays. *Nat. Genet. Suppl.* 32: 490, 2002.

Clark, D.P., and Russell, L.D. *Molecular Biology Made Simple and Fun*, 2nd edn. Cache River Press, Vienna, IL, 2000.

Comai, L. The advantages and disadvantages of being polyploid. *Nat. Rev. Genet.* 6: 836–846, 2005.

Cutler, C., and Antin, J.H. Peripheral blood stem cells for allogeneic transplantation: A review. *Stem Cells* 19: 108–117, 2001.

Dulbecco, R., and Ginsberg, H.S. *Virology*, 2nd edn. J.B. Lippincott Company, Philadelphia, PA, 1988.

Edgar, B.A., and Orr-Weaver, T.L. Endoreduplication cell cycles: More for less. *Cell* 105: 297–306, 2001.

Evans, M., and Kaufman, M. Establishment in culture of pluripotent cells from mouse embryos. *Nature* 292: 154–156, 1981.

Giarratana, M.C., Kobari, L., Lapillonne, H. et al. Ex vivo generation of fully mature human red blood cells from hematopoietic stem cells. *Nat. Biotechnol.* 23: 69–74, 2005.

Gilbert, S.F. *Developmental Biology*, 8th edn., chapter 9. Sinauer Associates, Stamford, CT, 2006.

Hacia, J.G., Fan, J.B., Ryder, O. et al. Determination of ancestral alleles for human single-nucleotide polymorphisms using high-density oligonucleotide arrays. *Nat. Genet.* 22: 164–167, 1999.

Haupt, Y., Bath, M.L., Harris, A.W., and Adams, J.M. Bmi-1 transgene induces lymphomas and collaborates with myc in tumorigenesis. *Oncogene* 8: 3161–3164, 1993.

Hwang, W.S., Roh, S.I., Lee, B.C. et al. Patient-specific embryonic stem cells derived from human SCNT blastocysts. *Science* 308: 1777–1783, 2005.

Keirstead, H.S., Nistor, G., Bernal, G. et al. Human embryonic stem cell-derived oligodendrocyte progenitor cell transplants remyelinate and restore locomotion after spinal cord injury. *J. Neurosci.* 25: 4694–4705, 2005.

Kennedy, D. Editorial retraction. *Science* 311(5759): 335, 2006.

Kikyo, N., and Wolffe, A.P. Reprogramming nuclei: Insights from cloning, nuclear transfer and heterokaryons. *J. Cell Sci.* 113: 11–20, 2000.

Klimanskaya, I., Chung, Y., Becker, S., Lu, S.J., and Lanza, R. Human embryonic stem cells derived without feeder cells. *Lancet* 365: 1636–1641, 2006.

Kohler, G., and Milstein, C. Continuous cultures of fused cells secreting antibody of predefined specificity. *Nature* 256: 495–497, 1975.

Kottler, M. Cytological technique, preconception and the counting of the human chromosomes. *Bull. Hist. Med.* 48: 465–502, 1974.

Kurinczuk, J.J. Safety issues in assisted reproduction technology: From theory to reality—Just what are the data telling us about ICSI offspring health and future fertility and should we be concerned. *Hum. Reprod.* 18: 925–931, 2003.

Langer, R., and Vacanti, J.P. Tissue engineering. *Science* 260: 920–926, 1993.

Lashkari, D.A., DeRisi, J.L., McCusker, J.H. et al. Yeast microarrays for genome wide parallel genetic and gene expression analysis. *Proc. Natl. Acad. Sci. USA* 94: 13057–13062, 1997.

Martin, G. Isolation of a pluripotent cell line from early mouse embryos cultured in medium conditioned by teratocarcinoma stem cells. *Proc. Natl. Acad. Sci. USA* 78: 7634–7638, 1981.

Martin, M.J., Muotri, A., Gage, F., and Varki, A. Human embryonic stem cells express an immunogenic non-human sialic acid. *Nat. Med.* 11: 228–232, 2005.

McCarthy, M. Bio-engineered tissues move towards the clinic. *Lancet* 348(9025): 466, 1996.

McFarland, D. Preparation of pure cell cultures by cloning. *Methods Cell Sci.* 22: 63–66, 2000.

McLaren, A. Cloning: Pathways to a pluripotent future. *Science* 288: 1775–1780, 2000.

Michael, W., Horn, M., and Holger, D. Fluorescence in situ hybridisation for the identification and characterization of prokaryotes. *Curr. Opin. Microbiol.* 6: 302–309, 2003.

Olson, W.P. *Separations Technology. Pharmaceutical and Biotechnology Applications.* Interpharm Press, Inc., Buffalo Grove, IL, 1995.

Painter, T.S. The spermatogenesis of man. *Anat. Res.* 23: 129, 1922.

Painter, T.S. Studies in mammalian spermatogenesis II: The spermatogenesis of humans. *J. Exp. Zoology* 37: 291–336, 1923.

Rundle, R.L. Cells 'tricked' to make skin for burn cases. *Wall Street J.*, 1994.

Yen, A.H., and Sharpe, P.T. Stem cells and tooth tissue engineering. *Cell Tissue Res.* 331: 359–372, 2008.

10 Nanobiotechnology

LEARNING OBJECTIVES

- Define nanotechnology
- Explain the significance of nanobiotechnology
- Discuss various applications of nanobiotechnology, including nanomedicine
- Explain the application of nanobiotechnology in food testing and water pollution
- Discuss nanoparticles and nanomaterials
- Discuss current research trends in nanobiotechnology

10.1 INTRODUCTION

Nanobiotechnology is the union of nanotechnology and biology. *Nanobiotechnology* is the branch of nanotechnology with biological and biochemical applications or uses. It often studies existing nanomaterials to fabricate new devices for medical and diagnostic applications. In the following two sections, we will examine both aspects to understand the significance of nanobiotechnology.

10.2 NANOTECHNOLOGY

The concept of nanotechnology was first envisioned as early as 1959 by the renowned physicist Richard Feynman, who said,

> I want to build a billion tiny factories, models of each other, which are manufacturing simultaneously. The principles of physics, as far as I can see, do not speak against the possibility of maneuvering things atom by atom. It is not an attempt to violate any laws; it is something, in principle, that can be done; but in practice, it has not been done because we are too big.

The term "nanotechnology" was defined by Tokyo Science University's Professor Norio Taniguchi in a 1974 paper as follows: "Nanotechnology mainly consists of the processing of, separation, consolidation, and deformation of materials by one atom or by one molecule." However, it was K. Eric Drexler who popularized the word "nanotechnology" in the 1980s by publishing a research paper in the journal *Proceeding of National Academy of Sciences of the United States of America*. To prove his passion for nanotechnology, Drexler spent many years in his lab describing and analyzing nanodevices based on the molecular scale. Later, as nanotechnology became an accepted concept, the meaning of the word shifted to encompass the simpler kinds of nanometer-scale technology.

The US National Nanotechnology Initiative (NNI) was created to fund nanotechnology research, and their definition includes anything smaller than 100 nanometers (nm) with novel properties. Nanotechnology is extremely diverse, ranging from novel extensions of conventional device physics, to completely new approaches based upon molecular self-assembly, to developing new materials with dimensions on the nanoscale—even to speculation on whether we can directly control matter on the atomic scale. Nanotechnology and nanoscience originated in the early 1980s with two major developments: the birth of cluster science and the invention of the scanning tunneling microscope (STM). This development led to the discovery of fullerenes in 1985 and carbon nanotubes a few years later. In another development, the synthesis and properties of semiconductor nanocrystals were studied. This led to a fast-increasing number of metal and metal oxide nanoparticles

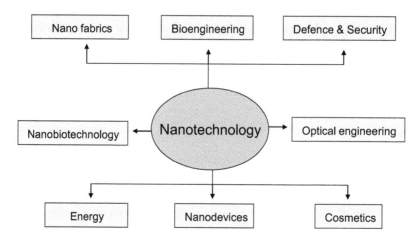

FIGURE 10.1   Applications of nanotechnology.

and quantum dots (qdots). The atomic force microscope was invented 6 years after the STM was invented. In 2000, the NNI was founded to coordinate federal nanotechnology research and development (R&D). The various applications of nanotechnology are illustrated in the Figure 10.1

10.3   NANOBIOTECHNOLOGY

Nanobiotechnology is a promising field of science and it is a marriage between the fields of microelectronics and biological sciences and combines the complementary strengths of biological molecules (to selectively bind with other molecules) and nanoelectronics (detection of the smallest electrical change caused by such binding). The integration of biological science with engineering is one of the most challenging and fastest growing sectors of nanobiotechnology. Nanotechnology is the creation and utilization of materials, devices, and systems through the control of matter on the nanometer-length scale (a nanometer is one billionth of a meter).

Nanobiotechnology is that branch of nanotechnology that deals with the study and application of biological and biochemical activities from elements of nature to fabricate new devices like biosensors. Nanobiotechnology is often used to describe the overlapping multidisciplinary activities associated with biosensors, particularly where photonics, chemistry, biology, biophysics, nanomedicine, and engineering converge. Measurement in biology using waveguide techniques such as the dual polarization interferometer is another example. One example of current nanobiotechnological research involves nanospheres coated with fluorescent polymers. Researchers are seeking to design polymers whose fluorescence are quenched when they encounter specific molecules. Different polymers would detect different metabolites. The polymer-coated spheres could become part of new biological assays, and the technology might someday lead to particles that could be introduced into the human body to track down metabolites associated with tumors and other health problems.

Nanobiotechnology holds considerable promise of advances in pharmaceuticals and healthcare. Initial work is being done to improve the various techniques and materials that are relevant to nanobiotechnology. It includes some of the physical forms of energy, such as nanolasers. Some of the technologies are scaling down, such as microfluidics to nanofluidic biochips. Some of the earliest applications are in molecular diagnostics. Nanoparticles, particularly qdots, are playing important roles. Various nanodiagnostics that have been reviewed will improve the sensitivity and extend the present limits of molecular diagnostics. An increasing use of nanobiotechnology by the pharmaceutical and biotechnology industries is anticipated. Nanotechnology can be applied at all stages of drug development, from formulations to diagnostic applications in clinical trials. Many of the assays based on nanobiotechnology will enable high-throughput screening of drug candidates.

Some nanostructures such as fullerenes are themselves drug candidates as they allow precise grafting of active chemical groups in three-dimensional orientations. Apart from offering a solution to solubility problems, nanobiotechnology provides intracellular delivery possibilities. Skin penetration is improved in transdermal drug delivery. A particularly effective application is on nonviral gene therapy vectors.

10.4 APPLICATIONS OF NANOBIOTECHNOLOGY

10.4.1 NANOMEDICINE

One of the major applications of nanotechnology is to develop diagnostic and therapeutic entities for better understanding and treatment of human diseases. The application of nanotechnology in medicine is called *nanomedicine*. The applications of nanomedicine range from the medical use of nanomaterials, to nanoelectronic biosensors, and even possible applications of molecular nanotechnology. Current challenges in the field of nanomedicine involve understanding the issues related to toxicity and the environmental impact of nanoscale materials. The application of nanomaterials has revolutionized drug discovery research. The journal *Nature Materials* showed that an estimated 130 nanotech-based drugs and delivery systems were being developed worldwide. The NNI expects new commercial applications in the pharmaceutical industry that may include advanced drug delivery systems, new therapies, and *in vivo* imaging. Neuroelectronic interfaces and other nanoelectronics-based sensors are another active goal of research. The speculative field of molecular nanotechnology believes that further down the line cell repair machines could revolutionize medicine and the medical field. Buckyballs may be used to trap free radicals generated during an allergic reaction and block the inflammation that results from an allergic reaction. Nanoshells may be used to concentrate the heat from infrared light to destroy cancer cells with minimal damage to surrounding healthy cells. Nanospectra Biosciences has developed such a treatment using nanoshells illuminated by an infrared laser that has been approved for a pilot trial with human patients. Nanoparticles, when activated by x-rays that generate electrons, cause the destruction of cancer cells to which they attach. This is intended to be used in place of radiation therapy, with much less damage to healthy tissue. Nanobiotix has released preclinical results for this technique. Aluminosilicate nanoparticles can more quickly reduce bleeding in trauma patients by absorbing water, causing blood in a wound to clot quickly. Z-Medica is producing medical gauze that uses aluminosilicate nanoparticles. Nanofibers can stimulate the production of cartilage in damaged joints.

Nanomedicine is a large industry, with nanomedicine sales reaching 6.8 billion dollars in 2004, and with over 200 companies and 38 products worldwide, a minimum of 3.8 billion dollars in nanotechnology R&D is being invested every year. As the nanomedicine industry continues to grow, it is expected to have a significant impact on the economy. The list of nanotechnology healthcare products is shown in Table 10.1

TABLE 10.1
Applications of Nanotechnology Products

Area of Application	Application
Agriculture-Nano modification of seeds/ fertilizers/pesticides	Pesticides, preservation,
Interactive smart food	Nanoemulsion, anti-caking agent, nanoencapsulation
Food fortification and modification	Interferons, interleukins, erythropoietin
Vaccines	Nutraceuticals, mineral and vitamin fortification, and water drinking purification system
Smart packaging and food tracking	UV protection, antimicrobials

10.4.1.1 Drug Delivery

In recent times, drug delivery has become an important field to improve the effectiveness of drug therapy, and efforts have been made to improve the drug delivery system to avoid pain and to enhance target-specific delivery with minimum side effects. We will first find out what kinds of drug delivery methods are available. *Drug delivery* is the method or process of administering a drug to achieve a therapeutic effect in humans. The most common methods of drug delivery include the preferred noninvasive peroral (through the mouth), topical (skin), transmucosal (nasal, buccal/sublingual, vaginal, ocular, and rectal) and inhalation routes. Many medications, such as peptides and proteins, antibodies, vaccines, and gene-based drugs in general, may not be delivered using these routes, because they might be susceptible to enzymatic degradation or cannot be absorbed into the systemic circulation efficiently enough to be therapeutically effective because of molecular size and charge issues; for this reason, many proteins and peptide drugs must be delivered by injection. For example, many immunizations are based on the delivery of protein drugs and are often administered by injection.

One of the main issues with the current drug delivery system is non-targeted drug delivery, which has several limitations that include not reaching the target site and drug action on normal cells or tissues. Constant efforts have been made in drug delivery, which include the development of targeted delivery in the target area of the body (e.g., in cancerous tissues) and sustained-release formulations in which the drug is released over a period of time in a controlled manner (Figure 10.2). Types of sustained release formulations include liposomes, drug-loaded biodegradable microspheres, and drug polymer conjugates. Nanomedical approaches to drug delivery center on developing nanoscale particles or molecules to improve the bioavailability of a drug (Figure 10.3). *Bioavailability* refers to the presence of drug molecules where they are needed in the body and where they will do the most good. Drug delivery focuses on maximizing bioavailability, both in specific places in the body and over a period. This will be achieved by molecular targeting by nanoengineered devices.

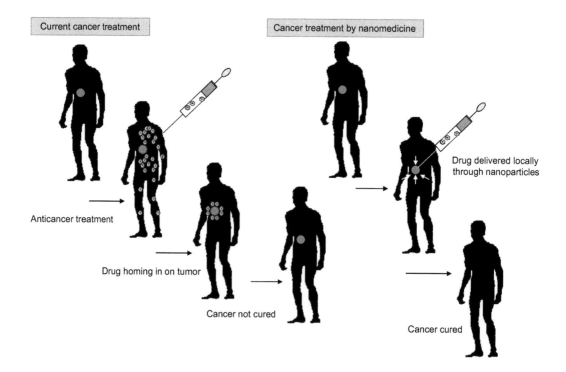

FIGURE 10.2   Application of nanomedicine in cancer treatment.

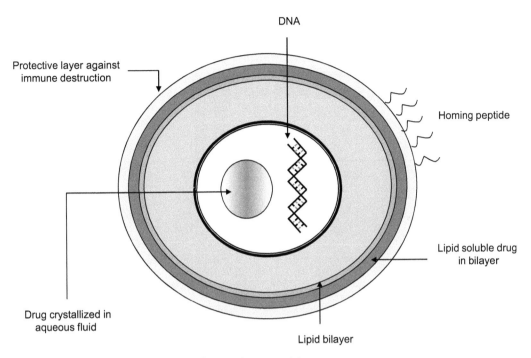

DNA

Protective layer against
immune destruction

Homing peptide

Lipid soluble drug
in bilayer

Drug crystallized in
aqueous fluid

Lipid bilayer

FIGURE 10.3 Drug delivery through liposomal nanoparticle.

It is all about targeting molecules and delivering drugs with cell precision. More than $65 billion are wasted each year due to poor bioavailability of drug molecules. *In vivo* imaging is another area where tools and devices are being developed. Using nanoparticle contrast agents, images such as ultrasound and magnetic resonance imaging (MRI) have a favorable distribution and improved contrast. The new methods of nanoengineered materials that are being developed might be effective in treating illnesses and diseases such as cancer. What nanoscientists will be able to achieve in the future is beyond current imagination. This will be accomplished by self-assembled biocompatible nanodevices that will detect, evaluate, treat, and report to the clinical doctor automatically.

Drug delivery systems, lipid- or polymer-based nanoparticles, can be designed to improve the pharmacological and therapeutic properties of drugs. The strength of drug delivery systems is their ability to alter the pharmacokinetics and biodistribution of the drug. Nanoparticles have unusual properties that can be used to improve drug delivery. Where larger particles would have been cleared from the body, cells take up these nanoparticles because of their size. Complex drug delivery mechanisms are being developed, including the ability to get drugs through cell membranes and into the cell cytoplasm. Efficiency is important, because many diseases depend upon processes within the cells and can only be impeded by drugs that make their way into the cell. A triggered response is one way for drug molecules to be used more efficiently. Drugs are placed in the body and only activate on encountering a particular signal. For example, a drug with poor solubility will be replaced by a drug delivery system where both hydrophilic and hydrophobic environments exist, improving the solubility. In addition, a drug may cause tissue damage, but regulated drug release with drug delivery systems can eliminate the problem. If a drug is cleared too quickly from the body, this could force a patient to use high doses, but clearance can be reduced by altering the pharmacokinetics of the drug with drug delivery systems. Poor biodistribution is a problem that can affect normal tissues through widespread distribution, but the particulates from drug delivery systems lower the volume of distribution and reduce the effect on nontarget tissue. Potential nanodrugs will work by very specific and well-understood mechanisms. One of the major impacts of nanotechnology and

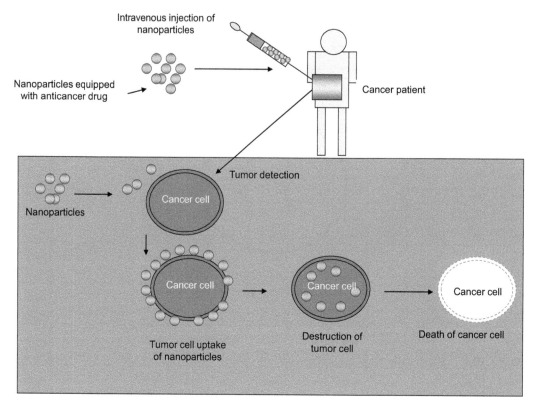

FIGURE 10.4 Use of nanoparticles in cancer treatment.

nanoscience will be in leading the development of completely new drugs with more useful performance and fewer side effects. The basic point in the use of drug delivery is based upon three facts:
(a) efficient encapsulation of the drugs, (b) successful delivery of these drugs to the targeted region
of the body, and (c) successful release of these drug in the targeted region.

Targeting ligands (such as monoclonal antibodies and sugar residues) can also be attached to
nanoparticles to achieve active targeting of these particles. The hepatocytes of the liver can be an
important target in the case of hepatitis and in other cases of gene therapy, when the administered
gene needs to be expressed in these cells of the liver. The nanoparticles having ligands for active
targeting should not only escape capture by the Kupffer cells of the liver, but also need to be small
enough less than 50 nm, or even less than 20 nm in size to be able to enter the hepatocytes with
the help of specific receptors on these surfaces and through fenestrations that are 100–150 nm in
diameter (Figures 10.4).

10.4.1.2 Cancer Diagnostics

The success of cancer treatment is primarily based on how quickly we detect the progression of
cancer cells so that effective treatment modalities can be implemented in patients. Over the past few
years, scientists have shown that nanoparticles are very useful diagnostic tools in oncology, particularly in imaging. Qdots (nanoparticles with quantum confinement properties, such as size-tunable
light emission), when used in conjunction with MRI, can produce exceptionally good images of
tumor sites. These nanoparticles are much brighter than organic dyes and only need one light source
for excitation. This means that the use of fluorescent qdots could produce a higher contrast image at
a lower cost than today's organic dyes used as contrast media. The downside, however, is that qdots
are usually made of quite toxic elements (Figure 10.5).

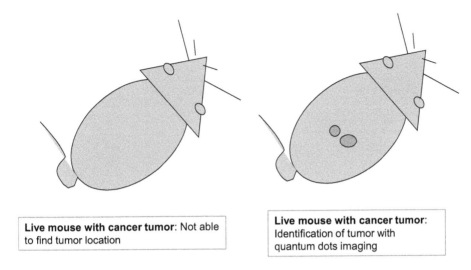

Live mouse with cancer tumor: Not able to find tumor location

Live mouse with cancer tumor: Identification of tumor with quantum dots imaging

FIGURE 10.5 Bio-imaging using carbohydrate-encapsulated quantum dots in live mouse.

Another property of nanoparticles is that they have a high surface-area-to-volume ratio, allowing many functional groups to be attached, which can be used to detect tumor cells. Additionally, the small size of nanoparticles (10–100 nm) allows them to preferentially accumulate at tumor sites, because tumors lack an effective lymphatic drainage system. A very exciting research question is how to make these imaging nanoparticles do more things for cancer. For instance, is it possible to manufacture multifunctional nanoparticles that would detect, image, and then proceed to treat a tumor? This question is under vigorous investigation, and the answer could shape the future of cancer treatment.

A promising new cancer treatment that may one day replace radiation and chemotherapy is edging closer to human trials. Sensor test chips containing thousands of nanowires, able to detect proteins and other biomarkers left behind by cancer cells, could enable the detection and diagnosis of cancer in the early stages from a few drops of a patient's blood. Headed by Prof. Jennifer West, researchers at Rice University have demonstrated the use of 120 nm diameter nanoshells coated with gold to kill cancer tumors in mice. The nanoshells can be targeted to bond to cancerous cells by conjugating antibodies or peptides to the nanoshell surface. By irradiating the area of the tumor with an infrared laser, which passes through flesh without heating it, the gold is heated sufficiently to cause the death of the cancer cells. Additionally, John Kanzius has invented a radio machine that uses a combination of radio waves and carbon or gold nanoparticles to destroy cancer cells. Nanoparticles of cadmium selenide (qdots) glow when exposed to UV light. When injected, they seep into cancer tumors. The surgeon can see the glowing tumor and use it as a guide for more accurate tumor removal (Figure 10.5).

One scientist, University of Michigan's James Baker, believes that he has discovered a highly efficient and successful way of delivering cancer-treatment drugs that is less harmful to the surrounding body. Baker has developed a nanotechnology that can locate and then eliminate cancerous cells. He looks at a molecule called a dendrimer. This molecule has more than 100 hooks on it that allow it to attach to cells in the body for a variety of purposes. Baker then attaches folic acid to a few of the hooks (folic acid, being a vitamin, is received by cells in the body). Cancer cells have more vitamin receptors than normal cells, so Baker's vitamin-laden dendrimer will be absorbed by the cancer cell. On the rest of the hooks on the dendrimer, Baker places anticancer drugs that will be absorbed with the dendrimer into the cancer cell, thereby delivering the cancer drug to the cancer cell and nowhere else (Figure 10.6).

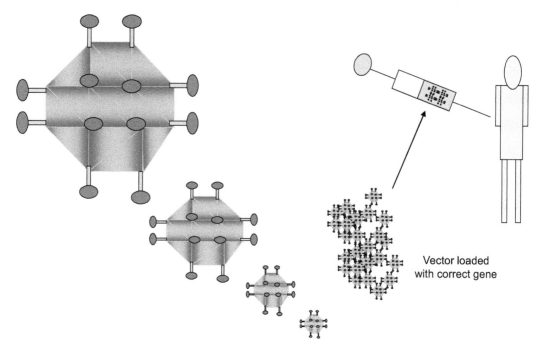

FIGURE 10.6   Polymer vector for gene therapy.

In photodynamic therapy, a particle is placed within the body and is illuminated with light from the outside. The light gets absorbed by the particle; if the particle is metal, energy from the light will heat the particle and surrounding tissue. Light may also be used to produce high-energy oxygen molecules that will chemically react with and destroy most organic molecules that are next to them (such as tumors). This therapy is appealing for many reasons. It does not leave a "toxic trail" of reactive molecules throughout the body (as in chemotherapy), because it is directed where only the light is made to shine and the particles exist. Photodynamic therapy has potential as a noninvasive procedure for dealing with diseases, growths, and tumors.

10.4.1.3   *In Vivo* Drug Imaging

Tracking the movement of drug molecules can help determine how well the drugs are being distributed or how substances are metabolized in the body. It is difficult to track a small group of cells throughout the body, so scientists previously dyed the cells. These dyes needed to be excited by light of a certain wavelength for them to light up. One of the main problems with this dye method was that it needs multiple light sources for visualization. This problem has been solved by using qdot nanoparticles that can attach to proteins that penetrate cell membranes. The dots can be random in size, can be made of bio-inert material, and they demonstrate the nanoscale property that color is size-dependent. As a result, sizes are selected so that the frequency of light used to make a group of qdots fluorescence is an even multiple of the frequency required to make another group incandesce; then both groups can be lit with a single light source. It has been widely observed that nanoparticles are promising tools for the advancement of drug delivery, medical imaging (Figure 10.7), and as diagnostic sensors. However, the biodistribution of these nanoparticles is mostly unknown, because of the difficulty in targeting specific organs in the body. Current research in the excretory systems of mice, however, shows the ability of gold composites to selectively target certain organs based on their size and charge. These composites are encapsulated by a dendrimer and assigned a specific charge and size. Positively-charged gold nanoparticles were found to enter the kidneys, while negatively-charged gold nanoparticles remained in the liver and spleen. It is suggested that

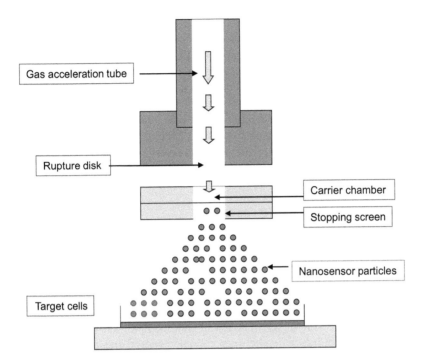

FIGURE 10.7   Bioimaging by nanosensor particles.

the positive surface charge of the nanoparticle decreases the rate of osponization of nanoparticles in the liver, thus affecting the excretory pathway. Even at a relatively small size of 5 nm, though, these particles can become compartmentalized in the peripheral tissues and will therefore accumulate in the body over time.

While the advancement of research proves that targeting and distribution can be augmented by nanoparticles, the dangers of nanotoxicity become an important next step in further understanding of their medical uses. Qdots may be used in the future for locating cancer tumors in patients and in the near term for performing diagnostic tests on samples. Invitrogen's website provides information about qdots that are available for both uses, although now, the "*in vivo*" use is limited to experiments with laboratory animals. Iron oxide nanoparticles can be used to improve MRI images of cancer tumors. The nanoparticles are coated with a peptide that binds to a cancer tumor. Once the nanoparticles are attached to the tumor, the magnetic property of the iron oxide enhances the images from the MRI scan.

Nanoparticles can attach to proteins or other molecules, allowing detection of disease indicators in a laboratory sample at a very early stage. There are several efforts to develop nanoparticle disease detection systems, and one system being developed by Nanosphere, Inc. uses gold nanoparticles. Nanosphere has clinical study results with their Verigene system involving its ability to detect four different nucleic acids. Another system, developed by T2 Biosystems, uses magnetic nanoparticles to identify specimens such as proteins, nucleic acids, and other materials.

10.4.1.4   Nanonephrology

Nanonephrology deals with (1) the study of kidney protein structures at the atomic level; (2) nano-imaging approaches to study cellular processes in kidney cells; and (3) nanomedical treatments that utilize nanoparticles to treat various kidney diseases. The creation and use of materials and devices at the molecular and atomic levels that can be used for the diagnosis and therapy of renal diseases are also parts of nanonephrology that will play a role in the management of patients with kidney disease in the future. Advances in nanonephrology will be based on discoveries in the preceding areas that can provide nanoscale information on the cellular molecular machinery involved in normal kidney processes and in pathological states. By understanding the physical and chemical properties of proteins and other macromolecules at the atomic level in various cells in the kidney, novel therapeutic approaches can be designed to combat major renal diseases. The nanoscale artificial kidney is a goal that many physicians dream of. Nanoscale engineering advances will permit programmable and controllable nanoscale robots to execute curative and reconstructive procedures in the human kidney at the cellular and molecular levels. Designing nanostructures compatible with the kidney cells that can safely operate *in vivo* is also a future goal. The ability to direct events in a controlled fashion at the cellular nano-level has the potential to significantly improve the lives of patients with kidney diseases.

10.4.1.5 Gene Therapy Using Nanotechnology

Nanoparticles have also been used in gene therapy. A pilot experiment involving gene therapy in human subjects has already been conducted for diseases like cystic fibrosis and muscular dystrophy. Three main types of gene-delivery systems have been described: (1) viral vectors; (2) nonviral vectors (particles and polymers); and (3) gene guns for direct injection of the genetic material into the target tissue. As viral vectors pose some serious problems, nonviral vectors are the gene-delivery system of choice. In this nonviral vector system, negatively charged plasmid DNA is condensed into nanoparticles that are 50–200 nm in size. The use of cationic lipids and cationic polymers gives a compact structure because of interaction between cationic material and anionic DNA. These compact structures also provide increased stability and uptake by the target cells. Some of the nonviral vectors for gene therapy based on nanoparticles are listed in Table 10.2. The targets for gene therapy using nanoparticles include liver hepatocytes, endothelial cells, the spleen, and lymph nodes, where some success has already been achieved.

10.4.1.6 Antimicrobial Techniques

One of the earliest nanomedicine applications was the use of nanocrystalline silver, which is an antimicrobial agent for the treatment of wounds. A nanoparticle cream has been shown to fight staph infections. The nanoparticles contain nitric oxide gas, which is known to kill bacteria. Studies in mice have shown that using the nanoparticle cream to release nitric oxide gas at the site of staph abscesses significantly reduced the infection. A welcome idea in the early study stages is the elimination of bacterial infections in a patient within minutes, instead of delivering treatment with antibiotics over a period of weeks.

10.4.1.7 Neural Differentiation

Nanoparticles have been extensively used to investigate their effects on the neuronal cells. These nanoparticles cause stem cells to differentiate into neuronal cells and promote neuronal cell

TABLE 10.2
Nanoparticles in Gene Therapy

Transfection Method	Description
Electroporation	**Delivery of genes into cells** Advantage: Higher transfection efficiency Disadvantage: Lower cell viability
Nanoparticles	**Delivery of genes into cells** Advantage: Higher cell viability Disadvantage: Lower transfection efficiency

survivability and neuronal cell growth and expansion. The nanoparticles have been tested both in *in vitro* and *in vivo* models. The nanoparticles with various shapes, sizes, and chemical compositions mostly produced stimulatory effects on neuronal cells, but there are a few that can cause inhibitory effects on the neuronal cells.

10.4.2   NEURO-ELECTRONIC INTERFACES

Neuro-electronic interfaces are a visionary goal dealing with the construction of nanodevices that will permit computers to be joined and linked to the nervous system. This technology will allow a computer to control neural activity in patients suffering from various neural disorders. This idea requires the building of a molecular structure that will permit control and detection of nerve impulses by an external computer. The computer will be able to interpret, register, and respond to signals the body gives off when it feels sensations. The demand for such structures is huge, because many diseases such as amyotrophic lateral sclerosis (ALS, also referred to as Lou Gehrig's Disease) and multiple sclerosis involve the decay of the nervous system. Also, many injuries and accidents may impair the nervous system, resulting in dysfunctional systems and paraplegia. If computers could control the nervous system through a neuro-electronic interface, problems that impair the system could be controlled to overcome the effects of diseases and injuries. Two approaches must be considered when selecting the power source for such applications: refuellable and non-refuellable strategies. A refuellable strategy implies that energy is refilled continuously or periodically with external sonic, chemical, tethered, magnetic, or electrical sources. A non-refuellable strategy implies that all power is drawn from an internal energy store, which would stop when all energy is drained. One limitation to this innovation is the fact that electrical interference is a possibility. Electric fields, electromagnetic pulses (EMPs), and stray fields from other *in vivo* electrical devices can all cause interference. Also, thick insulators are required to prevent electron leakage, and if high conductivity of the *in vivo* medium occurs there is a risk of sudden power loss and "shorting out." Finally, thick wires are also needed to conduct substantial power levels without overheating. Little practical progress has been made, even though research is ongoing. The wiring of the structure is extremely difficult, because the wires must be positioned precisely in the nervous system so that they are able to monitor and respond to nervous signals. The structures that will provide the interface must also be compatible with the body's immune system so that they will remain unaffected in the body for a long time. In addition, the structures must also sense ionic currents and be able to cause currents to flow backward. Although the potential for these structures is amazing, there is no timetable for when they will be available.

10.4.3   NANOROBOTS

It sounds like a science fiction movie; yes, soon, there is a possibility of creating nanorobots that will totally change the world of medicine once they are realized. According to Robert Freitas of the Institute for Molecular Manufacturing, a typical blood-borne medical nanorobot would be between 0.5 and 3 μm in size, because that is the maximum size possible because of capillary passage requirements. Carbon would be the primary element used to build these nanorobots owing to the inherent strength and other characteristics of some forms of carbon (diamond/fullerene composites), and nanorobots would be fabricated in desktop nanofactories specialized for this purpose. Nanodevices could be observed at work inside the body using MRI, especially if their components were manufactured using mostly ^{13}C atoms rather than the natural ^{12}C isotope of carbon, since ^{13}C has a nonzero nuclear magnetic moment. Medical nanodevices would first be injected into a human body and would then go to work in a specific organ or tissue mass. The doctor would monitor the progress and make sure that the nanodevices have gotten to the correct target treatment region. The doctor would also be able to scan a section of the body and see the nanodevices congregated neatly around their target (such as a tumor mass) so that he or she can be sure that the procedure was successful.

10.4.4 CELL REPAIR MACHINES

If there is a problem in the cells, we use drug molecules as a model to treat the problem, but it may be possible to treat dysfunctional cells with molecular machines in the future. A cell repair machine would be a device that has a set of minuscule arms and tools controlled by a nanocomputer. The whole system would be much smaller than a cell. The repair machine could work like a tiny surgeon, reaching into a cell, sensing damaged parts, repairing them, closing the cell, and moving on. By repairing and rearranging cells and surrounding structures, cell repair machines could restore tissues to health. Cells build and repair themselves using molecular machines, and cell repair machines will use the same principle. The main challenge will be to orchestrate these operations properly, once assemblers are able to build suitable tools. With molecular machines, there will be more direct repairs, as it would be possible to enter a diseased cell with great accuracy and precision. Not only are molecular machines capable of entering the cell, all specific biochemical interactions show that molecular systems can recognize other molecules by touch, build or rebuild every molecule in a cell, and disassemble damaged molecules. Finally, cells that replicate prove that molecular systems can assemble every system found in a cell. Therefore, because nature has demonstrated the basic operations needed to perform molecular-level cell repair, in the future, nanomachine-based systems will be built that are able to enter cells, sense differences from healthy ones, and make modifications to the structure. The possibilities of these cell repair machines are impressive. Comparable to the size of viruses or bacteria, their compact parts would allow them to be more complex. The early machines will be specialized.

10.4.5 NANOSENSORS

Nanosensors are any biological, chemical, or surgical sensory points used to convey information about nanoparticles to the macroscopic world. Their use is mainly for various medicinal purposes and as gateways to building other nanoproducts, such as computer chips that work at the nanoscale and nanorobots. Presently, there are several ways proposed to make nanosensors, including top-down lithography, bottom-up assembly, and molecular self-assembly. Medicinal uses of nanosensors mainly revolve around the potential of nanosensors to accurately identify particular cells or places in the body in need. By measuring changes in volume, concentration, displacement, and velocity, and changes in gravitational, electrical, and magnetic forces, pressure, or temperature of the cells in a body, nanosensors may be able to distinguish between and recognize certain cells (most notably, cancer cells) at the molecular level in order to deliver medicine to or monitor development in specific areas of the body.

In addition, nanosensors may be able to detect macroscopic variations from outside the body and communicate these changes to other nanoproducts working within the body. One example of nanosensors involves using the fluorescence properties of cadmium selenide qdots as sensors to uncover tumors within the body. By injecting a body with these qdots, a doctor could see where a tumor or cancer cell was by finding the injected qdots—an easy process because of their fluorescence. Developed nanosensor qdots would be specifically constructed to find only the particular cell to which the body was at risk. A downside to the cadmium selenide qdots, however, is that they are highly toxic to the body. As a result, researchers are working on developing alternate qdots made from a different, less toxic material, while still retaining some of the fluorescence properties. They have been investigating the specific benefits of zinc sulfide qdots, which, though not quite as fluorescent as cadmium selenide qdots, can be augmented with other metals, including manganese and various lanthanide elements.

In addition, these newer qdots become more fluorescent when they bond to their target cells. Potential predicted functions may also include sensors used to detect specific DNA to recognize explicit genetic defects, especially for individuals at high risk, and implanted sensors that can automatically detect glucose levels in diabetic subjects more simply than current detectors. DNA can also serve as a sacrificial layer for manufacturing a complementary metal-oxide semiconductor integrated circuit (CMOS IC), integrating a nanodevice with sensing capabilities. Therefore, using proteomic patterns and new hybrid materials, nanobiosensors can also be used to enable components to be configured into a hybrid semiconductor substrate as part of the circuit assembly. The development and miniaturization of nanobiosensors should provide interesting new opportunities. Other projected products most commonly involve using nanosensors to build smaller integrated circuits, as well as incorporating them into various other commodities made using other forms of nanotechnology for use in a variety of situations, including transportation, communication, improvements in structural integrity, and robotics. Nanosensors may also eventually be valuable as more accurate monitors of material states for use in systems where size and weight are constrained, such as in satellites and other aeronautic machines.

10.5 NANOPARTICLES

In terms of diameter, fine particles cover a range between 100 and 2500 nm, whereas ultra-fine particles are sized between 1 and 100 nm. Like ultrafine particles, nanoparticles are sized between 1 and 100 nm, though the size limitation can be restricted to two dimensions. Nanoparticles may or may not exhibit size-related properties that differ significantly from those observed in fine particles or bulk materials. Nanoclusters have at least one dimension between 1 and 10 nm and a narrow size distribution. Nanopowders are agglomerates of ultrafine particles, nanoparticles, or nanoclusters. Nanometer-sized single crystals or single-domain ultrafine particles are often referred to as *nanocrystals*. The term "nanocrystal" is a registered trademark of Elan Pharma International Limited (EPIL) used in relation to EPIL's proprietary milling process and nanoparticulate drug formulations. Nanoparticle research is currently an area of intense scientific research due to a wide variety of potential applications in the biomedical, optical, and electronic fields. The NNI has led to generous public funding for nanoparticle research in the United States.

Although nanoparticles are generally considered an invention of modern science, they have a very long history. Specifically, nanoparticles were used by artisans as far back as the ninth century in Mesopotamia to generate a glittering effect on the surface of pots. Even these days, pottery from the Middle Ages often retains a distinct gold- or copper-colored metallic glitter. This so-called luster is caused by a metallic film that was applied to the transparent surface of the glazing. The luster may still be visible if the film has resisted atmospheric oxidation and other weathering. The luster originates within the film itself, which contains silver and copper nanoparticles dispersed homogeneously in the glassy matrix of the ceramic glaze. These nanoparticles were created by the artisans by applying copper and silver salts and oxides together with vinegar, ochre, and clay on the surface of previously glazed pottery. The object was then placed in a kiln and heated to approximately 600°C in a reducing atmosphere. The glaze would soften in the heat, causing the copper and silver ions to migrate into the outer layers of the glaze. There, the reducing atmosphere reduced the ions back to metals, which then came together forming the nanoparticles that give the color and optical effects. Luster technique shows that craftsmen had a rather sophisticated empirical knowledge of materials. The technique originates in the Islamic world, as Muslims were not allowed to use gold in artistic representations; they had to find a way to create a similar effect without using real gold. The solution they found was using luster.

At the small end of the size range, nanoparticles are often referred to as clusters. Nanospheres, nanorods, nanofibers, and nanocups are just a few of the shapes that have been grown. Metal, dielectric, and semiconductor nanoparticles as well as hybrid structures (e.g., core-shell nanoparticles) have been formed. Nanoparticles made of semiconducting material may also be labeled qdots if they are small enough (typically <10 nm) for quantization of electronic energy levels to occur. Such nanoscale particles are used in biomedical applications as drug carriers or imaging agents. Semi-solid and soft nanoparticles have been manufactured. A liposome is a prototype nanoparticle of a semisolid nature. Various types of liposome nanoparticles are currently used clinically as delivery systems for anticancer drugs and vaccines.

The characterization of nanoparticles is necessary to establish an understanding and control of nanoparticle synthesis and applications. Characterization is done by using a variety of different techniques, which include electron microscopy, atomic force microscopy, dynamic light scattering, x-ray photoelectron spectroscopy, powder X-ray diffractometry, Fourier transform infrared spectroscopy (FTIR), matrix-assisted laser desorption/ionization time-of-flight mass spectrometry (MALDI TOF MS), and UV-visible spectroscopy.

10.5.1 NANOPARTICLES AND SAFETY ISSUES

Besides having so many beneficial applications, nanoparticles do cause health and environmental hazards. Most of these are because of the high surface-to-volume ratio, which can make the particles very reactive or catalytic. They can also pass through cell membranes in organisms, and their interactions with biological systems are relatively unknown. According to the *San Francisco Chronicle*,

> Animal studies have shown that some nanoparticles can penetrate cells and tissues, move through the body and brain, and cause biochemical damage. They also have been shown to cause a risk factor in men for testicular cancer. But whether cosmetics and sunscreens containing nanomaterials pose health risks remains largely unknown, pending completion of long-range studies recently begun by the FDA and other agencies.

Diesel nanoparticles have been found to damage the cardiovascular system in a mouse model. In October 2008, the Department of Toxic Substances Control within the California Environmental Protection Agency announced its intent to request information regarding "analytical test methods, fate and transport" in the environment, and other relevant information from manufacturers of carbon nanotubes. The purpose of this information request was to identify information gaps and to develop information about carbon nanotubes, an important emerging nanomaterial. The law places the responsibility to provide this information to the Department on those who manufacture and those who import the chemicals.

10.6 NANOTECHNOLOGY IN THE FOOD INDUSTRY

Nanotechnology is having an impact on several aspects of food science, from how food is grown to how it is packaged. Companies are developing nanomaterials that will make a difference not only in the taste of food, but also in food safety and the health benefits that food delivers. Nanoparticles are being developed that will deliver vitamins or other nutrients in food and beverages without affecting the taste or appearance. These nanoparticles encapsulate the nutrients and carry them through the stomach into the bloodstream. Researchers are using silicate nanoparticles to provide a barrier to gases (e.g., oxygen) or to moisture in the plastic film used for packaging. This could reduce the possibility of food spoiling or drying out. Zinc oxide nanoparticles can be incorporated into plastic packaging to block UV rays and provide antibacterial protection while improving the strength and stability of the plastic film.

Nanosensors are being developed that can detect bacteria and other contaminates (such as salmonella) at packaging plants. This will allow for frequent testing at a much lower cost than sending samples to a lab for analysis. This point-of-packaging testing, if conducted properly, has the potential to dramatically reduce the chance of contaminated food reaching grocery store shelves. Research is also being conducted to develop nanocapsules containing nutrients that would be released when nanosensors detect a vitamin deficiency in your body. Basically, this research could result in a super vitamin storage system in your body that delivers the nutrients you need, when you need them. "Interactive" foods are being developed by Kraft that would allow you to choose the desired flavor and color. Nanocapsules that contain flavor or color enhancers are embedded in the food, remaining inert until a hungry consumer triggers them. The method has not been published, so it will be interesting to see how this trick is accomplished. Researchers are also working with pesticides encapsulated in nanoparticles that only release the pesticide within an insect's stomach, minimizing the contamination of plants. Another development being pursued is a network of nanosensors and dispensers used throughout a farm field. The sensors recognize when a plant needs nutrients or water before there is any sign of their deficiency in the plant. The dispensers then release fertilizer, nutrients, or water as needed, optimizing the growth of each plant in the field.

10.7   WATER POLLUTION AND NANOTECHNOLOGY

Nanotechnology is being used to develop solutions to three very different problems in water quality. One challenge is the removal of industrial water pollution of ground water. Nanoparticles can be used to convert the contaminating chemical through a chemical reaction to make it harmless. Studies have shown that this method can be used successfully to reach contaminates dispersed in underground ponds and at much lower cost than methods that require pumping the water out of the ground for treatment. The second challenge is the removal of salt or metals from water. A deionization method using electrodes composed of nano-sized fibers shows promise for reducing the cost and energy requirements of turning salt water into drinking water. The third problem concerns the fact that standard filters do not work on virus cells. A filter only a few nanometers in diameter is currently being developed that should be capable of removing virus cells from water.

10.8   RESEARCH TRENDS

With a rapid increase in technological breakthroughs, it would be possible to create new cures and new methods of disease detection. One of the major issues regarding human diseases such as cancer, Parkinson's disease, and Alzheimer's disease is how early you can detect the disease. Nanotechnology has the potential to overcome the limitations of current approaches and thereby advance the diagnosis and treatment of life-threatening diseases. To make rapid, accurate, real-time detection possible, several advances will be required. The molecular signatures of diseases, also called *biomarkers*, are often present at concentrations that are too low to be measured by current technology, so new devices will be needed to improve sensitivity. One example of a promising new medical approach that uses nanobiotechnology is the bio-barcode assay for the detection of disease-related proteins or DNA in tissue samples. In its first application, the bio-barcode assay uses gold nanoparticles to amplify and detect amyloid beta-derived diffusible ligands (ADDLs), a molecular signature for early-stage Alzheimer's disease. This method is a million times more sensitive than the current diagnostic tests.

A key challenge for nanobiotechnology is the fabrication and assembly of nanoscale materials, devices, and systems. Naturally occurring nano-structured organisms such as diatoms have been converted from silica into a range of ceramic nanomaterials with new properties. Cells and viruses can also be engineered to manufacture or assemble nanomaterials. Certain bacteria that naturally produce magnetic nanoparticles have been engineered to produce nanoparticles coated with specific proteins. Researchers have used biological methods to discover viruses that can be used as scaffolds for selective attachment and growth of semiconductor nanoparticles to produce the first virus-assembled battery. Scientists are also learning to create complex inorganic nanostructures, including those with unique chemical, optical, and mechanical properties, and efforts are being made to create nanoparticles to quickly and accurately measure drug toxicity.

R&D is an important factor for the creation of novel products. Nanotechnology has the potential to profoundly change our economy and to improve our standard of living in a manner not unlike the impact made by advances in information technology over the last two decades. Although some commercial products are beginning to come to market, many major applications for nanotechnology are still 5–10 years away. Private investors look for shorter-term returns on investment, generally in the range of 1–3 years. Consequently, government support for basic R&D in its early stages needs to maintain a competitive position in the worldwide nanotechnology marketplace to realize nanotechnology's full potential. In the United States, federal funding for nanotechnology has increased from approximately $464 million in 2001 to nearly $1.5 billion for the fiscal year 2009. According to estimates, private industry is investing at least as much as the government. Besides the United States, the European Union and Japan have invested approximately $1.05 billion and $950 million, respectively, in nanotechnology. Korea, China, and Taiwan have invested $300 million, $250 million, and $110 million, respectively, in nanotechnology R&D.

PROBLEMS

Section A: Descriptive Type

Q1. What is nanobiotechnology?
Q2. How is nanobiotechnology used in medicine? Describe it by using an example.
Q3. How is qdots technology used in diagnosing cancer?
Q4. What is an *in vivo* drug imaging?
Q5. Is it possible to repair cells using a machine?
Q6. How are nanoparticles characterized?
Q7. Describe the future trends of nanobiotechnology.

Section B: Multiple Choice

Q1. As per the US NNI, what is the minimum size of a nanoparticle?
   a. 100 nm
   b. 100 pm
   c. 100 mm
   d. 100 cm
Q2. There are no US FDA directives to regulate nanobiotechnology. True/False
Q3. Bioavailability of a drug refers to the . . .
   a. Presence of the drug inside of the body
   b. Presence of the drug outside of the body
   c. Excretion of the drug
   d. Absorption of the drug
Q4. Nanoparticles have a high surface area, thus allowing many functional groups to attach. True/False
Q5. Dendrimer is a nanoparticle with . . .
   a. 100 hooks
   b. >100 hooks
   c. >500 hooks
   d. None of the above
Q6. Nanoparticles cause health and environmental hazards. True/False
Q7. ADDL is related to detection of . . .
   a. Parkinson's disease
   b. Huntington's disease
   c. Alzheimer's disease
   d. Mad cow disease

Section C: Critical Thinking

Q1. Do you agree that nanoparticles have the capability to diagnose the early onset of a disease? Explain.
Q2. What would happen to the human body if nanorobots worked as an artificial defense shield against all infections?

REFERENCES AND FURTHER READING

Allen, T.M., and Cullis, P.R. Drug delivery systems: Entering the mainstream. *Science* 303: 1818–1822, 2004.

Allman III, R.M. *Structural Variations in Colloidal Crystals.* MS thesis, UCLA, 1983. See Ref. 14 in Mangels, J.A., and Messing, G.L. (eds.). Forming of ceramics, microstructural control through colloidal consolidation, I.A. Aksay. *Adv. Ceram.* 9: 94. Proc. Amer. Ceramic Soc.

Books, B.R., Bruccoleri, R.E., Olafson, B.D. et al. CHARMM: A program for macromolecular energy, minimization, and dynamics calculations. *J. Comput. Chem.* 4: 187–217, 1983.

Buffat, P.H., and Burrel, J.P. Size effect on the melting temperature of gold particles. *Phys. Rev. A.* 13: 2287–2298, 1976.

Cavalcanti, A., Shirinzadeh, B., Freitas, R.A. Jr., and Hogg, T. Nanorobot architecture for medical target identification. *Nanotechnology* 19: 15, 2008.

Cavalcanti, A., Shirinzadeh, B., Freitas, R.A. Jr., and Kretly, L.C. Medical nanorobot architecture based on nanobioelectronics. *Recent Pat. Nanotechnol.* 1: 1–10, 2007.

Chemical information call-in web page. Department of Toxic Substances Control, 2008. www.dtsc.ca.gov/PollutionPrevention/Chemical_Call_In.cfm.

Choy, J.H., Jang, E.S., Won, J.H. et al. Hydrothermal route to ZnO nanocoral reefs and nanofibers. *Appl. Phys. Lett.* 84: 287–289, 2004.

Davidson, K. FDA urged to limit nanoparticle use in cosmetics and sunscreens. *San Francisco Chronicle.* www.sfgate.com/cgi-bin/article.cgi?file=/c/a/2006/05/17/MNGFHIT1161.DTL (retrieved on April 20, 2007).

Department of Toxic Substances Control [Report], 2008. www.dtsc.ca.gov/TechnologyDevelopment/Nanotechnology/index.cfm.

Editorial. Nanomedicine: A matter of rhetoric? *Nat. Mater.* 5: 243, 2006.

Fahlman, B.D. *Materials Chemistry*, Vol. 1. Springer, Mount Pleasant, MI, pp. 282–283, 2007.

Faraday, M. Experimental relations of gold (and other metals) to light. *Phil. Trans. Roy. Soc. London* 147: 145–181, 1857.

Freitas, R.A. Jr. *Nanomedicine, Volume I: Basic Capabilities.* Landes Bioscience, Georgetown, TX, 1999.

Freitas, R.A. Jr. What is nanomedicine? *Nanomedicine: Nanotech. Biol. Med.* 1: 2–9, 2005.

Hench, L.L., and West, J.K. The sol-gel process. *Chem. Rev.* 90: 33–72, 1990.

Khan, F.A., Almohazey, D., Alomari, M., and Almofty, S.A. Impact of nanoparticles on neuron biology: Current research trends. *International Journal of Nanomedicine* 13: 2767–2776, 2018. Published online May 9, 2018. PMCID: PMC5951135. PMID: 29780247. doi:10.2147/IJN.S165675.

LaVan, D.A., McGuire, T., and Langer, R. Small-scale systems for in vivo drug delivery. *Nat. Biotechnol.* 21: 1184–1191, 2003.

Loo, C., Lin, A., Hirsch, L. et al. Nanoshell-enabled photonics-based imaging and therapy of cancer. *Technol. Cancer Res. Treat.* 3: 33–40, 2004.

Minchin, R. Sizing up targets with nanoparticles. *Nat. Nanotechnol.* 3: 12–13, 2008.

Mnyusiwalla, A., Daar, A.S., and Singer, P.A. Mind the gap': Science and ethics in nanotechnology. *Nanotechnology* 14: R9–R13, 2003.

National Cancer Institute Alliance for nanotechnology in cancer. http://nano.cancer.gov/.

Onoda, G.Y. Jr., and Hench, L.L. *Ceramic Processing Before Firing.* Wiley & Sons, New York, 1979.

Roco, M.C., and Bainbridge, W.S. *Converging Technologies for Improving Human Performance: Nanotechnology, Biotechnology, Information Technology and Cognitive Science.* Kluwer Academic Publishers, Springer, Dordrecht and Boston, MA, 2003.

Shi, X., Wang, S., Meshinchi, S. et al. Dendrimer-entrapped gold nanoparticles as a platform for cancer-cell targeting and imaging. *Small* 3: 1245–1252, 2007.

Wagner, V., Dullaart, A., Bock, A.K., and Zweck, A. The emerging nanomedicine landscape. *Nat. Biotechnol.* 24(10): 1211–1217, 2006.

Whitesides, G.M. et al. Molecular self-assembly and nanochemistry: A chemical strategy for the synthesis of nanostructures. *Science* 254: 1312, 1991.

Ying, J. *Nanostructured Materials.* Academic Press, New York, 2001.

Zheng, G., Patolsky, F., Cui, Y., Wang, W.U., and Lieber, C.M. Multiplexed electrical detection of cancer markers with nanowire sensor arrays. *Nat. Biotechnol.* 23: 1294–1301, 2005.

11 Product Development in Biotechnology

LEARNING OBJECTIVES
- Define the product development process in the biotechnology industry
- Discuss scientific inventions and their commercialization
- Discuss various components of the biotechnology industry
- Explain the various phases of biotechnology product development
- Explain how a biotechnology company can be started
- Discuss the role of investment and financial implications in the biotechnology industry
- Discuss intellectual property rights in biotechnology
- Discuss the roles of Good Manufacturing Practices (GMPs), Good Laboratory Practice (GLP), the World Health Organization (WHO), and the US Food and Drug Administration (FDA) in biotechnology product development

11.1 INTRODUCTION

The biotechnology product development process is based on the kind of product you intend to develop or manufacture. For example, the product development process for medical applications is different from that for agricultural and food products. Likewise, the product development process for agricultural products is different from that for industrial products. In this chapter, we discuss product development pathways for all types of biotechnology products. We also describe in detail the development pathways for all products derived from the medical, agricultural, environmental, industrial, and nanobiotechnology sectors (Figure 11.1).

Biotechnology product development normally starts with a novel idea. A novel idea is basically a new way of thinking that can offer better products for human welfare. Let us find out the historical aspects of scientific inventions by using the scientific method. The *scientific method* refers to a body of techniques for investigating phenomena, acquiring new knowledge, or correcting and integrating previous knowledge. To be termed scientific, a method of inquiry must be based on gathering observable, empirical, and measurable evidence subject to specific principles of reasoning. A scientific method consists of the collection of data through observation and experimentation, and the formulation and testing of hypotheses. Although procedures vary from one field of inquiry to another, identifiable features distinguish scientific inquiry from other methodologies of knowledge.

Scientific researchers propose hypotheses as explanations of phenomena and then design experimental studies to test these hypotheses. These steps must be repeatable to dependably predict any future results. Theories that encompass wider domains of inquiry may bind many independently derived hypotheses together in a coherent, supportive structure. This in turn may help form new hypotheses or place groups of hypotheses into context. Since Ibn al-Haytham (Alhazen, 965–1039), one of the key figures in the development of the scientific method, the emphasis has been on seeking truth. Ibn al-Haytham discovered by experimentation that light travels in a straight line. Ibn al-Haytham's discovery was based on scientific experimentation, proved to be a milestone in scientific discovery, and continues to be of scientific relevance.

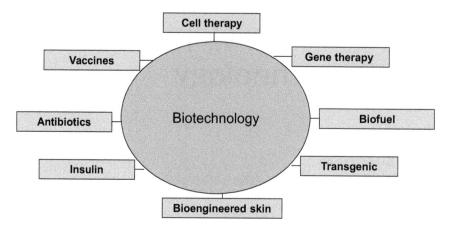

FIGURE 11.1 Products of biotechnology.

11.2 METHODS OF SCIENTIFIC ENQUIRY

There are different ways of outlining the basic method used for scientific inquiry. The scientific community and philosophers of science generally agree on a general classification of the method's components. These methodological elements and organization of procedures tend to be more characteristic of natural sciences than social sciences. Nonetheless, the cycle of formulating hypotheses, testing and analyzing the results, and formulating new hypotheses will resemble the sequence described below. This is also the expected format and outline of a scientific report.

The scientific method of investigation is based on four essential elements:

1. *Characterizations*—observations, definitions, and measurements of the subject of inquiry
2. *Hypotheses*—theoretical, hypothetical explanations of observations and measurements of the subject
3. *Predictions*—reasoning (including logical deduction) from the hypothesis or theory, or the identification of distinct and (ideally) mutually exclusive possible discernible outcomes
4. *Experiments*—tests of characterizations, hypotheses, and predictions

Each element of a scientific method is subject to peer review for possible mistakes. These activities do not describe all that scientists do but apply mostly to experimental sciences such as physics and chemistry. The elements above are often taught in the educational system (Figure 11.2).

The scientific method is not a recipe; it requires intelligence, imagination, and creativity. It is also an ongoing cycle, constantly developing more useful, accurate, and comprehensive models and methods. For example, when Einstein developed the special and general theories of relativity, he did not in any way refute or discount Newton's *Principia*. On the contrary, if the astronomically large, the vanishingly small, and the extremely fast are reduced out from Einstein's theories—all phenomena that Newton could not have observed—Newton's equations remain. Einstein's theories are expansions and refinements of Newton's theories, and thus increase our confidence in Newton's work.

11.2.1 CHARACTERIZATIONS OF SCIENTIFIC INVESTIGATION

The scientific method depends upon increasingly more sophisticated characterizations of the subjects of investigation (the *subjects* can also be called *unsolved problems* or the *unknowns*). The history of the discovery of the structure of DNA is a classic example of the elements of the scientific method. In 1950, it was known that genetic inheritance had a mathematical description, starting

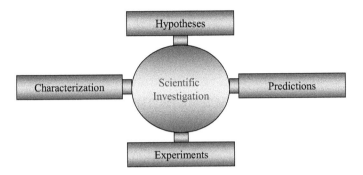

FIGURE 11.2   Methods of scientific investigation.

with the studies of Gregor Mendel, but the mechanism of the gene was unclear. Researchers in Bragg's laboratory at Cambridge University made x-ray diffraction pictures of various molecules, starting with crystals of salt and proceeding to more complicated substances. Using clues that were painstakingly assembled over the course of decades, beginning with the chemical composition of DNA, it was determined that it should be possible to characterize the physical structure of DNA, and x-ray images would be the vehicle. Linus Pauling proposed that DNA might be a triple helix. This hypothesis was also considered by Francis Crick and James Watson but discarded. When Watson and Crick learned of Pauling's hypothesis, they understood from existing data that Pauling was wrong, and that Pauling would soon admit his difficulties with that structure. Therefore, the race was on to figure out the correct structure (although, at the time, Pauling did not realize that he was in a race). James Watson, Francis Crick, and others hypothesized that DNA had a helical structure. This implied that DNA's x-ray diffraction pattern would be "x-shaped." This prediction followed from the work of Cochran, Crick, and Vand (and independently from the work by Stokes). The Cochran–Crick–Vand–Stokes theorem provided a mathematical explanation for the empirical observation that diffraction from helical structures produces x-shaped patterns. Also, in their first paper, Watson and Crick predicted that the double helix structure provided a simple mechanism for DNA replication, documenting "It has not escaped our notice that the specific pairing we have postulated immediately suggests a possible copying mechanism for the genetic material."

11.2.2   Scientific Inventions

The popular image of an inventor has been someone toiling away for months or years in a back shed and finally emerging with some wondrous gizmo that no one else had ever thought of. Another image is of a person in a white coat running out of their laboratory shouting "I've done it!" Sometimes, an invention is the result of the work of a team of people, particularly in the case of academic research or corporate R&D. Sometimes, Person A discovers a substance or process, while Person B later invents a way that it can be used or harnessed. Thus, while penicillin was "discovered" by Alexander Fleming, it was later isolated by Howard Florey and Ernst Chain in Cambridge, and it was Florey and his team who turned it into the practical medication that was used to save millions of lives. There were numbers of important people in the "invention" of penicillin as we know it, and Fleming, Chain, and Florey shared the 1945 Nobel Prize.

A patent application also includes one or more claims, although it is not always a requirement to submit these when first filing the application. The claims set out what the applicant is seeking to protect, in that they define what the patent owner has a right to exclude others from making, using, or selling. For a patent to be granted (i.e., to take legal effect in a country), the patent application must meet the patentability requirements of that country. Most patent organizations examine the application for compliance with these requirements. If the application does not comply, objections are communicated to the applicant or their patent agent or attorney, and one or more opportunities to respond to the objections to bring the application into compliance are usually provided. After rigorous technical evaluation, the patent agencies grant patent rights to the inventor, which in fact allow the inventor to manufacture the product.

11.3 COMMERCIALIZATION OF SCIENTIFIC DISCOVERY

Once scientific data is tested and proved by various techniques, the next step is to use the scientific information to create a product that may not be scientifically relevant but is economically viable. Most commercialization of a health product is largely dependent on the basic scientific data generated and published by the researchers working in the universities or federal research centers. Note that no company would like to invest in a project with no sound scientific basis, and at the same time they do not want to invest in noncommercial scientific projects. Let us take, for example, recombinant DNA (rDNA) insulin, which is basically synthetically manufactured insulin, a product of rDNA technology and used by most diabetic patients on a regular basis. One of the greatest breakthroughs in rDNA technology is the biosynthetic "human" insulin, the first medicine made via rDNA technology ever to be approved by the US Food and Drug Administration (US FDA). Due to the efforts of various scientists, rDNA technology has become one of the most important technologies in the modern time for generating various therapeutic proteins that are used in treating several diseases. Big companies had started investing billions of dollars to take advantage of this technology and to successfully produce large quantities of recombinant products.

11.4 BUSINESS PLAN

One of the basic requirements of any biotechnology product is to make a business plan, which includes work on the novel and economically feasible project. To ensure that the project is novel and economically feasible, a company consults with scientific and technical experts. The experts then review the project and examine it to ensure that the product is easy to use, safe for all users, and economical to manufacture. The design process is a cyclic, ongoing developmental process of generating ideas, testing these ideas, and selecting the ideas that work best. This developmental cycle begins with a range of possibilities and general ideas (conceptual designs) and gradually reaches a point where particular and fine details of the final product (final specifications) have been decided.

11.4.1 PROJECT FEASIBILITY

Before investing in a project, a company must know a lot of things to ensure its feasibility. The following are some of the questions that a company must know the answer to before investing in a project: Will the project work? Can it be done in time? Can it be done within budget? One of the biggest problems faced by any biotechnology company is, how to make a biotechnology product in large quantities. It may be possible to produce a product in small amounts or numbers, perhaps in a lab setting, but is it feasible to upscale this process for mass production? This is very important to know before a company can think of investing money in a biotechnology project. Biotechnology companies ask industry experts to prepare a feasibility report on the proposed project, which generally consists of the following components.

11.4.1.1 Market Research

The first thing that any biotechnology or healthcare company does is to carry out extensive research on existing products that could be potential competitors in the future. This kind of survey is conducted to identify the problems of existing products in terms of effectiveness of the treatment, side effects, cost, and—most important—the market demand for the product, such as how big a market there is for the product in terms of sales. Market research can provide information about customer preferences, identify gaps in the market, get customer feedback, and identify any social or cultural issues that may be involved in the production and sale of a product.

11.4.1.2 Significance of a Project

For the proposed project there must be a clear understanding of both the scientific and technical content, with each component mentioned in detail, and how the proposed project will be unique in terms of the quality of the products and treatment modalities in comparison with existing product lines.

11.4.1.3 Technical Outline

An outline specification lists key features and performance of a product. It lists what the product is expected to do and what materials, properties, approximate amounts, processes, equipment, and expertise are needed.

11.4.1.4 Time Plan

It is very important to know the deadlines for product R&D, manufacturing, formulation, and the preclinical and clinical testing phases. These deadlines help both the owner and company management to successfully launch the product on time.

11.4.1.5 Project Cost

The monetary requirement is very critical, and any company must know how much they must invest to manufacture the product. This cost estimate also helps a company to generate funds through various sources. To make a biotechnology product, approximately 500–700 million US dollars is primarily used for R&D, patent application, and preclinical and clinical trial costs. The money is either funded by the biotechnology company (or corporate investor) itself, or by a group of investors (venture investor). The venture capitalist can contribute at various phases by signing an agreement with the owner of the biotechnology product.

11.4.1.6 Legal and Regulatory Issues

Besides having scientific and technical information about a product, a company must know the intellectual property (IP), regulatory controls, environmental issues, and health and safety issues associated with the proposed product. This is to protect the company's ideas or to make sure that other companies do not transgress others' IP rights.

11.4.1.7 Quality of the Product

In product development terms, quality is a relative term that is defined by how well the product meets or exceeds what it was designed to do. A product's quality or fitness for purpose is judged by the degree to which it matches the desired specifications such as (1) measuring against specifications and (2) meeting customer requirements. Measuring against specifications requires a quality assurance system.

11.5 BIOTECHNOLOGY PRODUCT DEVELOPMENT

The field of biotechnology keeps expanding with new knowledge and advanced technology. Various components of biotechnology product development are illustrated in Figure 11.3. Even though such products may be very different and may use different biological processes for production, most biotechnology companies operate in a similar way. Industrial and environmental biotechnology make manufacturing processes cleaner and more efficient; create new materials, food ingredients, and other products; unlock cleaner and greener sources of energy; and reduce industrial waste. For example, biotechnology-based enzymes are used in such wide-ranging products as cheese, detergents, environmentally friendly plastics, and renewable fuels like cellulosic ethanol. We will now look at the

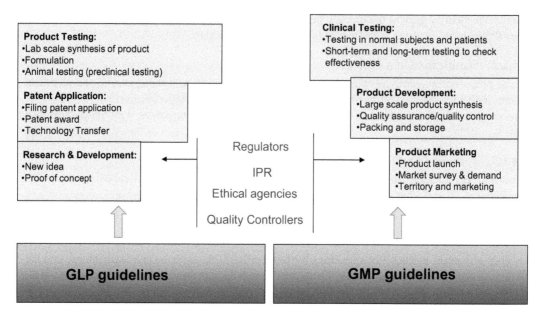

FIGURE 11.3 Components of biotechnology product development.

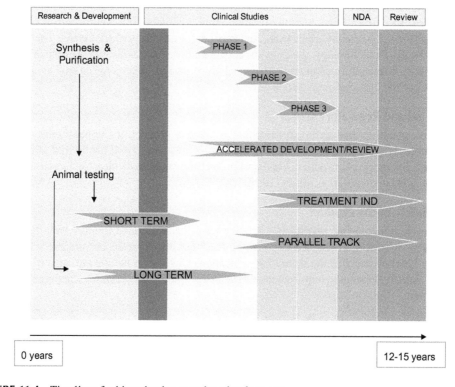

FIGURE 11.4 Timelines for biotechnology product development.

pharmaceutical industry process of the major functions involved in the discovery, development, and marketing of a new biotechnology product (Figure 11.4). Some of these processes occur in lab settings, some in the field, and some in manufacturing facilities. To prove that intended research work is novel and reproducible, the researchers must work in the approved laboratories or centers and carryout experiments to generate enough data under strict good laboratory practice (GLP) guidelines.

11.5.1 Infrastructure Requirements

One of the most important components of biotechnology product development is to set up the infrastructure required to manufacture the biotechnology products based on the GLP or good manufacturing practice (GMP) requirements and as per FDA guidelines. The manufacturing plant is generally divided into two major sections like GLP facilities that consist of R&D labs, animal house, bioequivalence labs, and clinical trial labs and GMP facilities that consist of manufacturing area, QA/QC labs, and formulation and packaging area (Figure 11.5). The infrastructure is primarily based on the kind of product required to be manufactured, but generally, any biotechnology manufacturing company has an R&D lab, animal testing lab, manufacturing area, quality assurance and quality control lab, and packaging and storage areas. This briefly describes the various components of any standard biotechnology manufacturing company.

11.5.1.1 Research and Development Facility

One of the first things any company does is to create a research facility and hire expert researchers and scientists to work on the project. The construction of research labs should be as per GLP requirements. In this section, we will discuss GLP for conducting research on biotechnology products.

A new product begins in the research laboratory, where scientists and laboratory technicians use biotechnology tools to learn about the causes of disease. Their discoveries lead to new ideas about how to combat diseases. For example, an antibody might be a cure for a particular disease. Many different antibodies are then tested to see which one works best. The laboratory facilities consist of a main working lab where researchers work on various research activities from DNA isolation to protein assays in well-defined designated areas. For example, all DNA- and RNA-related activities are mainly performed under sterile conditions, such as in the biosafety hood or laminar flow, to avoid contamination. To avoid contamination, nonclinical labs (which handle microbes and virus)

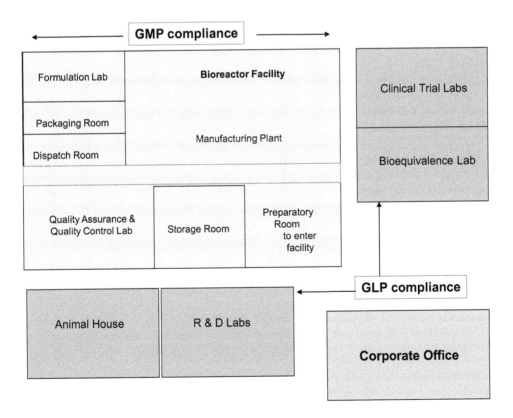

FIGURE 11.5 Biotechnology manufacturing company: Plant layout.

are separated from clinical labs (which handle human cells and DNA) by a proper barrier. The instruments (such as a PCR thermal cycler, spectrophotometer, laminar flow, centrifuge, and micro-pipettes) used in the labs should also be well calibrated, as per GLP requirements.

11.5.1.2 Animal Testing Facility or Preclinical Lab

Animal testing is used for countless products and applications. Everything from items in your home to products you use and medications you take have likely been tested on animals at some point prior to their distribution. Some of the products that commonly involve animal testing are cosmetics, drugs, food additives, food supplements, pesticides, and industrial chemicals. An animal facility has three major components: the breeding room, the surgical room, and the testing room. All these rooms are properly separated from each other to avoid contamination and infections. Aside from this, the temperature, light–dark cycle, and humidity are fully regulated as per international animal standards.

Preclinical testing, also known as *animal testing*, is a stage of research that begins before clinical trials (testing in humans) can begin. The main goal of preclinical studies is to determine a product's ultimate safety profile. Products may include new, iterated, or like-kind medical devices, drugs, or gene therapy solutions. Each class of product may undergo different types of preclinical research. For instance, drugs may undergo pharmacodynamics (PD); pharmacokinetics (PK); absorption, distribution, metabolism, and excretion (ADME); and toxicity testing through animal testing. This data allows researchers to allometrically estimate a safe starting dose of the drug for clinical trials in humans. Medical devices that do not have a drug attached will not undergo these additional tests and may go directly to GLP testing for the safety of the device and its components. Some medical devices will also undergo biocompatibility testing, which helps to show whether a component of the device or all its components are sustainable in a living model. Most preclinical studies must adhere to GLP guidelines issued by the International Conference on Harmonisation (ICH guidelines) to be acceptable for submission to regulatory agencies such as the US FDA. Typically, both *in vitro* and *in vivo* tests will be performed. Studies of a drug's toxicity include which organs are targeted by that drug and whether there are any long-term carcinogenic effects or toxic effects on mammalian reproduction.

The information collected from these studies is vital before safe human testing can begin. In drug development studies, animal testing typically involves two species. The most commonly used models are murine and canine, although primate and porcine are also used. The choice of species is based on which will give the best correlation to human trials. Differences in the gut, enzyme activity, circulatory system, or other considerations make certain models more appropriate, based on the dosage form, site of activity, or noxious metabolites. For example, canines may not be good models for solid oral dosage forms because the characteristic carnivore intestine is underdeveloped compared to that of omnivores and gastric emptying rates are increased. In addition, rodents cannot act as models for antibiotic drugs because the resulting alteration to their intestinal flora causes significant adverse effects. Depending on a drug's functional group, it may be metabolized in similar or different ways between species, which will affect both efficacy and toxicology.

11.5.1.3 Bioequivalence Lab

The US FDA has defined *bioequivalence* as

> the absence of a significant difference in the rate and extent to which the active ingredient in pharmaceutical equivalents or pharmaceutical alternatives becomes available at the site of drug action when administered at the same molar dose under similar conditions in an appropriately designed study.

In determining bioequivalence—for example, between two products such as a commercially-available brand product and a potential to-be-marketed generic product—PK studies are conducted, whereby each of the preparations is administered in a crossover study to volunteer subjects (generally, to healthy individuals, but occasionally to patients). Serum/plasma samples are obtained at regular intervals and assayed for parent drug (or occasionally metabolite) concentration. Occasionally, if blood concentration levels are neither feasible nor possible to compare the two products (such as inhaled corticosteroids), then PD endpoints rather than PK endpoints are used for comparison. For a PK comparison, the plasma concentration data are used to assess key PK parameters such as area-under-the-curve (AUC), peak concentration (Cmax), time to peak concentration (Tmax), and absorption lag time (Tlag). Testing should be conducted at several different doses, especially when the drug displays nonlinear PK. Bioequivalence cannot be claimed based only on *in vitro* testing or only on animal studies. Bioequivalence of human drugs must be determined in humans via established measures of bioavailability. By the same token, animal drugs must be tested for bioequivalence in the animal species for which the drug is intended.

11.5.1.4   Clinical Trial Center

Once a drug has completed preclinical trials with a great success rate, the drug molecule is now ready for testing in humans, starting with normal and healthy individuals, to check the toxicity or side effects of the drug molecule. If the drug does not show any apparent toxicity or side effects, it may now be tested in patients suffering from specific disease conditions. There are four phases of clinical trials: Phase I, Phase II, Phase III, and Phase IV. All these phases are conducted in specialized and FDA-approved clinical centers and hospitals. In healthcare, clinical trials are conducted to allow safety and efficacy data to be collected for new drugs or devices. These trials can only take place once satisfactory information has been gathered on the quality of the product and its nonclinical safety. Health authority or ethics committee approval is granted in the country where the trial is taking place. Depending on the type of product and the stage of its development, investigators enroll healthy volunteers and/or patients in small pilot studies initially, followed by larger-scale studies in patients that often compare the new product with the currently prescribed treatment. As positive safety and efficacy data are gathered, the number of patients is typically increased. Clinical trials can vary in size from a single center in one country to multicenter trials in multiple countries.

11.5.1.5   Manufacturing Plant

During the clinical phase, a company builds a manufacturing plant to synthesize the drug in great capacity. A manufacturing plant normally consists of several divisions, such as a raw material room, a bioreactor room, a formulation room, a packaging room, a quality assurance and quality control lab, and a storage room. Unlike a research laboratory, a manufacturing plant works round-the-clock and most personnel work in shifts, with assigned duties and responsibilities. A manufacturing plant strictly follows the GMP guidelines as prescribed by the FDA and other competent organizations.

11.5.1.6   Formulation Lab

After successful testing in laboratories, the desired product is synthesized at smaller levels to check its efficacy and toxicity in animals. The formulation lab is basically used to synthesize molecules in the final form where it can be delivered in tablet form, syrup form, injection form, or ointment form. Researchers and formulation teams discuss and decide which formulation will be effective in patients. The formulation team mainly consists of biochemists and chemists. *Formulation* in biotechnology is the process in which different chemical substances, including the active drug, are combined to produce a final medicinal product. Formulation studies involve developing a preparation of the drug that is not only effective and stable but is also palatable to the patient. For oral drugs, this usually involves incorporating the drug into tablet, capsule, or syrup form. It is important to appreciate that a tablet contains a variety of other substances apart from the drug itself, and studies must be carried out to ensure that the drug is compatible with these other substances. Before using it for formulation, the chemical's physical, chemical, and mechanical properties must be completely characterized to choose what other ingredients should be used in the preparation. In dealing with protein preformulation, the important aspect is to understand the solution behavior of a given protein under a variety of stress conditions such as freeze/thaw, temperature, and shear stress, among others, to identify mechanisms of degradation and therefore their mitigation. Formulation studies then consider other factors such as particle size, polymorphism, pH, and solubility, all of which can influence bioavailability and hence the activity of a drug. The drug must be combined with inactive additives by a method which ensures that the quantity of drug present is consistent in each dosage unit, such as in each tablet. The dosage should have a uniform appearance, with an acceptable taste, tablet hardness, or capsule disintegration. It is unlikely that formulation studies will be complete by the time clinical trials commence. This means that simple preparations are developed initially for use in Phase I clinical trials. These typically consist of hand-filled capsules containing a small amount of the drug and a diluent. Proof of the long-term stability of these formulations is not required, as they will be tested in a matter of days. Consideration must be given to what is called the *drug load*, the ratio of the active drug to the total content of the dose. A low drug load may cause homogeneity problems. A high drug load may pose flow problems or require large capsules if the compound has a low bulk density.

By the time Phase III clinical trials are reached, the formulation of the drug should have been developed to be close to the preparation that will ultimately be used in the market. Knowledge of stability is essential by this stage, and conditions must have been developed to ensure that the drug is stable during its preparation. If the drug proves unstable, it will invalidate the results from clinical trials, as it would be impossible to know what the administered dose was. Stability studies are carried out to test whether temperature, humidity, oxidation, or photolysis (UV light or visible light) have any effect, and the preparation is analyzed to see if any degradation products have been formed. It is also important to check whether there are any unwanted interactions between the preparation and the container. If a plastic container is used, tests are carried out to see whether any of the ingredients become adsorbed on to the plastic and whether any plasticizers, lubricants, pigments, or stabilizers leach out of the plastic into the preparation. Even the adhesives for the container label need to be tested to ensure they do not leach through the plastic container into the preparation.

11.5.1.7   Quality Assurance and Quality Control Lab

One of the important aspects of biotechnology product development is to ensure the quality of a product at all given times. Companies cannot afford to lose market share due to poor quality, hence the companies themselves create an internal quality assurance and quality control system. The people who work as quality controllers are basically trained biotechnology researchers whose role is to make sure that the company always produces high-quality and safe products.

11.6   PHASES OF BIOTECHNOLOGY PRODUCT DEVELOPMENT

The drug-development process will normally proceed through all four phases over many years. Drugs that successfully pass through Phases I, II, III, and IV are approved by the national regulatory authority for use by the general population. Phase IV covers "post-approval" studies. Before pharmaceutical companies start clinical trials on a drug, they conduct extensive preclinical studies (Figure 11.6).

11.6.1   Preclinical Studies

Preclinical studies involve *in vitro* (test tube) and *in vivo* (animal) experiments using a wide range of doses of the drug under study to obtain preliminary efficacy, toxicity, and PK information. Such tests help pharmaceutical companies to decide whether a drug candidate has scientific merit for further development as an investigational new drug.

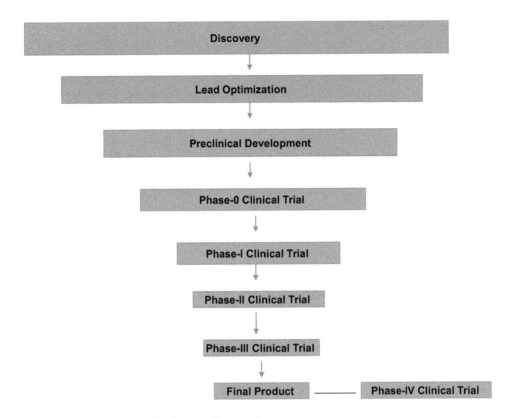

FIGURE 11.6   Phases of biotechnology product development.

11.6.2   PHASE 0

Phase 0 is a recent designation for exploratory, first-in-human trials conducted in accordance with the US FDA's 2006 Guidance on Exploratory Investigational New Drug (IND) Studies. Phase 0 trials are also known as *human microdosing studies* and are designed to speed up the development of promising drugs or imaging agents by establishing very early on whether the drug or agent behaves in human subjects as was expected from preclinical studies. Distinctive features of Phase 0 trials include the administration of single sub-therapeutic doses of the drug under study to a small number of subjects (usually ranging from 10 to 15) to gather preliminary data on the agent's PK (how the body processes the drug) and PD (how the drug works in the body).

11.6.3   PHASE I

Phase I trials are the first stage of testing in human subjects. Normally, a small group of healthy volunteers (usually from 20 to 80) will be selected. This phase includes trials designed to assess the safety, tolerability, PK, and PD of a drug. These trials are often conducted in an inpatient clinic, where the subject can be observed by a full-time staff. The subject who receives the drug is usually observed until several half-lives of the drug have passed. Phase I trials also normally include dose-ranging (also called *dose escalation*) studies so that the appropriate dose for therapeutic use can be found. The tested range of doses will usually be a fraction of the dose that causes harm in animal testing. Phase I trials most often include healthy volunteers; however, there are some circumstances when real patients are used, such as patients who have end-stage disease and lack other treatment options.

11.6.3.1 Single Ascending Dose

SAD studies are those in which small groups of subjects are given a single dose of the drug while they are observed and tested for a period. If they do not exhibit any adverse side effects and if the PK data is roughly in line with predicted safe values, the dose is escalated, and a new group of subjects is then given a higher dose. This is continued until either precalculated PK safety levels are reached or intolerable side effects start showing up, at which point the drug is said to have reached the maximum tolerated dose (MTD).

11.6.3.2 Multiple Ascending Doses

MAD studies are conducted to better understand the PK and PD of multiple doses of the drug. In these studies, a group of patients receives multiple low doses of the drug, while samples of blood and other fluids are collected at various time points and analyzed to understand how the drug is processed within the body. The dose is subsequently escalated for further groups, up to a predetermined level.

11.6.3.3 Food Effect

Food effect studies are short trials designed to investigate any differences in absorption of a drug by the body caused by eating before the drug is given. These studies are usually run as a crossover study, with volunteers being given two identical doses of the drug on two different occasions: one after having fasted, and one after being fed.

11.6.4 PHASE II

Once the safety of a drug under study has been confirmed in Phase I trials, Phase II trials are performed on larger groups (usually ranging from 20 to 300 participants). These trials are designed to assess how well the drug works and to continue Phase I safety assessments in a larger group of volunteers and patients. Phase II studies are sometimes divided into Phases IIA and IIB. *Phase IIA* is specifically designed to assess dosing requirements or how much drug should be given. *Phase IIB* is specifically designed to study efficacy or how well the drug works at the prescribed dose(s). Some trials combine Phase I and Phase II, and test both efficacy and toxicity. Some Phase II trials are designed as case series, demonstrating a drug's safety and activity in a selected group of patients. Other Phase II trials are designed as randomized clinical trials, where some patients receive the drug/device and others receive placebo/standard treatment. Randomized Phase II trials have far fewer patients than randomized Phase III trials.

11.6.5 PHASE III

Phase III studies are randomized controlled multicenter trials on large patient groups (usually ranging from 300 to 3000 or more, depending on the disease/medical condition studied) and are aimed at being the definitive assessment of how effective a drug is in comparison with the current "gold standard" treatment. Because of their size and comparatively long duration, Phase III trials are the most expensive, time-consuming, and difficult trials to design and run, especially in therapies for chronic medical conditions. It is common practice that certain Phase III trials will continue while the regulatory submission is pending at an appropriate regulatory agency. This allows patients to continue to receive possibly lifesaving drugs until the drug can be obtained by purchase. Other reasons for performing trials at this stage include attempts by the sponsor at "label expansion" (to show that the drug works for additional types of patients/diseases beyond the original use for which the drug was approved for marketing), to obtain additional safety data, or to support marketing claims for the drug. Studies in this phase are categorized as Phase IIIB studies by some companies.

Although not required in all cases, it is typically expected that there will be at least two successful Phase III trials demonstrating a drug's safety and efficacy in order to obtain approval from the appropriate regulatory agencies such as the FDA in the United States, the Therapeutic Goods Administration (TGA) in Australia, the European Medicines Agency (EMA) in the European Union, or the Central Drug Standard Control Organization (CDSCO) or the Indian Council of Medical Research (ICMR) in India. Once a drug has proved satisfactory after Phase III trials, the trial results are usually combined into a large document containing a comprehensive description of the methods and results of human and animal studies, manufacturing procedures, formulation details, and shelf life. This collection of information makes up the "regulatory submission" that is provided for review to the appropriate regulatory authorities in different countries. They will review the submission, and it is hoped that the sponsor will be given the approval to market the drug.

Most drugs undergoing Phase III clinical trials can be marketed under FDA norms with proper recommendations and guidelines. In case any adverse effects are reported anywhere, the drugs need to be recalled immediately from the market. While most pharmaceutical companies refrain from this practice, it is not abnormal to see many drugs undergoing Phase III clinical trials in the market.

11.6.6   PHASE IV

A Phase IV trial is also known as a *post-marketing surveillance trial*. Phase IV trials involve the safety surveillance and ongoing technical support of a drug after it receives permission to be sold. Phase IV studies may be required by regulatory authorities or may be undertaken by the sponsoring company for competitive (finding a new market for the drug) or other reasons (e.g., the drug may not have been tested for interactions with other drugs, or on certain population groups such as pregnant women, who are unlikely to subject themselves to trials). The safety surveillance is designed to detect any rare or long-term adverse effects over a much larger patient population and longer time period than was possible during the Phase I–III clinical trials. Harmful effects discovered by Phase IV trials may result in a drug being no longer sold or being restricted to certain uses. Recent examples involve cerivastatin (brand names Baycol and Lipobay), troglitazone (Rezulin), and rofecoxib (Vioxx).

11.7   BIOTECHNOLOGY ENTREPRENEURSHIP

During the last two decades of the twentieth century, many biotechnology companies came into existence. According to the US Biotechnology Industry Organization (BIO), there were 1379 biotechnology companies in the United States in 2001. Similarly, according to the Biotechnology Information Databank (BID) maintained at the University of Siena (Italy), there were 2104 biotechnology companies in Europe in the same year. Due to an anticipated growth in the biotechnology industry, biotechnology companies in both the United States and Europe also attracted huge investments in the stock market. While some of these companies achieved success, others met with failure and closure.

In the past, many scientists who started using biotechnology companies for commercial gains had no background or familiarity with the business world. Similarly, experts in the world of business were sometimes unfamiliar with the tools of biotechnology research that have an important bearing on the biotechnology business. Consequently, those who were in the biotechnology business learned through their mistakes. Furthermore, young entrepreneurs in this area often repeat these mistakes, since the biotechnology business is young, and the mistakes committed by others do not become quickly and widely known. In view of this, one would like to have knowledge of factors that influence the success of a biotechnology company. In this chapter, therefore, an effort is being made to familiarize readers with various aspects of biotechnology that have a bearing on the success of biotechnology businesses.

11.7.1 STARTING A BIOTECHNOLOGY COMPANY

To develop a successful biotechnology product, a company must have a sound economic plan. There are different ways a company can generate funds to develop a product. One way is to fund from the company's own source to maintain a market monopoly. This kind of investment is called *corporate investment*. If it is difficult for a company to find the required funds, they usually find investors, present their business proposals, and influence them to invest in the project. There are several options to obtain funding for a project.

11.7.1.1 Grants

Academics still in the research stages might qualify for government grants for equipment costs and staff (graduate students and technicians) salaries. There are grants available for academic collaborations with industry to facilitate invention commercialization—for example, in Canada, the Industrial Research Assistance Program (IRAP) funds many collaborative biotechnology projects. In the United States, funding from the National Institutes of Health (NIH) comes with certain data sharing policies that must be followed. Universities that have recognized the potential of their research programs have organizations to help commercialize the discoveries of their scientists.

11.7.1.2 Private Investors

Many startup companies rely on funding from private investors that have an interest and belief in a biotechnology product. This may come from friends and family or from acquaintances with money. These people might be the easiest to convince that your product is a viable investment, and they typically demand the least control over your company. However, if the company folds, you have the most to lose in terms of your relationships with them.

11.7.1.3 Angel Investors

Friends and family do not often provide substantial amounts of funding, but you might score bigger with "angel investors." These are individuals with money or capital who invest privately in new businesses. A typical angel investor will demand a larger share of the company than friends or family and thus more control over the company. However, you may benefit from their experience and advice. Angels know what they are up against, and you risk less in terms of personal relationships by taking this route.

11.7.1.4 Venture Capitalists

Like angel investors, venture capitalists will also demand a fair amount of control over your operations and decision making. However, the role of a venture capitalist is also to rally around the business, help with the management, promote the business, and provide contacts to protect their investment.

11.7.1.5 Bank Loans

Look into loans for new businesses and be sure to have a thorough business plan. It is usually easier to get a small business loan if you already have paying customers. If this is not an option, you can try for a personal loan (if it is enough to get you started). The downside of doing so is that, if the business fails, you must still repay the loan. Although the amount of funding you gain may be less than with investors as listed previously, by starting with "debt" financing (loans, lines of credit, and credit cards), you demonstrate to investors that you have faith in the company and are willing to take risks to make it work.

11.8 CAPITAL INVESTMENT IN BIOTECHNOLOGY

The first biotechnology company in the United States to make an initial public offering (IPO) in the capital market was Genentech, which made an offer of $35 million at the rate of $35 per share.

This share soared to $88 within the first 20 min and closed at the end of the day at $56 per share, thus giving the company a valuation of $400 million. Similarly, in March 1981, Cetus's gross IPO was $120 million, giving a valuation of approximately $500 million to this company. However, this trend in the United States did not continue later in the 1980s and 1990s. The average IPO raised by an individual company during 1980–2000 did not exceed $20–30 million/year, although the total capital raised due to biotechnology business improved significantly because of an increase in the number of biotechnology companies.

In the year 1986 (which was the best year in the biotechnology business in the 1980s), the US biotechnology industry raised $900 million (all companies together), which steadily improved over the years, reaching a level of several billion dollars per year during the 1990s. However, in general, the biotechnology industry did not attract investors very much during the 1990s, except towards the end of the twentieth century. For instance, in the year 1999, the share price of Tularik (a premier biotechnology company) improved from $11 to $13 per share in October 1999 to $90 in February 2000, and the company valuation improved from $500 million to $4 billion during the same period. Other companies, impressed by the performance of Tularik, suddenly began filing for IPOs, so that the year 2000 was a record year for biotechnology financing with 63 IPOs completed and $5.4 billion (with average IPO proceeds of an individual company rising from $30 million to $85 million) raised for the biotechnology industry. As many as 37 new biotechnology companies were floated in *genomics research* alone, although most of them may not be able to sustain growth and may therefore merge with or be acquired by other successful companies. The revival of the biotechnology industry was also witnessed in Europe, as suggested by several European biotechnological companies such as Neurotech, Transgene, NicOx, and Cytomix, which raised a total of $194 million in May 2001 alone.

11.9 MERGERS AND ACQUISITIONS OF BIOTECHNOLOGY COMPANIES

Another feature witnessed in the biotechnology industry during 1990–2000 was the merger and/or acquisitions of biotechnology companies. In most cases, this was because of the inability of several companies to retain their independent existence. For instance, Celera Genomics announced its acquisition of Axys Pharmaceuticals, while Lexicon Genetics announced its purchase of Coelacanth. Many more mergers, acquisitions, and collaborations took place in the first decade of the twenty-first century. In 2001, Sequinom Biotech Company merged with Gemini Genomics. Both companies specialized in information mining for the Human Genome Project. While Sequinom has the mass spectrometry-based genetic analysis system, Gemini Genomics specialized in clinical population genomics. The purpose of the merger was to combine human genetic resources with the rapid analytical system to generate a more powerful data-generating machine. This merger is seen as a synergistic effort, where two youthful and vigorous companies, which raised $250 million in IPO, decided to merge—not due to poverty, but due to a cleverly formulated strategy.

11.10 BIOTECHNOLOGY PRODUCTS AND INTELLECTUAL PROPERTY RIGHTS

IP is the term used to describe the branch of law that protects the application of thoughts, ideas, and information that are of commercial value. It thus covers the law relating to patents, copyrights, trademarks, trade secrets, and other similar rights. IP protection for biotechnology is currently in a state of flux. Although it used to be the case that living organisms were largely excluded from protection, attitudes are now changing, and biotechnology is increasingly receiving some form of protection. These changes have largely taken place in the United States and other industrialized countries, but other countries that wish to compete in the new biotechnological markets are likely to change their national laws to protect and encourage investment in biotechnology.

Now, there is no clear international consensus on how biotechnology should be treated. Although bodies such as the World Intellectual Property Organization (WIPO), the permanent body of the United Nations (UN) that is primarily responsible for international cooperation in IP, and OECD have conducted separate studies and produced various reports, these initiatives have only sought to make governments more aware of the potential problems and to offer some suggested solutions. In view of the highly controversial nature of providing IP protection for biotechnology, it is likely that in the short term developments will be at a national and regional level.

11.10.1 PATENTING, LICENSING, AND PARTNERSHIP IN THE BIOTECHNOLOGY INDUSTRY

The success of a biotechnology company also depends on its patent portfolio and licensing revenue; therefore, from the very beginning, a biotechnology company needs to carefully design its patent portfolio. Patents for biological inventions were not allowed till 1980 but doing so became necessary only with the advent of biotechnology. Consequently, patents are now allowed for biological inventions involving microorganisms, vectors, DNA/RNA, proteins, monoclonal antibodies and hybridoma, isolated antigens, vaccine compositions, and transgenic animals and plants. Patents have also been allowed for methods involving isolation or purification of biological material; gene cloning and production of proteins; diagnosis, treatment, and use of a product; and screening methods.

The development of the biotechnology industry partly depends on its licensing revenue, which stems from a good patent portfolio. Similarly, the success of a pharmaceutical company partly depends upon its revenue that is generated through marketing of compounds that are licensed by a biotechnology company. The revenue generated by pharmaceutical companies due to licensed compounds is estimated to have increased from 24% in 1992 to 35% in 2002. Similarly, the licensing revenue in the biotechnology industry rose from $5.7 billion in 1998 to more than $6.4 billion in 2000, which was expected to rise to the $7.8 billion mark in the year 2003. Biogen, for example, made about $600 million from its licensing activity during 1991–1995, which was a prerequisite for licensing its first product, *Avonex*, in the year 1996. In 2000, the licensing revenue of Biogen comprised approximately 18% of its total revenue. This example suggests that the biotechnology industry depends heavily on long-term deals with their pharmaceutical partners, which not only determines the total revenue, but also the company's share price in the stock market. For instance, the very news that Curagen (United States) has entered a deal for a drug development alliance with Bayer (Germany) sent Curagen's share price up by 35% to approximately $36. This illustrates that survival of biotechnology companies sometimes depends on successful deals with pharmaceutical and other companies, as does the success of the deals. The success of the deals also depends on finding the right partner and the post-deal governance.

11.10.2 INTELLECTUAL PROPERTY PROTECTION

There are currently two main systems of protection for biotechnology: rights in plant varieties and patents. Both systems provide exclusive, time-limited rights of commercial exploitation. Keeping biotechnology a "secret" is a valuable form of protection. National treatment of trade secrets is diverse, and all attempts to harmonize trade secret laws in Europe, for example, have failed. Most jurisdictions do provide some form of protection against those who steal or use others' trade secrets unfairly. However, the problem with this form of protection is that the secret generally becomes public once the biotechnology is used commercially and thus the protection is lost. It is conceivable that copyright laws could afford some protection for biotechnology. Lines of genetic code are analogous to some extent with computer program code, which has now been incorporated into the copyright systems of most industrialized countries. However, this route to protection is fraught with practical and conceptual difficulties and is generally thought to be unsuitable. There is as yet no recorded case of biotechnologists claiming copyright for their inventions. Trademarks are also unlikely to be of much use in protecting biotechnology, though they may of course prove important later when marketing products, processes, or services. An attempt to register the name of a plant or an animal as a trade mark is unlikely to be successful, as public policy would prevent it. For example, in England, registrations for names of varieties of roses have been removed from the Trademark Register for lack of distinctiveness and because of the likelihood of confusion.

11.10.3 PATENTS AND BIOTECHNOLOGY PRODUCTS

A *patent* is a grant of exclusive rights for a limited time with respect to a new and useful invention. The exact requirements for the grant of a patent, the scope of protection it provides, and its duration differ depending on national legislation. However, generally the invention must be patentable subject matter, novel (new), nonobvious (inventive), of industrial application, and sufficiently disclosed. A patent will provide a wide range of legal rights, including the right to possess, use, transfer by sale or gift, and to exclude others from similar rights. The duration will be for approximately 20 years (although for only 17 years in the United States). These rights are generally restricted to the territorial jurisdiction of the country granting the patent, and thus an inventor wishing to protect his or her invention in several countries will need to seek separate patents in each of those countries. Whilst most countries provide some form of patent protection, only a few provide patent protection for biotechnology. These include Australia, Bulgaria, Canada, Czechoslovakia, Hungary, Japan, Romania, the Soviet Union, and the parties to the European Patent Convention (EPC). The reasons for this may differ, but generally it has been because biotechnology has been thought inappropriate for patent protection, either because the system was originally designed for mechanical inventions, for technical or practical reasons, or for one or more ethical, religious, or social concerns. In all the national patent offices where patents are granted for biotechnology, there is a considerable backlog of pending applications. Even in those countries where patent protection is provided, the type and extent of that protection are different in nearly every national system.

It has largely been the United States that has broken new ground in providing the possibility of patent protection. Patents have been granted for plants under the Plant Patent Act since 1930. However, prior to 1980, the US Patent Office would not grant utility patents (separate from the Plant Patent Act) for living matter, because it deemed products of nature not to be within the terms of the utility patent statute. That was until the landmark decision of the US Supreme Court in *Diamond v. Chakrabarty* which held that a genetically engineered bacterium was statutory subject matter for a utility patent. This decision has been the basis upon which patents have been granted for higher life forms. Subsequently, it has been held that a utility patent may be granted for plants.

11.10.4 INTERNATIONAL TREATIES

There are three international IP treaties which are of importance for the protection of biotechnology: The Paris Convention for the Protection of Industrial Property (the Paris Convention), the Budapest Treaty on the International Recognition of the Deposit of Microorganisms for the Purposes of Patent Procedure (the Deposit Treaty), and the Patent Cooperation Treaty (PCT). The PCT simplifies the process of simultaneously filing patent applications in several countries. Under the PCT, a single application may be filed in one of the officials receiving offices (ROs), designating any number of PCT member countries, which can eventually result in a national patent (and/or a Europatent) being granted in each of the designated states. Unfortunately, the eventual outcome is not a "world patent," and there is no harmonization patent law under the PCT apart from the procedural aspects. The PCT is discussed in greater detail next.

11.10.4.1 Patent Cooperation Treaty

The PCT is an international patent law treaty concluded in 1970. It provides a unified procedure for filing patent applications to protect inventions in each of its contracting states. A patent application filed under the PCT is called an *international patent application*, or *PCT application*. A single filing of an international application is made with a RO in one language. It then results in a search performed by the relevant International Searching Authority (ISA), accompanied by a written opinion regarding the patentability of the invention that is the subject of the application. It is optionally followed by a preliminary examination, performed by the relevant International Preliminary Examining Authority (IPEA). Finally, the relevant national or regional authorities administer matters related to the examination of applications (if provided by national law) and issuance of the patent. The PCT does not provide for the grant of an "international patent," as such a multinational patent does not exist, and the grant of patent is a prerogative of each national or regional authority.

11.10.4.2   World Intellectual Property Organization

The World Intellectual Property Organization (WIPO) is one of the 16 specialized agencies of the UN. WIPO was created in 1967 "to encourage creative activity, to promote the protection of IP throughout the world." WIPO currently has 184-member states and is headquartered in Geneva, Switzerland. The current Director-General of WIPO is Francis Gurry, who took office on October 1, 2008. Almost all UN Members as well as the Holy See are Members of WIPO (nonmembers are the states of Kiribati, Marshall Islands, Micronesia, Nauru, Palau, Solomon Islands, Timor-Leste, Tuvalu and Vanuatu, as well as the entities of the Palestinian Authority, the Sahrawi Republic, and Taiwan).

11.10.4.3   Agreement on Trade-Related Aspects of Intellectual Property Rights

The Agreement on Trade-Related Aspects of Intellectual Property Rights (TRIPS) is an international agreement administered by the World Trade Organization (WTO) that sets down minimum standards for many forms of IP regulation as applied to nationals of other WTO members. It was negotiated at the end of the Uruguay Round of the General Agreement on Tariffs and Trade (GATT) in 1994. Specifically, TRIPS contains requirements that nations' laws must meet for copyright rights, including the rights of performers, producers of sound recordings, and broadcasting organizations; geographical indications, including appellations of origin; industrial designs; integrated circuit layout-designs; patents; monopolies for the developers of new plant varieties; trademarks; trade dress; and undisclosed or confidential information. TRIPS also specifies enforcement procedures, remedies, and dispute resolution procedures. Protection and enforcement of all IP rights must meet the objectives to contribute to the promotion of technological innovation and to the transfer and dissemination of technology, to the mutual advantage of producers and users of technological knowledge, and in a manner conducive to social and economic welfare and to a balance of rights and obligations.

11.10.4.4   Issues with Biotechnology Patents

TRIPS covers everything from pharmaceuticals to information technology, software and human gene sequences, and is emerging as a major issue dividing the North and the South. The TRIPS agreement is controversial in at least two areas. First, it threatens the right of poor countries to manufacture or to import cheap generic versions of patented drugs. The AIDS epidemic and other diseases kill millions every year because people in poor countries cannot afford the exorbitant prices the pharmaceutical giants are charging for the patented drugs. The existing TRIPS agreement also forces all countries to accept a medley of new biotechnology patents covering genes, cell lines, organisms, and living processes that turn living matter into commodities. Governments all over the world have been persuaded into accepting these "patents on life" before anyone understood the scientific and ethical implications. The patenting of life forms and living processes is covered under Article 27.3(b) of the TRIPS agreement.

11.10.4.5   Patent Infringements

Farmers may face legal liability issues in agricultural biotechnology and in legal disputes involving IP. More particularly, companies that create transgenic crops have IP rights, usually patents, on those crops and take legal action against farmers who grow the transgenic crops without the companies' permission. However, one should not forget that seed companies also protect their patents on non-transgenic seeds and plants. Patent infringement cases in agriculture are not unique to transgenic seeds and plants. Four cases for patent infringement regarding transgenic crops have resulted in written opinions by courts in Canada and the United States: *Monsanto Canada, Inc. v. Schmeiser* (2001); *Monsanto Company v. Trantham* (2001); *Monsanto Company v. McFarling* (2002); and *Monsanto Company v. Swann* (2003).

The Schmeiser Case: The Canadian courts found the following to be factually true. Mr. Schmeiser sprayed three or four acres of his canola in 1997 with Roundup (Monsanto Corp., St. Louis) herbicide because he thought his canola field contained Roundup Ready canola. Sixty percent of the sprayed area survived the herbicide, thereby showing herbicide tolerance. Mr. Schmeiser separately harvested and stored the canola seed from the sprayed acres. In 1998, Mr. Schmeiser decided to use the canola from the sprayed acres as his seed canola for the 1998 crop. When Monsanto Canada pursued Mr. Schmeiser for these actions in a lawsuit, the grow-out and DNA tests of Schmeiser's 1998 crop showed 95%–98% Roundup Ready (Monsanto Corp., St. Louis) canola from tests conducted by Monsanto, 95%–98% Roundup Ready canola on the 1998 crop from a Canadian laboratory, 63%–70% Roundup Ready canola from grow-out tests by Mr. Schmeiser himself on 1997 and 1998 crops, and 0%–98% on various samples from the 1997 and 1998 crops submitted by Mr. Schmeiser to the University of Manitoba, with the 1997 saved seed specifically testing in the 95%–98% range.

The Canadian courts held that Mr. Schmeiser infringed the Monsanto patents after consideration of two defenses. First, the courts ruled that the fact that Mr. Schmeiser did not use Roundup herbicide was irrelevant to the patent violation. The court stated that the Monsanto patent claims related to the gene and cells of the canola plants, and this particular patent had nothing to do with the herbicide. Considering the patent relevant to this lawsuit, Mr. Schmeiser could only grow these patented seeds if he had permission to do so. Mr. Schmeiser admitted that he had not signed a technology use agreement with Monsanto. Second, the court ruled that regardless of the origin of the 1997 seeds—for example pollen drift, seed spills from bags, or wind-blown from truck beds—Mr. Schmeiser infringed the patent when he knew or should have known that he was planting Roundup Ready canola in 1998. The court rejected Mr. Schmeiser's claim as an innocent grower, because his actions demonstrated that he was not innocent—that is, he knew or should have known that he was growing patented seeds. The court left undecided whether Monsanto would have an infringement claim against a truly innocent grower.

11.11   BIOTECHNOLOGY STOCK INVESTMENT: PROS AND CONS

Investing in biotechnology stocks is somewhat unlike investing in other stocks, because in valuing biotechnology stocks, it has always been difficult to use traditional net present value and discounted cash flow approaches, particularly for the clinical and preclinical-stage companies. Predicting the probability of a single product's success in the clinic depends on many variables such as clinical trial design, difficulty of indication, and quality of Phase II data. In addition, the company's financial well-being and corporate partnerships may further complicate the valuation analysis. The large cap and profitable biotechnology companies have had the broadest appeal to investors, but that is only a handful of companies. Investors in biotechnology stocks take a long-term approach to investing. There are stocks that can significantly appreciate overnight if a trial is successful; conversely, they can also drop by 30%–70% in value with disappointing results. Biotechnology companies' stocks tend to be heavily influenced by favorable or unfavorable news regarding the development or testing of a product. In this section, we discuss some of the critical factors that need to be considered before investing in biotechnology companies.

11.11.1 Products in the Pipeline

Look for companies with at least two drugs in clinical trials, because if for some reason the product proves to lack efficacy, then at least the company has something to fall back on. Another approach is to look for companies diversified around a specific disease class or those that have a niche technology that can be used as a platform for a range of different drugs.

11.11.2 Collaboration and Merger

Companies that fail to link up with a corporate or academic partner can have trouble surviving. To ensure survival or lower risk, biotechnology companies will engineer several collaborative agreements with various pharmaceutical companies for research or marketing. Look for substantial milestone payments and cash commitments when the deal is announced, not just "talk" about a research alliance. For example, Abgenix has made several deals with biotechnology and pharmaceuticals for licensing its XenoMouse technology for making humanized monoclonal antibodies that are worth hundreds of millions of dollars.

11.11.3 Experienced Management

Early-phase companies may or may not have someone in senior management with a proven track record of taking a drug through the regulatory hurdles and/or to the marketplace; even so, look at their financials and go to the section on management to see who is working there and what they have accomplished in the past.

11.11.4 Cash Flow

For many biotechnology companies, the release of a commercial product is often many years away and requires millions of dollars. Thus, a company's burning of cash in ongoing R&D, or "burn rate," is a critical measure of a company's longevity. Look for companies that have a minimum of 2 years cash reserves.

11.12 ROLE OF REGULATORS IN BIOTECHNOLOGY PRODUCT DEVELOPMENT

11.12.1 World Health Organization

The World Health Organization (WHO) is a specialized agency of the UN that acts as a coordinating authority on international public health. Established on April 7, 1948, and headquartered in Geneva, Switzerland, the agency inherited the mandate and resources of its predecessor, the Health Organization, which had been an agency of the League of Nations. As well as coordinating international efforts to monitor outbreaks of infectious diseases such as SARS, malaria, and AIDS, the WHO also sponsors programs to prevent and treat such diseases. The WHO supports the development and distribution of safe and effective vaccines, pharmaceutical diagnostics, and drugs. After over two decades of fighting smallpox, the WHO declared in 1980 that the disease had been eradicated—the first disease in history to be eliminated by human effort. The WHO aims to eradicate polio within the next few years. The organization has already endorsed the world's first official HIV/AIDS Toolkit for Zimbabwe (from October 3, 2006), making it an international standard. In addition to its work in eradicating disease, the WHO also carries out various health-related campaigns—for example, to boost the consumption of fruits and vegetables worldwide and to discourage tobacco use. Experts met at the WHO headquarters in Geneva in February 2007 and reported that their work on pandemic influenza vaccine development had achieved encouraging progress. More than 40 clinical trials have been completed or are ongoing. Most have focused on healthy adults. Some companies, after completing safety analyses in adults, have initiated clinical trials in the elderly and in children. So far, all vaccines appear to be safe and well-tolerated in all the age groups tested.

11.12.2 INTERNATIONAL CONFERENCE FOR HARMONIZATION

The birth of the ICH took place in April 1990 at a meeting hosted by the European Federation of Pharmaceutical Industries and Associations (EFPIA) in Brussels. Representatives of the regulatory agencies and industry associations of Europe, Japan, and the United States met primarily to plan an international conference, but the meeting also discussed the ICH's wider implications and terms of reference. The ICH Steering Committee, which was established at that meeting, has since met at least twice a year, with the location rotating between the three regions. The ICH of Technical Requirements for Registration of Pharmaceuticals for Human Use is a unique project that brings together the regulatory authorities of Europe, Japan, and the United States and experts from the pharmaceutical industry in the three regions to discuss scientific and technical aspects of product registration. The purpose is to make recommendations on ways to achieve greater harmonization in the interpretation and application of technical guidelines and requirements for product registration in order to reduce or obviate the need to duplicate the testing carried out during the R&D of new medicines. The objective of such harmonization is a more economical use of human, animal, and material resources, and the elimination of unnecessary delay in the global development and availability of new medicines, while maintaining safeguards on quality, safety, and efficacy, and regulatory obligations to protect public health.

11.12.3 UNITED STATES FOOD AND DRUG ADMINISTRATION

The United States Food and Drug Administration (US FDA) is an agency within the US Public Health Service, which is a part of the Department of Health and Human Services. The FDA regulates over $1 trillion worth of products, which account for 25 cents of every dollar spent annually by American consumers and touches the lives of virtually every American every day. It is the FDA's job to see that food is safe and wholesome, cosmetics will not hurt people, medicines and medical devices are safe and effective, and that radiation-emitting products such as microwave ovens will not cause harm. Feed and drugs for pets and farm animals also come under FDA scrutiny. The FDA ensures that all of these products are labeled truthfully with the information that people need to use them properly. The FDA requires that both prescription and over-the-counter drugs be proven safe and effective. In deciding whether to approve new drugs, the FDA does not itself conduct research, but rather examines the results of studies done by the manufacturer. The FDA must determine that the new drug produces the benefits it is supposed to without causing side effects that would outweigh those benefits.

The FDA tests food samples to see if any substances, such as pesticide residues, are present in unacceptable amounts. If contaminants are identified, the FDA takes corrective action. The FDA also sets labeling standards to help consumers know what is in the foods they buy. The FDA also ensures that medicated feeds and other drugs given to pets and animals raised for food are not threatening to health. FDA investigators examine blood bank operations, from record keeping to testing for contaminants. The FDA also ensures the purity and effectiveness of biological products (medical preparations made from living organisms and their products) such as insulin and vaccines. These are classified and regulated by the FDA according to their degree of risk to the public. Devices that are life-supporting, life-sustaining, or implanted (such as pacemakers) must receive agency approval before they can be marketed. The FDA can have unsafe cosmetics removed from the market. Dyes and other additives used in drugs, foods, and cosmetics are also subject to FDA scrutiny. The agency must review and approve these chemicals before they can be used.

FDA investigators and inspectors collect domestic and imported product samples for scientific examination and for label checks. If a company is found violating a law that the FDA enforces, the FDA can encourage the firm to voluntarily correct the problem or to recall a faulty product from the market. When a company cannot (or will not) correct a public health problem with one of its products voluntarily, the FDA has legal sanctions it can bring to bear. The FDA can go to court to force a company to stop selling a product and to have items already produced seized and destroyed. When warranted, criminal penalties—including prison sentences—are sought against manufacturers and distributors. About 3000 products a year are found to be unfit for consumers and are withdrawn from the marketplace, either by voluntary recall or by court-ordered seizure. In addition, approximately 30,000 import shipments a year are detained at the port of entry because the goods appear to be unacceptable. Evidence to back up FDA legal cases is prepared by FDA laboratory scientists. Some analyze samples to see, for instance, if products are contaminated with illegal substances. Other scientists review test results submitted by companies seeking agency approval for drugs, vaccines, food additives, coloring agents, and medical devices.

11.12.4   GOOD LABORATORY PRACTICE

Good Laboratory Practice (GLP) generally refers to a quality system of management controls for laboratories and research organizations that regulate how nonclinical safety studies are planned, performed, monitored, recorded, reported, and archived. It ensures the consistency and reliability of results for submissions to the US FDA, the OECD, and other national organizations. GLP practices are intended to promote the quality and validity of test data. GLP is a regulation that goes beyond good analytical practice. Good analytical practice is important, but it is not enough. For example, the laboratory must have a specific organizational structure and procedures to perform and document lab work. The objective is not only quality of data, but also traceability and integrity of data. However, the biggest difference between GLP and non-GLP work is the type and amount of documentation. A GLP inspector normally examines the documentation process to find out who has done a study, how the experiment was carried out, which procedures have been used, whether there has been any problem, and if so, how it has been solved.

11.12.5   GOOD MANUFACTURING PRACTICE

GMP regulations promulgated by the US FDA under the authority of the Federal Food, Drug, and Cosmetic Act has the force of law and require that manufacturers, processors, and packagers of drugs, medical devices, food, and blood take proactive steps to ensure that their products are safe, pure, and effective. GMP regulations require a quality approach to manufacturing, enabling companies to minimize or eliminate instances of contamination, mix-ups, and errors. This in turn protects the consumer from purchasing a product that is not effective or may even be dangerous. Failure of firms to comply with GMP regulations can result in very serious consequences, including recall, seizure, fines, and jail time. GMP regulations address issues including record keeping, personnel qualifications, sanitation, cleanliness, equipment verification, process validation, and complaint handling. Most GMP requirements are very general and open-ended, allowing each manufacturer to decide individually how best to implement the necessary controls. This provides much flexibility, but also requires that the manufacturer interpret the requirements in a manner that makes sense for each individual business. GMP is also sometimes referred to as Current Good Manufacturing Practice (cGMP). The "c" stands for "current," reminding manufacturers that they must employ technologies and systems that are up-to-date to comply with the regulation. In the United States, the phrase "current GMP" appears in 501(B) of the 1938 Food, Drug, and Cosmetic Act (21 USC 351). US courts may theoretically hold that a drug product is adulterated even if there is no specific regulatory requirement that was violated, if the process was not performed according to industry standards.

PROBLEMS

Section A: Descriptive Type

Q1. What are the different methods of scientific enquiry?
Q2. Describe a scientific invention and provide an example.
Q3. How do companies commercialize scientific discovery? Explain.
Q4. Explain the significance of a bioequivalence lab in biotechnology product testing.
Q5. Why are quality assurance and quality control essential for biotechnology companies?
Q6. What is the PCT?
Q7. What are some of the challenges in filing patents in biotechnology?
Q8. Explain how a biotechnology company can be started.

Section B: Multiple Choice

Q1. Ibn al-Haytham (Alhazen) is known for the . . .
  a. Development of the scientific method
  b. Development in the field of medicine
  c. Development in the field of chemistry
Q2. GLP deals with . . .
  a. Clinical testing
  b. Research laboratories
  c. Manufacturing plants
  d. None of the above
Q3. A biosafety hood or laminar hood is used to . . .
  a. Process tissues
  b. Avoid contamination
  c. Prepare reagents
Q4. To avoid contamination, nonclinical labs are separated from clinical labs by a proper barrier. True/False
Q5. Bioequivalence tests deal with . . .
  a. Animal testing
  b. *In vitro* testing
  c. Human testing
  d. None of the above
Q6. Preclinical testing is done after clinical testing. True/False
Q7. Phase IV trials are also known as post-marketing surveillance trials. True/False
Q8. Which of the following is the first biotechnology company in the United States to make an IPO?
  a. Genzyme
  b. Biogen
  c. Genentech
  d. Immunogen
Q9. The PCT is a regional patent law treaty. True/False
Q10. WIPO is affiliated with . . .
  a. WHO
  b. UNESCO
  c. UN
  d. US FDA
Q11. Both the US FDA and ICH are drug regulatory bodies. True/False
Q12. This is a systematic preventive approach to food safety and pharmaceutical safety.

    a. US FDA
    b. ICH
    c. HACCP

Section C: Critical Thinking

Q1. Why are market research and analysis important in biotechnology product development?
Q2. Why can biotechnology products not be sold at "discount rates" like other products?

FIELD TRIP

Organize a field trip to any pharmaceutical or biotechnology manufacturing bases in your city. Learn how its biotechnology product is manufactured and submit your observations in the form of a field report to your class instructor.

REFERENCES AND FURTHER READING

Angell, M. *The Truth About Drug Companies*. Random House Trade Paperbacks, New York, 2004, p. 30.
Angell, M. Drug companies & doctors: A story of corruption. *New York Rev. Books* 56(1): 8–12, 2009.
Arcangelo, V.P., and Peterson, M.A. *Pharmacotherapeutics for Advanced Practice: A Practical Approach*. Lippincott Williams & Wilkins, Philadelphia, PA, 2005. ISBN: 0781757843.
Brater, C.D., and Daly, W.J. Clinical pharmacology in the middle ages: Principles that presage the 21st century. *Clin. Pharmacol. Ther.* 67: 447–450, 2000.
Cancellation of FED-STD-209E—Institute of Environmental Sciences and Technology, November 2001. www.iest.org/i4a/pages/index.cfm?pageid=3480 (retrieved on March 12, 2007).
Chronicle of the World Health Organization. World Health Organization, 1948, p. 54. http://whqlibdoc.who.int/hist/chronicles/chronicle_1948.pdf (retrieved on July 18, 2007).
Cleanroom Classification/Particle Count/FS209E/ISO TC209/. http://cleanroom.byu.edu/particlecount.phtml.
David, W.T. Arab roots of European medicine. *Heart Views* 4: 2, 2003.
Demain, A.L., and Davies, J.E. *Manual of Industrial Microbiology and Biotechnology*. ASM Press, Washington, DC, 1999.
FAO/WHO. FAO/WHO guidance to governments on the application of HACCP in small and/or less-developed food businesses (PDF).
Food Safety Research Information Office. A focus on hazard analysis and critical control points. Created June 2003, Updated March 2008.
Glossary of Clinical Trial Terms. NIH Clinicaltrials.gov/ct/info/glossary (retrieved on March 18, 2008).
Guidance for Industry 2006. Investigators, and reviewers exploratory IND studies. Food and Drug Administration. www.fda.gov/cder/guidance/7086fnl.htm (retrieved on January 5, 2007).
ICH guideline for good clinical practice: Consolidated guidance. www.vichsec.org/en/GL48-st4.doc.
International Conference on Harmonization. www.ich.org/cache/compo/276-254-1.html.
International Conference on Harmonization of Technical Requirements for Registration of Pharmaceuticals for Human Use. www.ich.org.
International HACCP Alliance. International HACCP Alliance (PDF). http://haccpalliance.org/alliance/HACCPall.pdf (retrieved on October 12, 2007).
Iriye, A. *Global Community: The Role of International Organizations in the Making of the Contemporary World*. University of California Press, Berkeley, CA, 2002.
ISO 14644-1 Scope 2007. www.iest.org/i4a/pages/index.cfm?pageid=3322#ISO_14644-1.
Patents: Frequently asked questions. *World Intellectual Property Organization*. www.wipo.int (retrieved on February 22, 2009).
Sox, H.C., and Rennie, D. Seeding trials: Just say "no". *Ann. Intern. Med.* 149: 279–280, 2008.
Toby, E.H. *The Rise of Early Modern Science: Islam, China, and the West*. Cambridge University Press, Cambridge and New York, p. 218, 2003.
What is informed consent? US National Institutes of Health. www.Clinicaltrials.gov;en.wikipedia.org/wiki/Informed_consent (retrieved on May 18, 2011).
World Health Organization. Workers' health: Global plan of action. *Sixtieth World Health Assembly*, May 23, 2007 (retrieved on September 15, 2008).

12 Trends in Biotechnology

CHAPTER'S OBJECTIVES
- Discuss trends in biotechnology
- Major trends in biotechnology
  - Forensic Science
  - Regenerative Medicine
  - Biosimilars
  - Synthetic Biology
  - Bioinformatics

12.1 INTRODUCTION

Biotechnology is a highly interdisciplinary field where integration of various fields and science merged to bring new technology and products which are beneficial to all of us. In this chapter we will discuss some major emerging topics in biotechnology that have enormously affected our lives. These emerging fields are directly or indirectly associated with biotechnology. In this chapter, we will briefly describe important trends in the biotechnology field.

12.2 FORENSIC SCIENCE

Forensic science is the application of science to mainly investigate the criminal or civil law cases in which evidence is being used for criminal procedures in the court. During ancient time, criminal investigations and court trials heavily relied on forced confessions and witness testimony. The success of trial was purely based on the availability of witnesses, if witnesses turned blind or did not show up to the courtroom, criminals could easily get released with no punishment due to lack of evidence. In the twentieth century, due to the technological revolution, various tools, machines, and chemicals were discovered that can assist the investigator in framing charges against any crime preparator and this evidence-based method has become the main tools for the investigating agencies around the world. The first written account of using medicine to solve criminal cases is attributed to the book written by Song Ci in the year 1248. The author was a director of justice, jail, and supervision during the Song Dynasty. At that time, the book provided knowledge to distinguish between genuine suicide or pretend suicide cases. Here is a brief story about solving the case in which a man was murdered with a steel rod. To solve the murder mystery, the investigator instructed the suspect to bring his rod to one location. During trial, the investigator used flies to identify the rod that had been used to kill the person—now here is science as flies were attracted to the smell of human blood. As soon as the suspect showed his rod, the flies attracted by the smell of blood eventually gathered on the rod. Considering this test, the murderer finally confessed the crime. What we learn here is that crime can be solved by proving facts with the evidence. This marked the beginning of evidence-based investigations. Thereafter various methods have been used to solve criminal cases in which investigators have used saliva and examination of the mouth and tongue to determine innocence or guilt. Later, this evidence-based field became advanced with the advent of new tools and new technology. Evidence-based science is now referred to as forensic science.

12.2.1 Classifications of Forensic Science

The field of forensic science keeps growing with the emergence of new technology and methodology; with a view to better understand forensic science, we have classified forensic science into various sub-fields and each field has its own significance and application.

12.2.1.1 Forensic Anthropology

The application of science to identify human origins or to identity by using the human skeleton in criminal cases is known as forensic anthropology. To identify a victim who died due to major accident or fire where the victim's body is completely mutilated and beyond recognition, in that situation the investigator uses the victim's parts of the body (bone, hair, or tissue) to solve the mystery. The investigator who examined the human body or tissue to solve the mystery is called a forensic anthropologist.

12.2.1.2 Forensic Accounting

Forensic accounting is the specialty of accountancy that defines engagements that result from actual or anticipated disputes or litigation related to financial theft and forgery. In banking and financial organizations, currency theft or forgery are the main problems and to tackle these problems, the organizations need the help of investigators to solve the theft or forgery problem. These investigators are called forensic accountants or forensic auditors.

12.2.1.3 Forensic Archaeology

Forensic archaeology is the application of archaeological philosophies, procedures, and methodologies to solve a crime or mystery by excavating the earth. When war, volcanic eruption, or natural disasters, which could have happened hundreds of years earlier, cause the death of many people, to find out how many people died and their identity the investigators dig into the earth and locate human or treasure remains. The science of locating and identifying the subject (human or treasure) from such a site is called forensic archaeologist.

12.2.1.4 Forensic Fingerprinting

Fingerprinting is a human finger impression and the science of using human finger friction ridges for solving crimes is called forensic fingerprinting. It is known that each human being has a distinctive and unique impression of finger ridges that are different in everyone. To solve the crime, investigators normally use impressions of human finger's left on a glass, door handle, or cigarette bud at the crime site. These unique ridges are used to identify an individual's identity.

12.2.1.5 Forensic Dentistry

The evaluation of a victim's tooth as evidence to solve the crime is called forensic dentistry. In some cases, victim's teeth are used as evidence-based proof to solve the crime in the court of law. The method of forensic dentistry usually involves dental radiographs, post-mortem photographs, and analysis of DNA obtained from the victim's dental tissues.

12.2.1.6 Forensic Pathology

The branch of pathology that determines the cause of death by examination of the victim's body tissues is called forensic pathology. The post-mortem autopsy is performed by the forensic pathologist at the request of a medical doctor or examiner. The forensic pathologist processes the tissues using biochemical and microscopic techniques to determine the cause of death. The biochemical and cytological data obtained from the victim's body are compared with those of normal tissues to conclude what was the cause of the death. Additionally, forensic pathologists write a report to approve the identity of a victim.

12.2.1.7 Forensic Serology

To study human fluid to solve a crime or mystery is called forensic serology. In forensic serology, the victim's bodily fluids (such as fecal matter, blood, and semen) and the suspect's bodily fluids (such as fecal matter, blood, and semen) are analyzed and compared to find out any relationship. For example, the presence of a suspect's semen inside a victim's body can confirm the crime and is considered as evidence-based proof to prosecute the suspect. The investigators who examine such fluids are called forensic serologist. Additionally, a forensic serologist is involved in DNA analysis.

12.2.1.8 Forensic Toxicology

Forensic toxicology is the use of toxicology to aid medical or legal investigation of a death that is caused by chemical usage, poisoning, or drug use. In some cases, where death occurred due to chemical leakage, food contamination, or drug abuse, such types of cases are referred to the forensic toxicology department for analysis. The investigators who examine such samples are known as forensic toxicologist. The primary role of forensic toxicologist is to analyze the samples and submit a report of any foul play.

12.2.1.9 Forensic Facial Reconstruction

Forensic facial reconstruction is the method of reconstructing the face of an individual from their skeletal remains by using artistic images. The artist uses anatomy and the structure of the skeleton to reconstruct the face. This reconstructed face normally is used by the police department to identify suspects or killers. Besides it benefits, this technique is one of the most controversial techniques in the field of forensic anthropology. Despite this controversy, facial reconstruction has proved successful to identify culprits and suspects.

12.2.1.10 Forensic Firearm Examination

The forensic process of examining the characteristics of firearms, cartridges, or bullets left behind at a crime scene is called forensic firearm examination. The forensic specialists usually analyze the bullets and cartridges of weapons used by criminals. To determine who is the owner of the weapon, obliterated serial numbers can be identified and recorded from the record. Moreover, examiners can also look for fingerprints on the weapon and cartridges, and then feasible prints can be processed through fingerprint databases for a probable match.

12.2.1.11 Digital Forensics

It is a branch of forensic science about the recovery of material found in digital devices. Digital forensics was originally used for computer forensics but now has expanded to cover investigation of all devices capable of storing digital data. As there is a growing number of softwares that are being used by most of the banking and financial organizations, it became very critical for such organizations to have digital forensics experts to protect and safeguard the enormous amount of digital data.

12.2.1.12 DNA Fingerprinting

DNA profiling is done to identify individuals on the basis of their respective DNA and genetic profiles. DNA profiles are encoded sets of numbers that mirror a person's DNA makeup. DNA profiling is sequencing but that should not be confused with a full genome sequencing. Although 99.9% of human DNA sequences are the same in every person, there are sequence that are different in each individual. The unique sequencing can be measured by *variable number tandem repeats* (VNTR) in the DNA. It has been found that in closely related individuals VNTRs loci are very similar, whereas VNTRs are variable in unrelated individuals. The DNA profiling technique was first reported in 1985 by Sir Alec Jeffreys, University of Leicester, England. *DNA fingerprinting* is a technique to determine the possibility that genetic material came from a particular individual or group. In the case of plants such as grapes, researchers compared the similarities between different species of grapes and were able to piece together parent subspecies that could have contributed to the present prize-winning varieties.

There are many applications of DNA fingerprinting, and these fall into three main categories. By using this technique, we can find out certain issues such as (1) to know where we came from; (2) to discover what we are doing at present; and (3) to predict where we are going. Regarding where we came from, DNA fingerprinting is commonly used to examine the heredity of people, as people inherit the arrangement of their base pairs from their parents. Comparing the banding patterns of a child and a parent can be done to generate a probability of relatedness. If the two patterns are similar then they may be family. However, DNA fingerprinting cannot distinguish between identical twins, as their banding designs are the same.

DNA fingerprinting can also tell us about present-day conditions. Conceivably the best known use of DNA fingerprinting is in forensic medicine. DNA samples gathered at a crime scene can be associated with the DNA of a suspect to prove whether or not he or she was present at the scene. Databases of DNA fingerprints are only available from known offenders, so it is not yet possible to fingerprint the DNA from a crime scene and then pull out names of probable matches from the general public. However, this may happen in the future if DNA fingerprints replace more traditional and forgeable forms of identification. In a real case, trading standards agents found that 25% of caviar is bulked up with roe from different categories, the high-class equivalent of cheating the consumer by not filling the metaphorical pint glass all the way up to the top. DNA fingerprinting confirmed that the "suspect" (inferior) caviar was present at the crime scene.

Finally, genetic fingerprinting can help us forecast our future health. DNA fingerprinting is often used to track down the genetic basis of inherited diseases. If a particular pattern turns up time and time again in different patients, scientists can narrow down which gene(s) or at least which part of the DNA might be involved. Since knowing the genes involved in disease susceptibility gives clues about the underlying physiology of the disorder, genetic fingerprinting aids in developing therapies.

Prenatally, genetic fingerprinting can also be used to screen parents and fetuses for the presence of inherited abnormalities, such as Huntington's disease or muscular dystrophy. The process begins with a sample of an individual's DNA. The most desirable method of collecting a reference sample is the use of a buccal swab as this reduces the possibility of contamination. When this is not available, other procedures may need to be used to collect a sample of blood, saliva, semen, or other appropriate fluid or tissue from a toothbrush or razor or from stored samples. Samples obtained from biological relatives can provide a sign of an individual's profile, as could human remains that have been previously profiled.

12.3   REGENERATIVE MEDICINE

Regenerative medicine is the science of making artificial organs, tissues, and cells under controlled conditions. It has been suggested that regenerative medicine helps the natural process of healing by using special materials that can be regrown on the missing or damaged tissues. Recently, doctors or physicians use regenerative medicine to speed up healing of the damaged tissues and cells that cannot heal on their own. Regenerative therapies have been tested both in laboratory experiments and clinical conditions in which broken bones, bad burns, blindness, deafness, heart damage, nerve damage, Parkinson's disease, and other conditions can be restored by regenerative medicine. Lately, stem cell research has created quite a lot of interest among patients suffering from Parkinson's disease, because stem-cell-based therapy may provide treatment for these patients. Although it may be too early to comment on the effectiveness of such a novel therapy, trials on animals have proved to be quite interesting. The field of cell and tissue engineering has emerged as a multidisciplinary field by combining different fields such as biology, medicine, and engineering. In addition to therapeutic applications, tissue engineering can also be used in diagnostic assays where the cells and tissues can be used for testing new drug molecules or candidates.

12.3.1 Stem Cells

Stem cells are prominent cells that have two important characteristics. First, they are highly unspecialized cells capable of renewing themselves through cell division and second, under certain physiologic or experimental conditions, these cells can be induced to become tissue- or organ-specific cells with special functions. In a few organs, like the gut and bone marrow, these stem cells regularly grow and divide to repair and replace damaged tissues. In other human organs, such as the pancreas and the heart, stem cells only divide under special conditions.

12.3.1.1 Adult Stem Cells

Adult stem cells (ASCs) are undifferentiated cells that are found throughout the body. These adult stem cells multiply by cell division to replenish dying or damaged cells or tissues. These cells are somatic stem cells, because they can be found in young as well as adult stages. One of the main characteristics of ACSs is their ability to divide or self-renew for many passages and upon differentiation, they can generate all the cell types of the organ from which they originate. Unlike embryonic stem cells, ACSs are considered safe for clinical uses without any ethical concern. There are several types of adult stem cells.

12.3.1.1.1 Dental Pulp-Derived Stem Cells

These stem cells can be successfully recovered in the perivascular niche of dental pulp and these cells have the same cellular markers and differential abilities as *mesenchymal stem cells* (MSCs). The deciduous baby teeth can be shed naturally, and stem cells can be harvested through noninvasive and painless methods.

12.3.1.1.2 Hematopoietic Stem Cells

Hematopoietic stem cells (HSCs) are multipotent stem cells that form all types of blood cells, including neutrophils, basophils, eosinophils, monocytes and macrophages, erythrocytes, megakaryocytes, platelets, dendritic cells, T-cells, B-cells, and NK-cells. The hematopoietic tissue committed multipotent, oligopotent, and unipotent progenitor cells and possess long-term and short-term regeneration capacities.

12.3.1.1.3 Mammary Stem Cells

The stem cells that are derived from the mammary gland are called mammary stem cells and they can be isolated during puberty and gestation. These stem cells play a critical role in carcinogenesis of the breast. These mammary stem cells can also be isolated from human and mouse tissues and many cell lines have been established from mammary glands. It has been found that single mammary stem cells can form both luminal and myoepithelial cell types of the gland, and an entire organ can be regenerated in the mice model.

12.3.1.1.4 Mesenchymal Stem Cells

Mesenchymal stem cells (MSCs) are multipotent stem cells that can differentiate into a variety of cell types; MSCs can be differentiated into chondrocytes, myocytes, osteoblasts, adipocytes, endothelium, and beta-pancreatic islets cells under *in vitro* conditions. These MSCs have been extensively used in the *in vitro* drug testing and cell transplantation studies.

12.3.1.1.5 Neural Stem Cells

The cells in the adult brain that have been shown to possess neurogenesis capabilities (the birth of new neurons) are called neural stem cells (NSCs). These NSCs were first reported in adult mice and later they were also found in rats, songbirds, primates, and also in humans. It has been found that adult neurogenesis is mostly restricted in two brain regions, (1) the subventricular zone, which lines the lateral ventricles, and (2) the dentate gyrus of the hippocampal formation. These NSCs can be cultured *in vitro* as commonly called as *neurospheres*, floating heterogeneous aggregates of cells that contain a large proportion of stem cells.

12.3.1.1.6 Olfactory Adult Stem Cells

The stem cells that can be harvested from adult human olfactory mucosa cells are called olfactory adult stem cells (OSCs). These OSCs are mostly found in the lining of the nose and are involved in the sense of smell. Under appropriate conditions, these cells possess the ability like embryonic stem cells to develop into many different cell types. It has been found that OSCs hold potential applications in various therapeutic and regenerative medicines. Another advantage of these OSCs is that they can be isolated with ease without any harm to the patients, which means that they can be easily harvested from people of all the ages.

12.3.1.1.7 Applications of Adult Stem Cells

For many years, adult stem cells have been successfully used to treat leukemia and related bone and blood cancers by using a patient's own bone marrow. It has been shown that transplantation of neural adult stem cells into a dog's brain was effective in treating cancerous tumors. It's not possible to treat such conditions by using traditional techniques. The research team from Harvard Medical School has induced intracranial tumors in rodents and then injected human neural stem cells (NSCs). Post-transplantation analysis showed the NSCs were present in the brain and then migrated to the cancerous area and produced an enzyme (cytosine deaminase) that converts a pro-drug into a chemotherapeutic agent. After few weeks, they found that animals with NSCs transplantation had tumor mass reduced by 81%.

A team of Korean researchers in 2004 reported that they had successfully transplanted multi-potent adult stem cells derived from umbilical cord blood into a patient who was suffering from a spinal cord injury. After transplantation, the patient could walk on her own without difficulty. The patient previously could not stand for 19 years and for this clinical trial, the scientists isolated adult stem cells from umbilical cord blood and then injected them into the damaged part of the spinal cord.

In another clinical trial in 2005, doctors in the UK transplanted corneal stem cells into the cornea of a patient who was blinded due to acid attack. The cornea, which is the transparent window of the eye, is a particularly appropriate site for transplantation. In fact, the first successful human cell transplant was done in the cornea because the human cornea has an extraordinary characteristics that it does not contain any blood vessels, making it relatively easy to transplant. The majority of corneal transplants carried out today are because of a degenerative disease called keratoconus. The University Hospital of New Jersey, USA. claims that the success rate when growing new cells from transplanted stem cells varies from 25% to 70%. In 2009, researchers at the University of Pittsburgh Medical Center, USA demonstrated that stem cells isolated from human corneas can restore transparency in mice with corneal damage.

12.3.1.2 Embryonic Stem Cells

Embryonic stem cells (ESCs) are those which are isolated from the inner cell mass of a 5-day-old embryo which is known as a *blastocyst*. These inner cell masses consist of 50–150 cells. One unique property of ESCs is *pluripotency*. This means that ESCs can differentiate into all three primary germ layers which are ectoderm, endoderm, and mesoderm. ESCs can also differentiate into more than 220 cell types present in the human body. ESCs can maintain pluripotency for many cell divisions. The presence of pluripotent adult stem cells remains a subject of scientific debate, although research has demonstrated that pluripotent stem cells may have therapeutic potential. Because of their plasticity and potentially unlimited capacity for self-renewal, ESC therapies have been proposed for many diseases such as genetic diseases, cancers, juvenile diabetes; Parkinson's disease; blindness, and spinal cord injuries.

12.3.1.3 Bioengineered Skin

Skin is maybe the only organ that can be artificially produced from *in vitro* cell culture and can be used for grafting when skin is severely damaged due to severe burns/damage. The keratinocytes comprise 90% of the epidermis of skin and they are responsible for giving rise to the dead cells making the external cornified layer of the skin, and their proliferation is facilitated by the fibroblasts that are found in the dermis layer of the skin. Since fibroblasts are useful for culturing keratinocytes, fibroblast cells called 3T3 cells were used to cover the bottom of a vessel before adding epidermal cells for culturing. Culture media was added with epidermal growth factors, cholera toxin, and a mixture of other growth factors. It has been found that only 1% to 10% of epidermal cells proliferate, others having already started the process of differentiation. These cells form colonies, are separated again, and are transferred to fresh culture to allow better growth. The process of separating cells from colonies and reculturing them is continued to discourage stratification of cell layers and permit the cell colonies to become confluent, forming a sheet of pure epithelium. The cells of this sheet of epithelium are linked by desmosomes. This cultured epithelium can be detached from the vessel using the enzyme dispase, washed, and brought to the hospital to be used for grafting on patients with severe burns.

12.3.1.4 Bioengineered Organ Transplantation

The first full transplant of a grown human organ was done in 2008 by Paolo Macchiarini, Hospital Clínic of Barcelona on Claudia Castillo, a Columbian female adult whose trachea had collapsed due to tuberculosis. A team of researchers from the University of Padua, the University of Bristol, and Politecnico di Milano harvested a section of trachea received from a donor and stripped off the cells that could cause an immune reaction, leaving a grey trunk of cartilage. This section of trachea was then seeded with stem cells taken from Ms. Castillo's bone marrow, and a new section of trachea was grown in the laboratory over 4 days. The new section of trachea was then transplanted into the left main bronchus of the patient. Since the stem cells were harvested from the patient's own bone marrow, Professor Macchiarini did not think it was necessary for her to be given any immunosuppressive medication, and when the procedure was reported 4 months later in *The Lancet*, the patient's immune system displayed no signs of rejecting the transplants.

12.3.1.5 Biomaterials in Tissue Engineering

Several biomaterials have been utilized in tissue engineering and these may range from synthetic polymers with biological characteristics to biologically derived polymers. They can be used to repair damaged or diseased tissues and to create entirely new tissues for transplantation. They can also serve as matrices to guide tissue regeneration and to release growth factors. Biomaterials can also block antibody permeation in transplanted tissue, which could otherwise lead to transplant rejection. Recombinant DNA technology has also been used to generate self-assembly of biomaterials, which are designed to mimic natural tissue matrices that are potentially useful for engineering tissue for injured bones.

12.3.1.5.1 Tissue Regeneration

Usually tissue does not regenerate except under healthy conditions, hence biomaterials can be designed that not only regenerate the tissues but also acquire the desired architecture as in the healthy condition. While designing biomaterials, the density of the adsorbed proteins may be so accustomed that it should support cell spreading, receptor clustering, and cytoskeleton organization. This approach was actually tried in patients with skin wounds and cartilage damage during the 1970s and 1980s. In the US, a living skin cell has already been approved for this purpose and is being used for treatment of diabetic ulcers, skin cancer, and severe burns. Cartilage is another tissue that is being produced artificially and used for orthopedic, craniofacial, and urological applications.

Since cartilage has low nutrient needs, it does not require new blood vessels. In the US, approval has been granted to a company (Genzyme Tissue Repair) to engineer tissues (derived from a patient's own cells) for the repair of knee cartilage damage. Full regeneration in this case takes 12–18 months.

12.3.1.5.2 Bone Grafts

Every year millions of orthopedic operations are performed in many parts of the world and often these procedures involve repair of injured bones. Autogenous bone grafts are usually used for repair of these injured bones. In these autogenous grafts, bone parts are taken from one part of the body such as a non-essential part of a bone, like the brim of the pelvis, and transplanted to repair bones that are essential for weight bearing or other functions such as leg bones. In this procedure of autologous grafts, the patient suffers the pain and potential complications due to harvest from one bone to repair another bone. Groundbreaking alternatives to autologous bone grafts are now becoming available and two skin products were also created by the company, known by the name Advanced Tissue Sciences. These are known by the trade names Transcyte and Dermagraft. Transcyte consists of functional dermis and is alive only until frozen for shipment and until taken for off-the shelf use; it does not carry any living cells when used by the patient.

12.4 BIOINFORMATICS

Bioinformatics, also known as computational biology, is the science where biological data is used to develop software or models to study the biological function or properties of cells or molecules. To understand the complex structure of the human cell, DNA, or proteins requires efficient computing systems to store and analyze such data and with the advent of high-speed computers it is now possible to store and analyze the biological data. The biological data generated from molecular biology, genomics, and proteomics are mostly used in computational biology. It now becomes easy to not only store the biological data but also to share it with other researcher using high-speed internet connection. The field of Computational Biology or Bioinformatics started in the early 1970s and it was considered the science of biological systems through computation. The computational biology platform made it easy for researchers to analyze and compare biological data with great ease and efficiency. Till 1982, biological information was being shared amongst researchers using punch cards and it began to grow exponentially by the end of the 1980s. It was difficult to analyze biological data with low configured computers, and hence thereafter efforts were made to build high-speed computers with large storage capacities. In the 1990s and 2000s, computational biology became an important part of biological research and discovery for sequencing the human genome, creating accurate models of the human brain, and modeling biological systems.

12.4.1 SIGNIFICANCE OF BIOINFORMATICS

Biological data must be combined to form a comprehensive picture of biological activities in the computer to study how normal functions are altered in different disease conditions. Therefore, bioinformatics mainly involves the analysis and interpretation of various types of biological data which includes nucleotide and amino acid sequences, protein domains, and protein structures. The real process of analyzing and interpreting biological data is known as computational biology.

12.4.1.1 Amino Acid Sequencing

Protein biosynthesis is most commonly performed by ribosomes in human cells and protein primary structures can be directly sequenced or inferred from DNA sequences. Protein primary structure is the linear sequence of amino acids in a peptide or protein. The primary structure of a protein is reported starting from the amino-terminal (N) end to the carboxyl-terminal (C) end.

12.4.1.2   Protein Domain

A protein domain is a conserved part of a given protein sequence and (tertiary) structure that can evolve, function, and exist independently of the rest of the protein chain. Each domain forms a compact three-dimensional structure and often can be independently stable and folded. Many proteins consist of several structural domains. One domain may appear in a variety of different proteins. Molecular evolution uses domains as building blocks and these may be recombined in different arrangements to create proteins with different functions. The shortest domains such as zinc fingers are stabilized by metal ions or disulfide bridges. Domains often form functional units, such as the calcium-binding EF hand domain of calmodulin. Because they are independently stable, domains can be swapped by genetic engineering between one protein and another to make chimeric proteins. The primary goal of bioinformatics is to increase the understanding of biological processes. What sets it apart from other methods, however, is its focus on developing and applying computationally intensive techniques such as pattern recognition, data mining, machine learning algorithms, and visualization to achieve this goal. Major research efforts in the field include sequence alignment, gene finding, genome assembly, drug design, drug discovery, protein structure alignment, protein structure prediction, prediction of gene expression and protein–protein interactions, genome-wide association studies, the modeling of evolution and cell division/ mitosis. Based on the application of biological data, bioinformatics is classified into various subtypes, which are discussed as below:

12.4.2   FUNCTIONAL BIOINFORMATICS

12.4.2.1   Sequencing

Computers became essential in molecular biology when protein sequences became available after Frederick Sanger determined the sequence of insulin in the early 1950s. Comparing multiple sequences manually turned out to be impractical and laborious. A pioneer in the field was Margaret Oakley Dayhoff, who has been associated with David Lipman, director of the National Center for Biotechnology Information. Dayhoff compiled one of the first protein sequence databases, initially published as books. Another early contributor to the bioinformatics field was Elvin A. Kabat, who pioneered biological sequence analysis in 1970 with his comprehensive volumes of antibody sequences released between 1980 and 1991.

12.4.2.2   Computational Anatomy

Computational anatomy (CA) is a discipline within medical imaging focusing on the study of anatomical shape and form at the visible or gross anatomical scale of morphology. It comprises the development and application of computational, mathematical data for modeling and simulation of biological structures and organisms. Its emphasis is on the anatomical structures being imaged, rather than the medical imaging devices. It is like computational linguistics, a discipline that focuses on the linguistic structures rather than the sensor acting as the transmission and communication medium. With the advent of dense three-dimensional measurements through technologies such as MRI, CA has emerged as a subfield of medical imaging and bioengineering for mining anatomical coordinate systems. In CA, the diffeomorphism group is used to study different coordinate systems via coordinate transformations as generated via the Lagrangian and Eulerian velocities of flow from one anatomical configuration to another. CA intersects the study of Riemannian manifolds where groups of diffeomorphisms are the central focus, intersecting with emerging high-dimensional theories of shape emerging from the field of shape statistics. The metric structures in CA are related in spirit to morphometrics, with the distinction that CA focuses on an infinite-dimensional space of coordinate systems transformed by a diffeomorphism, hence the central use of the terminology diffeomorphometry, the metric space study of coordinate systems via diffeomorphisms. At CA's heart is the comparison of shape by recognizing in one shape the other. This connects it to D'Arcy Wentworth Thompson's developments in *On Growth and Form* which has led to scientific explanations of morphogenesis, the process by which patterns are formed in biology.

12.4.2.3 Computational Biomodeling

Computational biomodeling is a field concerned with building computer models of biological systems. Computational biomodeling aims to develop and use visual simulations in order to assess the complexity of biological systems. This is accomplished through the use of specialized algorithms and visualization software. These models allow for prediction of how systems will react under different environments. This is useful for determining if a system is robust. A robust biological system is one that "maintain their state and functions against external and internal perturbations," which is essential for a biological system to survive. Computational biomodeling generates a large archive of such data, allowing for analysis from multiple users. While current techniques focus on small biological systems, researchers are working on approaches that will allow for larger networks to be analyzed and modeled. A majority of researchers believe that this will be essential in developing modern medical approaches to creating new drugs and gene therapy. A useful modeling approach is to use Petri nets via tools such as esyN.

12.5 SYNTHETIC BIOLOGY

Synthetic biology is a scientific discipline that combines knowledge of science with engineering tools in order to design and build novel biological tissues and systems. The application includes the design and construction of new biological parts of the human body, devices to be used in the human body, and systems to perform human body function. Synthetic biology has been known to produce diagnostic tools for various diseases such as HIV and hepatitis viruses. Furthermore, the term synthetic biology was first used for genetically engineered bacteria that were created by using recombinant DNA technology and synthetic biology is used as a means to design human body parts using biomimetic chemistry, where organic molecules are used to generate artificial molecules that mimic natural molecules like enzymes. Recently, the engineering community has been seeking to extract components from the biological systems to be reassembled in a way that can copycat the living body. Additionally, the biological information can be used to design products that can be used in various ways, for example DNA, which is basically located in the nucleus of a cell, consists of double-stranded anti-parallel strands each having four various nucleotides assembled from bases, sugars, and phosphates, and these combinations can easily be synthetically designed in the lab. In contrast, this is not true for the protein molecule, which is normally more complex and sometimes unpredictable.

12.5.1 Significance of Synthetic Biology

Recent advancements in recombinant DNA technology have revolutionized the biological and medical sciences including the synthetic biology field. These technologies refer to techniques by which DNA molecules that code for a protein of interest are synthesized using the blueprint of a known genetic sequence. Thereafter, by employing a variety of enzymatic techniques, these genetic sequences or genes are transferred into another organism and this bioengineered organism then uses its own genetic information together with the inserted gene to produce the protein of interest for human use. The whole concept of using genetic engineering techniques is to enhance the production of biomolecules outside of the human body.

The main reasons synthetic biology came into existence is the approval of human insulin, which was the first medicine made by using recombinant DNA technology, and later the development of the first recombinant vaccine. In addition, researchers were also able to successfully synthesize the entire genome of *Mycoplasma mycoides* based on the known sequence of the microbe, and could replace the DNA from the bacterium *Mycoplasma capricolum* with this synthetic genome and produce functional bacterial cells similar to natural *Mycoplasma mycoides*. There are various applications of synthetic biology which create new organisms for biofuel production, to degrade bio-wastes, and for enhancement of the productivity of agriculture and food production.

Furthermore, over the past six decades, several crucial advances have been made that completely transformed the life science field, which include the discovery of the DNA structure, decoding of the genetic information, the growth of recombinant DNA technology, and the decoding of the human genome. In addition, researchers have also made it possible to design and construct human or animal body components by using biological information integrated with the engineering field. The field of synthetic biology has also been revolutionized after creating synthetic DNA and the application of synthetic biology includes the creation of bioengineered microorganisms that can produce useful products such as pharmaceutical chemicals, identify toxic chemicals, break down pollutants and toxins, repair defective genes, and kill cancerous cells. Over a period of a few years, the synthetic biology field is considered an engineering discipline, where human body parts or components are designed and manufactured.

The field of synthetic biology was only limited to laboratory levels till 2004, and there was an international attempt to bring all synthetic biologists to one platform by conducting the first synthetic biology international conference conducted at the Massachusetts Institute of Technology (MIT), Cambridge, USA. The whole purpose of organizing such an event was to bring together all researchers who are working to design and build biological parts, devices, and integrated biological systems that are used for human benefits. This conference was a milestone in establishing synthetic biology as a separate and independent field of research as various universities started interdisciplinary research towards synthetic biology and teaching the life sciences along with engineering subjects. Moreover, there are many technologies which are related to synthetic biology that have existed for several years; for example, the application of metabolic engineering of bacteria used for natural product synthesis was first realized in the year 1970, and engineered bacterial plasmids for biotechnology were established during the year 1980. Moreover, the main difference between genetic engineering and synthetic biology is that though the former involves the transfer of individual genes from one organism to another, the latter comprises the assembly of distinct microbial genomes from a set of standardized genetic parts and these components can be natural genes that are being applied for a new application. Furthermore, natural genes that are redesigned can proficiently work.

12.5.2 Applications of Synthetic Biology

There are various applications of synthetic biology that we have briefly described.

12.5.2.1 Genome Design and Construction

The most important application of synthetic biology is to restructure the genomes of existing microbes to make them more effective in carrying out new functions. It has been reported that researchers were able to simplify the genome of a bacteriophage called T-7 by separating overlapping genes and removing nonessential DNA sequences to facilitate future modifications. Moreover, another group of researchers has redesigned the bacterium *Mycoplasma genitalium*, which has the smallest known bacterial genome studied so far, to have all of the biochemical pathways needed to metabolize, grow, and reproduce. Additionally, efforts have also been made to develop a synthetic version of the *Mycoplasm genitalium* genome that has the minimum number of genes required to support independent life. Furthermore, the primary goal of this minimum genome project was to build a basic microbial platform to which new genes can be added, and also create synthetic organisms with known characteristics and functionality.

12.5.2.2 Synthetic Genome

Synthetic genome basically refers to the set of technologies that makes it possible to construct any specified gene from short strands of synthetic DNA called oligonucleotides, which are normally produced chemically and are 50 and 100 base pairs in length. In August 2002, the virologist Eckard Wimmer announced that over a period of several months his research team had developed live, infectious poliovirus from customized oligonucleotide viral genome available online and similarly in the year 2003, researchers at the Center Institute developed a quick method for genome assembly by using synthetic oligonucleotides to construct a bacteriophage called lambda X174 in two weeks' time. Additionally, in 2005, scientists at the US Centers for Disease Control and Prevention synthesized the Spanish influenza virus that was responsible for the 1918 flu pandemic that killed 50 to 100 million people worldwide. Considering the various benefits of synthetic biology, it's also possible to reconstruct any existing virus for which the complete DNA sequence is known. The arrival of high throughput DNA synthesis machines has sharply reduced the cost of DNA and in the year 2000, the price of customized oligonucleotides was as low as $1.0 per base pairs; in 2004, researchers from Harvard Medical School, USA invented a new multiplex DNA synthesis technique that has finally reduced the cost of DNA synthesis to 20,000 base pairs for $1.0.

12.5.3. APPLIED PROTEIN DESIGN

In addition to synthetically designed genes and DNA molecules, scientists are also working to make synthetic proteins, as this idea was first proposed in 1980 by Genex Corporation, which had suggested that synthetic proteins can be achieved by systematic alteration of the genes that code for certain proteins. After that, efforts have been made to develop protein-engineering technology to improve catalytic efficiency or to alter substrate specificity, or develop proteins which can tolerate high temperatures. It has been suggested that engineered enzymes are being utilized as laundry detergents and in various industrial applications. Furthermore, researchers have also been working to design proteins that are present in the human body, and these bioengineered proteins can be used to treat various diseases. Moreover, scientists have decoded the genetic code of an organism to specify unnatural amino acids, which can be replaced by natural proteins to change their strength as well as their catalytic and binding properties. Additionally, this technique has made it possible to design protein-based drugs that can fight the rapid degradation of proteins in the body.

12.5.3.1 Natural Product Synthesis

Molecular cloning by using recombinant DNA technology has been in use for making multiple copies of a single gene, and these multiple copies of genes are later used to make specific types of proteins. It has been observed that useful proteins such as human insulin can be produced by this technique. Presently, with the advent of synthetic biology, scientists can bioengineer microbes to perform complex multistep syntheses of natural products by assembling components of animal or plant genes that code for all of the enzymes in a synthetic pathway. Furthermore, synthetic biology techniques have been used to program yeast cells to manufacture the drug artemisinin, a natural product that is extremely effective in treating malaria. Currently, this compound can be extracted chemically from the sweet wormwood plant. The extraction of artemisinin is difficult and costly, which reduces its availability and affordability in developing countries of the world. Interestingly, researchers have found a method to reduce the cost of the drug by using a bioengineered metabolic pathway of yeast for the synthesis of precursor (artemisinic acid). Additionally, scientists have also assembled a group of several genes from sweet wormwood that code for the series of enzymes needed to make artemisinic acid, and inserted this cassette into the yeast *Saccharomyces cerevisiae*. Later on, the scientists analyzed the expression levels of each gene so that the entire multi-enzyme pathway can function competently. Once the engineered yeast cells have been coaxed into producing high yields of the artemisinin precursor, it becomes conceivable to manufacture this compound economically in large quantities by using fermentation technology. Moreover, a similar approach can be used to mass produce other drugs that are currently available in limited quantities from natural sources, such as anticancer drugs and the anti-HIV compounds.

12.5.3.2 Creation of Standardized Biological Parts and Circuits

In addition to designing DNA and protein molecules, researchers have been working to create biological parts and circuits to simulate biological functions. For instance, researchers have made biological components which are called Bio-Bricks (short pieces of DNA) that contain functional genetic components. The Bio-Bricks are basically a promoter sequence that initiates the transcription of DNA into messenger RNA and a terminator sequence that stops RNA transcription. In 2006, the Bio-Bricks registry contained 167 basic biological parts, including sensors, actuators, input and output devices, and regulatory elements. In addition to these parts, Bio-Bricks registries contained 421 composite parts, and an additional 50 fragments were being created. These synthetic product registries can be accessed through a public website and the website also invited interested researchers to use the existing knowledge and contribute to it. The main goal of such effort is to develop a methodology for the assembly of Bio-Bricks into circuits with practical applications. Interestingly, these Bio-Bricks have been gathered into a few simple genetic circuits and one such circuit condenses a film of bacteria that are highly sensitive to light so that it can capture an image like a photographic negative. Furthermore, it is anticipated that the long-term goal of this kind of work is to convert bio-engineered cells into tiny programmable chips or computers, so that it become possible to direct their operations by means of chemical signals or light.

12.5.4 Health Risk and Ethical Issues Associated with Synthetic Biology

In recent times, synthetic biologists have had a great deal of success in a short period of time, but major difficulties remain to be overcome before the actual applications of the technology can be realized. Furthermore, there is one problem in that the behavior of bioengineered systems remains unpredictable and it has been suggested that genetic circuits can also tend to mutate quickly and become non-functional over time. It has been hypothesized that synthetic biology will not achieve its potential until investigators can accurately predict how a new genetic circuit will perform inside a living cell or living organism. Another issue which needs to be considered is that most of the bioengineered biological systems remain costly, unreliable, and unplanned because scientists do not understand the molecular processes of cells well enough to manipulate them reliably.

Furthermore, there are also risks involved in synthetic biology research that need to be discussed, because engineered microorganisms are self-replicating and capable of evolution, they can be associated with a different risk category than toxic chemicals or radioactive materials. Additionally, there are three main areas of risk in synthetic biology: first, synthetically designed microorganisms might escape from a research laboratory facility and cause environmental damage or threaten the public; second, synthetically designed microorganism developed for some applied purpose might cause destructive side effects after being intentionally released into the open environment; and third, fanatics and militants might exploit synthetic biology for hostile or hateful purposes.

12.5.4.1 The Risk of Testing in the Open Environment

One of the major applications of synthetic biology is the development of synthetic microorganisms to use in biosensing and cleaning up soil contaminated with toxic chemicals, and these synthetic microorganisms can also cause harmful effects on the natural flora and fauna. This problem is mainly associated with genetically modified organisms or crops, where it has been cautioned that extensive use of genetically modified organisms or crops may lead to health and environmental problems. Moreover, environmental safety agencies require a severe risk assessment before it can approve genetically modified organisms. Few genetically modified organisms have been synthe-sized for this purpose, including genetically modified strains of the soil bacterium *Pseudomonas putida* and in that experiment, scientists used and monitored bioengineered bacteria while they degraded a toxic chemical. Hypothetically, there could be three types of negative effects associated with synthetic biology products such as a synthetic microorganism when released into the natural environment could interrupt the local environment by means of competition or infection that, in the worst case, could lead to the extinction of one or more natural species of organism. The second negative effect would be, once a synthetic organism has well colonized a locality, it might become widespread and thus impossible to eradicate. Finally, the synthetic organism might disrupt some aspect of the habitat into which it was first presented, upsetting the natural balance and causing the degradation or destruction of the local environment.

12.5.4.2 The Risk of Deliberate Misuse

In 1972, use and development of biological war weapons was banned in the Biological War Convention (BWC) held in the United States. In addition, the BWC prohibits the synthesis of known or novel microorganisms for intimidating purposes. Additionally, if synthetic organisms are designed to pro-duce chemical toxins, then the development and production of these poisons for weapons purposes can be prohibited as per the regulations of Biological War Convention. Nevertheless, one potential misuse of synthetic biology would be to recreate known pathogens such as the "Ebola virus" in the laboratory for intimidation and terror activity. The viability of collecting an entire infectious viral genome from synthetic oligonucleotides has already been established for poliovirus and the Spanish influenza virus.

12.5.4.3 On-line Selling of Oligonucleotides

With a rapid advancement of information technology tools, it became possible to buy genetically engineered molecules such as oligonucleotides and genes by online shopping. These oligonucle-otides producers emerged in several countries around the world and started selling these oligonucle-otides. A US-based company voluntarily uses special software to screen all oligonucleotide and gene orders for the presence of DNA sequences from select agents of bio-terrorism and when such a sequence is detected, the request is denied. However, oligonucleotide suppliers are currently under no legal obligation to screen their orders and because many clients value privacy, companies might face problems in doing so. There are two possible solutions to this problem. First, the US Congress could pass a law requiring suppliers to check all oligonucleotide and gene orders for the presence of pathogenic DNA sequences. Otherwise, suppliers could agree among themselves to screen orders willingly.

12.5.5 Challenges in Synthetic Biology

In addition to various useful applications of synthetic biology, there are technical problems associ-ated with synthetic biology-based products. There are reports that suggest that many synthetically designed biological parts have not been fully tested and characterized and especially their perfor-mances are not tested for long periods of time in order to know their longevity. In this direction, efforts have been made to optimize lactose production by using synthetically designed microbes, but showed little success without much documentation. Moreover, it has been reported that about 1500 registry parts have been known to be in working condition, but there are 50 parts which are failing while performing the activities. In order to improve the quality and performances of the synthetically designed products, the registry has been stepping up efforts to encourage contributors to include documentation on part function and performances. Meanwhile, a new project is proposed to reduce some of the variability in measurements from different laboratories and some researchers have found that it could eliminate half of the variation arising from experimental conditions and instruments. Measurements are tricky to standardize; however, in mammalian cells, for example, genes introduced into a cell integrate unpredictably into the cell's genome and neighboring regions often affect expression, and it has been suggested that this type of complexity is very difficult to capture by standardized characterization. One of the major problems with synthetically designed parts is that these parts don't show any function when they are assembled together and synthetic biologists are often caught in a laborious process of testing as these synthetically designed bio-parts are not predictable like other modern engineering products. Computer modeling could help reduce this guesswork. In 2009, Collins and his colleagues created several slightly different versions of two promoters and they used one version of each to create a genetic timer, a system that would cause cells to switch from expressing one gene to another gene after a certain interval time. Later on, they tested the timer, and fed the results back into a computer system and predicted how timers built from other versions would behave. By using such modeling techniques, researchers could improve computationally rather than test every version of a network. Nonetheless, designs cannot work perfectly all the time, but they can be eventually refined and this refining process is called directed evolution. It has been suggested that this directed evolution is generally involved in mutating DNA sequences, screening their performance, selecting the best candidates, and repeating the process until the system is optimized. Another issue related to synthetically designed parts is their complexity and physical size, as during construction the circuits get bigger and the process of constructing and testing them becomes more daunting. A system was developed that uses about a dozen genes to produce a precursor of the antimalarial compound artemisinin in microbes. It has been estimated that it will take roughly 150 years of work to uncover genes involved in the pathway and to develop or refine parts to control their expression. It has been further stated that researchers had to test many part variations before they found a configuration that sufficiently increased production of an enzyme needed to consume a toxic intermediate molecule. Additionally, another research group had created a system to combine genetic parts through automation as these parts have predefined flanking sequences, verbalized by a set of guidelines called the "BioBrick standard," and can be assembled by robots. Another research group had developed a system that engineered *E. coli* cells, called assembler cells, that are equipped with enzymes that can cut and stitch together DNA parts. Moreover, *E. coli* cells can be engineered to act as selective cells that can sort out the completed products from the surplus parts.

12.6 BIOSIMILARS

12.6.1 INTRODUCTION

A biosimilar is a biologic medical product that is almost an identical copy of an original product that is manufactured by a different company. Biosimilars are officially approved versions of original patented products and can be manufactured when the original product's patent expires. In 2009, the US Congress, through the Biologics Price Competition and Innovation Act (BPCI Act), created an abbreviated licensure pathway for biological products that are demonstrated to be biosimilar to or interchangeable with an FDA-approved biological product. This pathway was created as a way to provide more treatment options, increase access to lifesaving medications, and potentially lower health care costs through competition.

In contrast to generic drugs of the more common small molecule type, biologics generally exhibit high molecular complexity and may be quite sensitive to changes in manufacturing processes. In spite of that heterogeneity, all biopharmaceuticals, including biosimilars, must maintain consistent quality and clinical performance throughout their lifecycle. Follow-on manufacturers do not have access to the originator's molecular clone and original cell bank, to the exact fermentation and purification process, or to the active drug substance, but they have access to the commercialized innovator product. Overall, it is harder to ascertain fungibility between generics and innovators among biologics than it is among totally synthesized and semi-synthesized drugs. That is why the name "biosimilar" was coined to differentiate them from small-molecule generics.

12.6.2 SIGNIFICANCE OF BIOSIMILARS

After the termination of the patent of approved recombinant drugs such as insulin, human growth hormone, interferons, erythropoietin, or monoclonal antibodies, any other biotechnology company can develop and market these biologics which are called as biosimilars. Every biological or biopharmaceutical product displays a certain degree of variability, even between different batches of the same product, which is due to the inherent variability of the biological expression system and the manufacturing process. Any kind of reference product has undergone numerous changes in its manufacturing processes, and such changes in the manufacturing process (ranging from a change in the supplier of the cell culture media to new purification methods or new manufacturing sites) was substantiated with appropriate data and was approved by the EMA (European Medicines Agency). In contrast, it is mandatory for biosimilars to take both a nonclinical (*in vitro* and animal testing) and clinical test that the most sensitive clinical models are asked to show to enable detection of differences between the two products in terms of human pharmacokinetics (PK) and pharmacodynamics (PD), efficacy, safety, and immunogenicity.

Usually, once a drug is released in the market by the US FDA, it has to be reevaluated for its safety and efficacy once every six months for the first and second years. Afterward, reevaluations are conducted yearly, and the result of the assessment should be reported to authorities such as the FDA. Biosimilars are required to undergo pharmacovigilance (PVG) regulations as its reference product. Thus biosimilars approved by the EMA are required to submit a risk management plan (RMP) along with the marketing application and have to provide regular safety update reports after the product is on the market. The RMP includes the safety profile of the drug and proposes the prospective pharmacovigilance studies.

Numerous PK studies, such as studies conducted by the Committee for Medicinal Products for Human Use (CHMP), have been conducted under various ranges of conditions: antibodies from an originator's product versus antibodies from a biosimilar, combination therapy and monotherapy, various diseases, etc. for the purpose of verifying comparability in PK of the biosimilar with the reference medicinal product in a sufficiently sensitive and homogeneous population. Importantly, provided that structure and functions, PK profiles, and PD effects and/or efficacy can be shown to be comparable for the biosimilar and the reference product, those adverse drug reactions which are related to exaggerated pharmacological effects can also be expected at similar frequencies.

12.6.3 BIOSIMILARS MARKET

To be able to make a biosimilar drug requires $75–$250 million, which includes the cost of legal requirements, FDA approval, and manufacturing costs. This market entry barrier affects not only the companies willing to produce them but could also delay availability of inexpensive alternatives for public healthcare institutions that subsidize treatment for their patients. Even though the biosimilars market is rising, the price drop for biological drugs at risk of patent expiration will not be as great as for other generic drugs; in fact, it has been estimated that the price for biosimilar products will be 65%–85% of that of their originators. Biosimilars are drawing the market's attention as there is an upcoming patent cliff, which will put nearly 36% of the $140 billion market for biologic drugs at risk as of 2011, and this is considering only the top 10 selling products. The global biosimilars market was $1.3 billion in 2013 and is expected to reach $35 billion by 2020 driven by the patent expiration of an additional ten blockbuster biologic drugs. The list of US approved biosimilar products is shown in table on next page.

12.6.4   LIST OF US APPROVED BIOSIMILARS

List of US FDA Approved Biosimilars

Date of FDA Approval	Biosimilar Product	Original Product
March 6, 2015	filgrastim-sndz/Zarxio	filgrastim/Neupogen
April 5, 2016	infliximab-dyyb/Inflectra	infliximab/Remicade
August 30, 2016	etanercept-szzs/Erelzi	etanercept/Enbrel
September 23, 2016	adalimumab-atto/Amjevita	adalimumab/Humira
April 21, 2017	infliximab-abda/Renflexis	infliximab/Remicade
August 25, 2017	adalimumab-adbm/Cyltezo	adalimumab/Humira
September 14, 2017	bevacizumab-awwb/Mvasi	bevacizumab/Avastin
December 1, 2017	trastuzumab-dkst/Ogivri	trastuzumab/Herceptin
December 13, 2017	infliximab-qbtx/Ixifi	infliximab/Remicade
May 15, 2018	epoetin alfa-epbx/Retacrit	epoetin alfa/Procrit
June 4, 2018	pegfilgrastim-jmdb/Fulphila	pegfilgrastim/Neulasta

12.6.5   DEFINITION OF BIOSIMILAR PRODUCT

A biosimilar is a biological product that is highly similar to and has no clinically meaningful differences from an existing FDA-approved reference product. Minor differences between the reference product and the proposed biosimilar product in clinically inactive components are acceptable. A manufacturer developing a proposed biosimilar demonstrates that its product is highly similar to the reference product by extensively analyzing (i.e., characterizing) the structure and function of both the reference product and the proposed biosimilar. State-of-the-art technology is used to compare characteristics of the products, such as purity, chemical identity, and bioactivity. The manufacturer uses results from these comparative tests, along with other information, to demonstrate that the biosimilar is highly similar to the reference product. Minor differences between the reference product and the proposed biosimilar product in clinically inactive components are acceptable. For example, these could include minor differences in the stabilizer or buffer compared to what is used in the reference product. Any differences between the proposed biosimilar product and the reference product are carefully evaluated by the FDA to ensure the biosimilar meets the FDA's high approval standards. As mentioned previously, slight differences (i.e., acceptable within-product variations) are expected during the manufacturing process for biological products, regardless of whether the product is a biosimilar or a reference product. For both reference products and biosimilars, lot-to-lot differences (i.e., acceptable within-product differences) are carefully controlled and monitored. A manufacturer must also demonstrate that its proposed biosimilar product has no clinically meaningful differences from the reference product in terms of safety, purity, and potency (safety and effectiveness). This is generally demonstrated through human pharmacokinetic (exposure) and pharmacodynamic (response) studies, an assessment of clinical immunogenicity, and, if needed, additional clinical studies.

12.6.6   DIFFERENCE BETWEEN A BIOSIMILAR AND AN INTERCHANGEABLE PRODUCT

An interchangeable product is a biosimilar product that meets additional requirements outlined by the BPCI Act. As part of fulfilling these additional requirements, information is needed to show that an interchangeable product is expected to produce the same clinical result as the reference product in any given patient. Also, for products administered to a patient more than once, the risk in terms of safety and reduced efficacy of switching back and forth between an interchangeable product and a reference product will have been evaluated. An interchangeable product may be substituted for the reference product without the involvement of the prescriber. The FDA's high standards for approval should assure health care providers that they can be confident in the safety and effectiveness of an interchangeable product, just as they would be for an FDA approved reference product.

A manufacturer of a proposed interchangeable product will need to provide additional information to show that an interchangeable product is expected to produce the same clinical result as the reference product in any given patient. Also, for a product that is administered to a patient more than once, a manufacturer will need to provide data and information to evaluate the risk, in terms of safety and decreased efficacy, of alternating or switching between the products. As a result, a product approved as an interchangeable product means that the FDA has concluded that it may be substituted for the reference product without consulting the prescriber. For example, say a patient self-administers a biological product by injection to treat their rheumatoid arthritis. To receive the biosimilar instead of the reference product, the patient may need a prescription from a health care prescriber written specifically for that biosimilar. However, once a product is approved by the FDA as interchangeable, the patient may be able to take a prescription for the reference product to the pharmacy and, depending on the state, the pharmacist could substitute the interchangeable product for the reference product without consulting the prescriber. Note that pharmacy laws and practices vary from state to state. The FDA undertakes a rigorous and thorough evaluation to ensure that all products, including biosimilar and interchangeable products, meet the Agency's high standards for approval.

12.6.7 BIOSIMILARS VERSUS GENERIC DRUGS

Biosimilars are drugs that offer more affordable treatment choices to patients. Both biosimilars and generics drugs are approved through different pathways to avoid duplicating costly clinical trials. But biosimilars are not generic drugs, and there are important differences between biosimilars and generic drugs. For example, the active ingredients of generic drugs are the same as those of brand name drugs. In addition, the manufacturer of a generic drug must demonstrate that the generic is bioequivalent to the brand name drug. By contrast, biosimilar manufacturers must demonstrate that the biosimilar is highly similar to the reference product, except for minor differences in clinically inactive components. Biosimilar manufacturers must also prove that there are no clinical differences between the biosimilar and the reference product in terms of safety and efficacy.

PROBLEMS

Section A: Descriptive Type

Q1. What is forensic science?
Q2. What is synthetic biology?
Q3. What is bioinformatics?
Q4. What is regenerative medicine?
Q5. What are biosimilars?

Section B: Critical Thinking

Q1. How does forensic science help to solve criminal cases?
Q2. Is it possible to reconstruct human organs?
Q3. Explain the role of bioinformatics in drug discovery.

ASSIGNMENTS

Make a poster based on the various applications of forensic science, synthetic biology, bioinformatics, regenerative medicine, and biosimilars. Display it in the classroom and discuss it with your colleagues.

REFERENCES AND FURTHER READING

Abeyewickremea, A., Kwoka, A., McEwanb, J.R., and Jayasinghe, S.N. Bio-electrospraying embryonic stem cells: Interrogating cellular viability and pluripotency. *Integrative Biology* 1: 260–266, 2009.

Adachi, T., Osako, Y., Tanaka, M., Hojo, M., and Hollister, S.J. Framework for optimal design of porous scaffold microstructure by computational simulation of bone regeneration. *Biomaterials* 27(21): 3964–3972, 2006.

Ang, T.H., Sultana, F.S., Hutmacher, D.W., Wong, Y.S., Fuh, J.Y., Mo, X.M., Loh, H.T., Burdet, E., and Teoh, S.H. Fabrication of 3D chitosan-hydroxyapatite scaffolds using a robotic dispersing system. *Materials Science and Engineering* C20: 35–42, 2002.

Baerlocher, G.M., Vultro, I., de Jong, G., and Lansdorp, P.M. Flow cytometry and FISH to measure the average length of telomeres. *Nature Protocols* 1(5): 2365–2376, 2006.

Bailey, S.M., and Goodwin, E.H. DNA and telomeres: Beginnings and endings. *Cytogenetic and Genome Research* 104: 109–115, 2004.

Barry III, R., Shepherd, R., Hanson, J., Nuzzo, R., Wiltzius, P., and Lewis, J. Direct-write assembly of 3D hydrogel scaffolds for guided cell growth. *Advanced Materials* 21(23): 2407–2410, 2009.

Bing, D.H., Boles, C., Rehman, F.N., Audeh, M., Belmarsh, M., Kelley, B., and Adams, C.P. Bridge amplification: A solid phase PCR system for the amplification and detection of allelic differences in single copy genes. Genetic Identity Conference Proceedings, Seventh International Symposium on Human Identification, 1996. www.promega.com/geneticidproc/ussymp7proc/0726.html.

Brtolo, P.J. State of the art of solid freeform fabrication for soft and hard tissue engineering. In: *Design and Nature III: Comparing Design in Nature with Science and Engineering*. WIT Press, Southampton, UK, 2006, pp. 233–243.

Brtolo, P.J., Almeida, H.A., and Laoui, T. Rapid prototyping and manufacturing for tissue engineering scaffolds. *International Journal of Computer Applications in Technology* 36: 1–9, 2009.

Brtolo, P.J., Almeida, H.A., Rezende, R.A., Laoui, T., Bidanda, B., and Brtolo, P. (eds). Advanced processes to fabricate scaffolds for tissue engineering. In: *Virtual Prototyping & Bio Manufacturing in Medical Applications*. Springer Verlag, Berlin, 2007, p. 299.

Brtolo, P.J., and Mitchell, G. Stereo-thermal-lithography. *Rapid Prototyping Journal* 9: 150–156, 2003.

Bucklen, B., Wettergreen, W., Yuksel, E., and Liebschner, M. Bone-derived CAD library for assembly of scaffolds in computer-aided tissue engineering. *Virtual and Physical Prototyping* 3(1): 13–23, 2008.

Cai, S., and Xi, J. A control approach for pore size distribution in the bone scaffold based on the hexahedral mesh refinement. *Computer-Aided Design* 40(10–11): 1040–1050, 2008.

Canela, A., Vera, E., Klatt, P., and Blasco, M.A. High-throughput telomere length quantification by FISH and its application to human population studies. *Proc. Natl. Acad. Sci. USA* 104(13): 5300–5305, 2007.

Carson, S.A., Gentry, W.L., Smith, A.L., and Buster, J.E. Trophectoderm microbiopsy in murine blastocysts: Comparison of four methods. *J. Assist. Reprod. Genet.* 10(6): 427–433, 1993. doi:10.1007/BF01228093. PMID 8019091.

Chan, G., and Mooney, D.J. New materials for tissue engineering: Towards greater control over the biological response. *Trends Biotechnol* 26(7): 382–392, 2008.

Cheah, C.M., Chua, C.K., Leong, K.F., Cheong, C.H., and Naing, M.W. Automatic algorithm for generating complex polyhedral scaffold structures for tissue engineering. *Tissue Engineering* 10(3–4): 595–610, 2004.

Cheng, S., Fockler, C., Barnes, W.M., and Higuchi, R. Effective amplification of long targets from cloned inserts and human genomic DNA. *Proc. Natl. Acad. Sci. USA* 91(12): 5695–5699, 1994. doi:10.1073/pnas.91.12.5695. PMID 8202550.

Chien, A., Edgar, D.B., and Trela, J.M. Deoxyribonucleic acid polymerase from the extreme thermophile thermus aquaticus. *J. Bacteriol.* 174(3): 1550–1557, 1976. PMID 8432.

Chu, T.M., Halloran, J.W., Hollister, S.J., and Feinberg, S.E. Hydroxyapatite implants with designed internal architecture. *Journal of Materials Science: Materials in Medicine* 12(6): 471–478, 2001.

Chua, C.K., Leong, K.F., and Tan, K.H. Specialized fabrication processes: Rapid prototyping. In: *Biomedical Materials*, R. Narayan (ed.). Springer Science, Berlin, 2009, pp. 493–523.

Chua, C.K., Leong, K.F., Tan, K.H., Wiria, F.E., and Cheah, C.M. Development of tissue scaffolds using selective laser sintering of polyvinyl alcohol/hydroxyapatite biocomposite for craniofacial and joint defects. *Journal of Materials Science: Materials in Medicine* 15(10): 1113–1121, 2004.

Cooke, M.N., Fisher, J.P., Dean, D., Rimnac, C., and Mikos, A.G. Use of stereolithography to manufacture critical-sized 3D biodegradable scaffolds for bone ingrowth. *Journal of Biomedical Materials Research Part B: Applied Biomaterials* 64B: 65–69, 2002.

Coutelle, C., Williams, C., Handyside, A., Hardy, K., Winston, R., and Williamson, R. Genetic analysis of DNA from single human oocytes: A model for preimplantation diagnosis of cystic fibrosis. *BMJ* 299(6690): 22–24, 1989. doi:10.1136/bmj.299.6690.22. PMID 2503195.

Crump, S.S. Apparatus and method for creating three-dimensional objects. Patent US5121329, June 9, 1992.

Cui, X., and Boland, T. Human microvasculature fabrication using thermal inkjet printing technology. *Biomaterials* 30(31): 6221–6227, 2009.

David, F., and Turlotte, E. An isothermal amplification method. *C.R. Acad. Sci Paris, Life Science* 321(1): 909–914, 1998.

Demko, Z., Rabinowitz, M., and Johnson, D. Current methods for preimplantation genetic diagnosis. *Journal of Clinical Embryology* 13(1): 6–12, 2010. www.genesecurity.net/wp-content/uploads/2010/04/Current-Methods-for-Preimplantation-Genetic-Diagnosis.pdf.

Engel, E., Michiardi, A., Navarro, M., Lacroix, D., and Planell, J.A. Nanotechnology in regenerative medicine: The materials side. *Trends in Biotechnology* 26: 39–47, 2007.

Fagagna, F., Hande, M.P., Tong, W.M., Roth, D., Lansdorp, P.M., Wang, Z.Q., and Jackson, S.P. Effects of DNA nonhomologous end-joining factors on telomere length and chromosomal stability in mammalian cells. *Current Biology* 15: 1192–1196, 2001.

Falconer, E., Chavez, E.A., Henderson, A., Poon, S.S.S., McKinney, S., Brown, L., Huntsman, D.G., and Lansdorp, P.M. Identification of sister chromatids by DNA template strand sequences. *Nature* 463: 93–98, 2010.

Gardner, R.L., and Edwards, R.G. Control of the sex ratio at full term in the rabbit by transferring sexed blastocysts. *Nature* 218(5139): 346–349, 1968. doi:10.1038/218346a0. PMID 5649672.

Gianaroli, L., Magli, M.C., Ferraretti, A.P., and Munné, S. Preimplantation diagnosis for aneuploidies in patients undergoing in vitro fertilization with a poor prognosis: Identification of the categories for which it should be proposed. *Fertil. Steril.* 72(5): 837–844, 1999. doi:10.1016/S0015-0282(99)00377-5.

Griffith, M.L., and Halloran, J.W. Free-form fabrication of ceramics via stereolithography. *Journal of the American Ceramic Society* 79: 2601–2608, 1996.

Handyside, A.H., Lesko, J.G., Tarín, J.J., Winston, R.M., and Hughes, M.R. Birth of a normal girl after in vitro fertilization and preimplantation diagnostic testing for cystic fibrosis. *N. Engl. J. Med.* 327(13): 905–909, 1992.

He, J., Li, D., Liu, Y., Gong, H., and Lu, B. Indirect fabrication of microstructured chitosan-gelatin scaffolds using rapid prototyping. *Virtual and Physical Prototyping* 3(3): 159–166, 2008.

Hemann, M.T., Strong, M.A., Hao, L.Y., and Greider, C.W. The shortest telomere, not average telomere length, is critical for cell viability and chromosome stability. *Cell* 107: 67–77, 2001.

Herman, J.G., Graff, J.R., Myöhänen, S., Nelkin, B.D., and Baylin, S.B. Methylation-specific PCR: A novel PCR assay for methylation status of CpG islands. *Proc. Natl. Acad. Sci. USA* 93(13): 9821–9826, 1996.

Holding, C., and Monk, M. Diagnosis of beta-thalassaemia by DNA amplification in single blastomeres from mouse preimplantation embryos. *Lancet* 2(8662): 532–535, 1989.

Hollister, S.J. Porous scaffold design for tissue engineering. *Natural Materials* 4(7): 518–524, 2005.

Innis, M.A., Myambo, K.B., Gelfand, D.H., and Brow, M.A. DNA sequencing with thermus aquaticus DNA polymerase and direct sequencing of polymerase chain reaction-amplified DNA. *Proc. Natl. Acad. Sci. USA* 85(24): 9436–9440, 1988.

Isenbarger, T.A., Finney, M., Ríos-Velázquez, C., Handelsman, J., and Ruvkun, G. Miniprimer PCR, a new lens for viewing the microbial world. *Applied and Environmental Microbiology* 74(3): 840–849, 2008.

Jayasinghe, S. Bio-electrosprays: The development of a promising tool for regenerative and therapeutic medicine. *Biotechnology Journal* 2(8): 934–937, 2007.

Kahraman, S., Bahçe, M., Samli, H. et al. Healthy births and ongoing pregnancies obtained by preimplantation genetic diagnosis in patients with advanced maternal age and recurrent implantation failure. *Hum. Reprod.* 15(9): 2003–2007, 2000.

Khan, Z., Poetter, K., and Park, D.J. Enhanced solid phase PCR: Mechanisms to increase priming by solid support primers. *Analytical Biochemistry* 375(2): 391–393, 2008.

Kim, S.S., Utsunomiya, H., Koski, J.A., Wu, B.M., Cima, M.J., Sohn, J., Mukai, K., Griffith, L.G., and Vacanti, J.P. Survival and function of hepatocytes on a novel three-dimensional synthetic biodegradable polymer scaffold with an intrinsic network of channels. *Annals of Surgery* 228(1): 8–13, 1998.

Kleppe, K., Ohtsuka, E., Kleppe, R., Molineux, I., and Khorana, H.G. Studies on polynucleotides. XCVI: Repair replications of short synthetic DNA's as catalyzed by DNA polymerases. *J. Mol. Biol.* 56(2): 341–361, 1971.

Klitzman, R., Zolovska, B., Folberth, W., Sauer, M.V., Chung, W., and Appelbaum, P. Preimplantation genetic diagnosis on in vitro fertilization clinic websites: Presentations of risks, benefits and other information. *Fertil. Steril.* 92(4): 1276–1283, 2009.

Lam, C.X., Mo, X.M., Teoh, S.H., and Hutmacher, D.W. Scaffold development using 3D printing with a starch-based polymer. *Materials Science and Engineering* 20: 49–56, 2002.

Lam, C.X., Olkowski, R., Swieszkowski, W., Tan, K., Gibson, I., and Hutmacher, D. Mechanical and in vitro evaluations of composite PLDLLA/TCP scaffolds for bone engineering. *Virtual and Physical Prototyping* 3(4): 193–197, 2008.

Lan, P.X., Lee, J.W., Seol, Y.J., and Cho, D.W. Development of 3D PPF/DEF scaffolds using micro-stereo-lithography and surface modification. *Journal of Materials Science: Materials in Medicine* 20(1): 271–279, 2009.

Lawyer, F.C., Stoffel, S., Saiki, R.K., Chang, S.Y., Landre, P.A., Abramson, R.D., and Gelfand, D.H. High-level expression, purification, and enzymatic characterization of full-length Thermus aquaticus DNA polymerase and a truncated form deficient in 5' to 3' exonuclease activity. *PCR Methods Appl.* 2(4): 275–287, 1993. PMID 8324500.

Lee, G., and Barlow, J.W. Selective laser sintering of bioceramic materials for implants. 96 SFF Symposium, Austin, TX, August 12–14, 1996.

Leong, K.F., Cheah, C.M., and Chua, C.K. Solid freeform fabrication of three-dimensional scaffolds for engineering replacement tissues and organs. *Biomaterials* 24(13): 2363–2378, 2003.

Leong, K.F., Chua, C., Sudarmadji, N., and Yeong, W. Engineering functionally graded tissue engineering scaffolds. *Journal of the Mechanical Behavior of Biomedical Materials* 1(2): 140–152, 2008.

Levy, R.A., Chu, T.M., Halloran, J.W., Feinberg, S.E., and Hollister, S. CT-generated porous hydroxyapatite orbital floor prosthesis as a prototype bioimplant. *AJNR American Journal of Neuroradiology* 18(8): 1522–1525, 1997.

Lewis, C.M., Pinêl, T., Whittaker, J.C., and Handyside, A.H. Controlling misdiagnosis errors in pre-implantation genetic diagnosis: A comprehensive model encompassing extrinsic and intrinsic sources of error. *Hum. Reprod.* 16(1): 43–50, 2001.

Li, M., DeUgarte, C.M., Surrey, M., Danzer, H., DeCherney, A., and Hill, D.L. Fluorescence in situ hybridization reanalysis of day-6 human blastocysts diagnosed with aneuploidy on day 3. *Fertil. Steril.* 84(5): 1395–1400, 2005.

Liu, V.A., and Bhatia, S.N. Three dimensional patterning of hydrogels containing living cells. *Biomedical Microdevices* 257–266, 2002.

Liu, Y.G., and Whittier, R.F. Thermal asymmetric interlaced PCR: Automatable amplification and sequencing of insert end fragments from P1 and YAC clones for chromosome walking. *Genomics* 25(3): 674–681, 1995.

Marcondes, A.M., Bair, S., Rabinovitch, P.S., Gooley, T., Deeg, H.J., and Risques, R. No telomere shortening in marrow stroma from patients with MDS. *Annals of Hematology* 88: 623–628, 2009.

Matsumoto, T., and Mooney, D.J. Cell instructive polymers. *Advances in Biochemical Engineering and Biotechnology* 102: 113–137, 2006.

McArthur, S.J., Leigh, D., Marshall, J.T., de Boer, K.A., and Jansen, R.P. Pregnancies and live births after trophectoderm biopsy and preimplantation genetic testing of human blastocysts. *Fertil. Steril.* 84(6): 1628–1636, 2005.

Melchels, F.P.W., Feijen, J., and Grijpma, D.W. A poly(D,L-lactide) resin for the preparation of tissue engineering scaffolds by stereolithography. *Biomaterials* 30(23–24): 3801–3809, 2009.

Mironov, V., Boland, T., Trusk, T., Forgacs, G., and Markwald, R.R. Organ printing: Computer-aided jet-based 3D tissue engineering. *Trends in Biotechnology* 21(4): 157–161, 2003.

Mironov, V., Gentile, C., Brakke, K., Trusk, T., Jakab, K., Forgacs, G., Kasyanov, V., Visconti, R., and Markwald, R. Designer 'blueprint' for vascular trees: Morphology evolution of vascular tissue constructs. *Virtual and Physical Prototyping* 4(2): 63–74, 2009. [informaworld].

Mironov, V., Prestwich, G., and Forgacs, G. Bioprinting living structures. *Journal of Materials Chemistry* 17(20): 2054–2060, 2007.

Mironov, V., Visconti, R., Kasyanov, V., Forgacs, G., Drake, C., and Markwald, R. Organ printing: Tissue spheroids as building blocks. *Biomaterials* 30(12): 2164–2174, 2009.

Mistry, A.S., and Mikos, A.G. Tissue engineering strategies for bone regeneration. *Advances in Biochemical Engineering and Biotechnology* 94: 1–22, 2005.

Montag, M., van der Ven, K., Dorn, C., and van der Ven, H. Outcome of laser-assisted polar body biopsy and aneuploidy testing. *Reprod. Biomed.* Online 9(4): 425–429, 2004.

Mota, C., Chiellini, F., Brtolo, P.J., and Chiellini, E. A novel approach to the fabrication of polymeric scaffolds for tissue engineering applications. VII Convegno Nazionale Scienza e Tecnologia Dei Materiali, Tirrenia, Italy, June 2009.

Mota, C., Mateus, A., Brtolo, P.J., Almeida, H., and Ferreira, N. Processo e equipamento de fabrico rapido por bioextrusao/Process and equipment for rapid fabrication through bioextrusion—Portuguese Patent Application, 2009.

Mueller, P.R., and Wold, B. In vivo foot printing of a muscle specific enhancer by ligation mediated PCR. *Science* 246(4931): 780–786, 1988.

Mullis, K. The unusual origin of the polymerase chain reaction. *Scientific American* 262(4): 56–61, 64–65, 1990.

Mullis, K. *Dancing Naked in the Mind Field.* Pantheon Books, New York, 1998. ISBN: 0-679-44255-3.

Munné, S., Cohen, J., and Sable, D. Preimplantation genetic diagnosis for advanced maternal age and other indications. *Fertil. Steril.* 78(2): 234–236, 2002.

Munné, S., Dailey, T., Sultan, K.M., Grifo, J., and Cohen, J. The use of first polar bodies for preimplantation diagnosis of aneuploidy. *Hum. Reprod.* 10(4): 1014–1020, 1995. PMID 7650111. http://humrep.oxford-journals.org/cgi/pmidlookup?view=long&pmid=7650111.

Munné, S., Magli, C., Cohen, J. et al. Positive outcome after preimplantation diagnosis of aneuploidy in human embryos. *Hum. Reprod.* 14(9): 2191–2199, 1999.

Myrick, K.V., and Gelbart, W.M. Universal fast walking for direct and versatile determination of flanking sequence. *Gene* 284(1–2): 125–131, 2002. doi:10.1016/S0378-1119(02)00384-0. PMID 11891053.

Nair, L.S., and Laurencin, C.T. Polymers as biomaterials for tissue engineering and controlled drug delivery. *Advances in Biochemical Engineering and Biotechnology* 102: 47–90, 2006.

Navidi, W., and Arnheim, N. Using PCR in preimplantation genetic disease diagnosis. *Hum. Reprod.* 6(6): 836–849, 1991.

Newton, C.R., Graham, A., Heptinstall, L.E., Powell, S.J. et al. Analysis of any point mutation in DNA: The amplification refractory mutation system (ARMS). *Nucleic Acids Research* 17(7): 2503–2516, 1989.

Nietzel, A., Rocchi, M., Starke, H., Heller, A. et al. A new multi-colour FISH approach for the characterization of marker chromosomes: Centromere-specific multicolor-FISH (cenM-FISH). *Human Genetics* 108: 199–204, 2001.

Ochman, H., Gerber, A.S., and Hartl, D.L. Genetic applications of an inverse polymerase chain reaction. *Genetics* 120(3): 621–623, 1988.

Park, D.J. 3'RACE LaNe: A simple and rapid fully nested PCR method to determine 3'-terminal cDNA sequence. *Biotechniques* 36(4): 586–588, 590, 2004.

Park, D.J. A new 5' terminal murine GAPDH exon identified using 5'RACE LaNe. *Molecular Biotechnology* 29(1): 39–46, 2005. doi:10.1385/MB:29:1:39.

Pavlov, A.R., Pavlova, N.V., Kozyavkin, S.A., and Slesarev, A.I. Recent developments in the optimization of thermostable DNA polymerases for efficient applications. *Trends Biotechnol.* 22(5): 253–260, 2004. doi:10.1016/j.tibtech.2004.02.011.

Pavlov, A.R., Pavlova, N.V., Kozyavkin, S.A., and Slesarev, A.I. Thermostable DNA polymerases for a wide spectrum of applications: Comparison of a robust hybrid TopoTaq to other enzymes. In: *DNA Sequencing II: Optimizing Preparation and Cleanup*, J. Kieleczawa (ed.). Jones & Bartlett, Burlington, MA, pp. 241–257, 2006.

Pehlivan, T., Rubio, C., Rodrigo, L. et al. Impact of preimplantation genetic diagnosis on IVF outcome in implantation failure patients. *Reprod. Biomed.* Online 6(2): 232–237, March 2003.

Pernthaler, A., Pernthaler, J., and Rudolf, A. Fluorescence in situ hybridization and catalyzed reporter deposition for the identification of marine bacteria, applied and environmental microbiology. *Appl. Environ. Microbiol.* 68(6): 3094–3101, 2002. doi:10.1128/AEM.68.6.3094-3101.2002.

Pierce, K.E., and Wangh, L.J. Linear-after-the-exponential polymerase chain reaction and allied technologies real-time detection strategies for rapid, reliable diagnosis from single cells. *Methods Mol. Med.* 132: 65–85, 2007.

Ramanath, H.S., Chandrasekaran, M., Chua, C.K., Leong, K.F., and Shah, K.D. Modeling of extrusion behavior of biopolymer and composites in fused deposition modeling. *Key Engineering Materials* 334–335: 1241–1244, 2007.

Ramanath, H.S., Chua, C.K., Leong, K.F., and Shah, K.D. Melt flow behaviour of poly-epsilon-caprolactone in fused deposition modelling. *Journal of Materials Science: Materials in Medicine* 19(7): 2541–2550, 2008.

Rath, S., Cohn, D., and Hutmacher, D. Comparison of chondrogenesis in static and dynamic environments using a SFF designed and fabricated PCL-PEO scaffold. *Virtual and Physical Prototyping* 3(4): 209–219, 2008. [informaworld].

Rychlik, W., Spencer, W.J., and Rhoads, R.E. Optimization of the annealing temperature for DNA amplification in vitro. *Nucl. Acids Res.* 18(21): 6409–6412, 1990.

Sachlos, E., Reis, N., Ainsley, C., Derby, B., and Czernuszka, J.T. Novel collagen scaffolds with predefined internal morphology made by solid freeform fabrication. *Biomaterials* 24(8): 1487–1497, 2003.

Sachs, E.M., Haggerty, J.S., Cima, M.J., and Williams, P.A. Three-dimensional printing techniques. Patent US5204055, April 20, 1993.

Saiki, R.K., Gelfand, D.H., Stoffel, S., Scharf, S.J., Higuchi, R. et al. Primer-directed enzymatic amplification of DNA with a thermostable DNA polymerase. *Science* 239(4839): 487–491, 1988.

Sambrook, J., and Russel, D.W. In vitro amplification of DNA by the polymerase chain reaction. In: *Molecular Cloning: A Laboratory Manual*, 3rd edn, chapter 8. Cold Spring Harbor Laboratory Press, Cold Spring Harbor, NY, 2001. ISBN: 0-87969-576-5.

Sandalinas, M., Sadowy, S., Alikani, M., Calderon, G., Cohen, J., and Munné, S. Developmental ability of chromosomally abnormal human embryos to develop to the blastocyst stage. *Hum. Reprod.* 16(9): 1954–1958, 2001.

Sarrate, Z., Vidal, F., and Blanco, J. Role of sperm fluorescent in situ hybridization studies in infertile patients: Indications, study approach, and clinical relevance. *Fertil. Steril.* 93(6): 1892–1902, 2010.

Sharkey, D.J., Scalice, E.R., Christy, K.G. Jr., Atwood, S.M., and Daiss, J.L. Antibodies as thermolabile switches: High temperature triggering for the polymerase chain reaction. *Bio-Technology* 12: 506–509, 1994. doi:10.1038/nbt0594-506.

Shkumatov, A., Kuznyetsov, V., Cieslak, J., Ilkevitch, Y., and Verlinsky, Y. Obtaining metaphase spreads from single blastomeres for PGD of chromosomal rearrangements. *Reprod. Biomed.* Online 14(4): 498–503, 2007.

Sieben, V.J., Debes-Marun, C.S., Pilarski, L.M., and Backhouse, C.G. An integrated microfluidic chip for chromosome enumeration using fluorescence in situ hybridization. *Lab on a Chip* 8(12): 2151–2156, 2008.

Sieben, V.J., Debes Marun, C.S., Pilarski, P.M., Kaigala, G.V. et al. FISH and chips: Chromosomal analysis on microfluidic platforms. *IET Nanobiotechnology* 1(3): 27–35, 2007.

Staessen, C., Platteau, P., Van Assche, E. et al. Comparison of blastocyst transfer with or without preimplantation genetic diagnosis for aneuploidy screening in couples with advanced maternal age: A prospective randomized controlled trial. *Hum. Reprod.* 19(12): 2849–2858, 2004. doi:10.1093/humrep/deh536.

Stemmer, W.P., Crameri, A., Ha, K.D., Brennan, T.M., and Heyneker, H.L. Single-step assembly of a gene and entire plasmid from large numbers of oligodeoxyribonucleotides. *Gene.* 164(1): 49–53, 1995. doi:10.1016/0378-1119(95)00511-4.

Stender, H. PNA-FISH: An intelligent stain for rapid diagnosis of infectious diseases. *Expert Review in Molecular Diagnostics* 5: 649–655, 2003.

Summers, P.M., Campbell, J.M., and Miller, M.W. Normal in-vivo development of marmoset monkey embryos after trophectoderm biopsy. *Hum. Reprod.* 3(3): 389–393, 1988.

Uhrig, S., Schuffenhauer, S., Fauth, C., Wirtz, A., Daumer-Haas, C. et al. Multiplex-FISH for pre- and postnatal diagnostic applications. *American Journal of Human Genetics* 65: 448–462, 1999.

Verlinsky, Y., Ginsberg, N., Lifchez, A., Valle, J., Moise, J., and Strom, C.M. Analysis of the first polar body: Preconception genetic diagnosis. *Hum. Reprod.* 5(7): 826–829, 1990.

Verlinsky, Y., Rechitsky, S., Schoolcraft, W., Strom, C., and Kuliev, A. Preimplantation diagnosis for Fanconi anemia combined with HLA matching. *JAMA* 285(24): 3130–3133, 2001. doi:10.1001/jama.285.24.3130.

Vincent, M., Xu, Y., and Kong, H. Helicase-dependent isothermal DNA amplification. *EMBO Reports* 5(8): 795–800, 2004.

Yang, S., Leong, K., Du, Z., and Chua, C. The design of scaffolds for use in tissue engineering. Part II: Rapid prototyping techniques. *Tissue Engineering* 8(1): 1–11, 2002.

Yeong, W.Y., Chua, C.K., and Leong, K.F. Indirect fabrication of collagen scaffold based on inkjet printing technique. *Rapid Prototyping Journal* 12(4): 229–237, 2006.

Yeong, W.Y., Chua, C.K., Leong, K.F., and Chandrasekaran, M. Rapid prototyping in tissue engineering: Challenges and potential. *Trends in Biotechnology* 22(12): 643–652, 2004.

Yeong, W.Y., Chua, C.K., Leong, K.F., Chandrasekaran, M., and Lee, M.W. Comparison of drying methods in the fabrication of collagen scaffold via indirect rapid prototyping. *Journal of Biomedical Materials Research B Applied Biomaterials* 82(1): 260–266, 2007.

Yildirim, E., Besunder, R., Guceri, S., Allen, F., and Sun, W. Fabrication and plasma treatment of 3D polycaprolactane tissue scaffolds for enhanced cellular function. *Virtual and Physical Prototyping* 3(4): 199–207, 2008. [informaworld].

Yuan, D., Lasagni, A., Shao, P., and Das, S. Rapid prototyping of microstructured hydrogels via laser direct-write and laser interference photopolymerisation. *Virtual and Physical Prototyping* 3(4): 221–229, 2008. [informaworid].

Zeltinger, J., Sherwood, J.K., Graham, D.A., Meller, R., and Griffith, L.G. Effect of pore size and void fraction on cellular adhesion, proliferation, and matrix deposition. *Tissue Engineering* 7(5): 557–572, 2001.

Zietkiewicz, E., Rafalski, A., and Labuda, D. Genome fingerprinting by simple sequence repeat (SSR)-anchored polymerase chain reaction amplification. *Genomics* 20(2): 176–183, 1994. doi:10.1006/geno.1994.1151.

13 Industrial Biotechnology

LEARNING OBJECTIVES
- Discuss the significance of industrial biotechnology
- Explain the large-scale production of biomolecules using bioreactors
- Discuss types of fermenters and their applications
- Discuss upstream and downstream processes for biotechnology products
- Explain biosafety issues and automation of industrial plants
- Discuss product validation and regulation of biotechnology products
- Discuss industrial applications of microbes, plants, and mammalian cells in biotechnology product development

13.1 INTRODUCTION

With the rapid increase in the human population that created an increased demand for healthcare products, manufacturing large quantities of healthcare products (such as vaccines and antibiotics) became essential. This could only be possible by using large-sized bioreactors or fermenters. When we talk about industrial biotechnology, the first thing that comes to mind is large-scale production. In the broadest sense, industrial biotechnology is concerned with all aspects of business that relate to microbiology (Figure 13.1). In a more restricted sense, industrial microbiology is concerned with employing microorganisms to produce a desired product and preventing microbes from diminishing the economic value of various products. Some of the commercial products made by microbes are antibiotics, steroids, human protein, vaccines, vitamins, organic acids, amino acids, enzymes, alcohols, organic solvents, and synthetic fuels. The potential of microbes was also realized in the recovery of metals from ores through bioleaching, recovery of petrol, and single-protein production. The term fermentation in industrial microbiology is used in a wider sense to include any chemical transformation of organic compounds carried out by using microbes and their enzymes. Production methods in industrial microbiology bring together raw materials (substrates), microorganisms (specific strains or microbial enzymes), and a controlled favorable environment (created in a fermenter) to produce the desired substance.

13.2 FERMENTER OR BIOREACTOR

A *bioreactor* is a device in which a substrate of low value is utilized by living cells or enzymes to generate a product of higher value. Bioreactors are extensively used for food processing, fermentation, waste treatment, etc. Based on the agent used, bioreactors are grouped into two broad classes: those based on living cells and those employing enzymes. All bioreactors deal with heterogeneous systems having two or more phases (liquid, gas, solid); therefore, optimal conditions for fermentation necessitate efficient transfer of mass, heat, and momentum from one phase to the other. A bioreactor normally has the following functions: (1) agitation, for mixing of cells and medium; (2) aeration, such as aerobic fermenters for oxygen supply; (3) regulation of factors such as temperature, pH, pressure, aeration, nutrient feeding, and liquid level; (4) sterilization and maintenance of sterility; and (5) withdrawal of cells or medium, such as for continuous fermenters. Modern fermenters are usually integrated with computers for efficient process monitoring and data acquisition. However, in terms of process requirements, bioreactors can be classified into four categories:

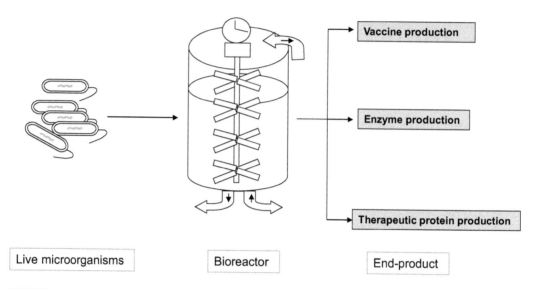

FIGURE 13.1 Application of live microorganisms in the development of various useful products.

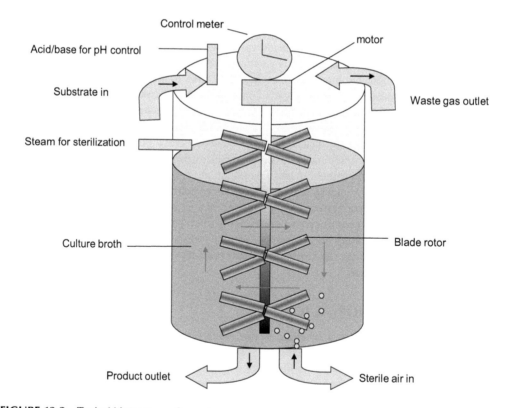

FIGURE 13.2 Typical bioreactor or fermenter.

aerobic, anaerobic, solid-state, and immobilized cell bioreactors. The size of fermenters ranges from 1.0 to 500,000 L or occasionally even more; fermenters of up to 1.2 million liters have been used. Generally, 20%–25% of fermenter volume is left unfilled with medium as "head space" to allow for splashing, foaming, and aeration. The design of fermenters varies greatly, depending on the type of fermentation for which it is used (Figure 13.2).

13.3 PRINCIPLE OF FERMENTATION

In a general sense, *fermentation* is the conversion of a carbohydrate such as sugar into an acid or an alcohol. More specifically, fermentation can refer to the use of yeast to change the sugar into alcohol or the use of bacteria to create lactic acid in certain foods. Given the right conditions, fermentation occurs naturally in many different foods, and humans have intentionally made use of it for thousands of years. The earliest uses of fermentation were to create alcoholic beverages such as mead, wine, and beer. These beverages were created as far back as 7000 BCE in parts of the Middle East. Fermentation of foods such as milk and various vegetables took place some thousands of years later in both the Middle East and China. Although the general principle of fermentation is the same across all drinks and foods, the precise methods of achieving it and the end results differ. Pickling foods such as cucumbers can be accomplished by submerging the vegetable one wants to pickle in a saltwater solution with vinegar added. Over time, bacteria create the lactic acid that gives the food its distinctive flavor and helps to preserve it. Other foods can be pickled simply by packing them in dry salt and allowing the natural fermentation process to occur. There are two types of fermentation processes: *aerobic fermentation* and *anaerobic fermentation*.

13.3.1 AEROBIC FERMENTATION

The main feature of aerobic fermentation is the provision of adequate aeration. In some cases, the amount of air needed per hour is about 60 times the medium volume; therefore, bioreactors used for aerobic fermentation have a provision for adequate supply of sterile air, which is generally sparged into the medium. In addition, these fermenters may have a mechanism for stirring and mixing the medium and cells. Aerobic fermenters may either be of the stirred-tank type, in which mechanical motor-driven stirrers are provided, or of the air-lift type, in which no mechanical stirrers are used and the agitation is achieved by the air bubbles generated by the air supply. Generally, these bioreactors are of closed or batch type, but continuous flow reactors are also used. Such reactors provide a continuous source of cells and are also suitable for product generation when the product is released into the medium. There are two major types of aerobic fermentation processes: *submerged culture method* and *semisolid* or *solid-state method*.

13.3.1.1 Submerged Culture Method

In this process, an organism is grown in a liquid medium that is vigorously aerated and agitated in large tanks (fermenters). The fermenter could be either an open tank or a closed tank, may be a batch type or a continuous type, and is generally made of a noncorrosive type of metal or is glass lined or made of wood. In batch fermentation, the organism is grown in a known amount of culture medium for a defined period of time, after which the cell mass is separated from the liquid before further processing in the continuous culture. The culture medium is withdrawn depending on the rate of product formation and the inflow of fresh medium. Most fermentation industries today use the submerged process for the production of microbial products.

13.3.1.2 Semisolid/Solid-State Methods

In this method, the culture medium is impregnated in a carrier such as bagasse, wheat bran, potato pulp, etc. and the organism is allowed to grow on this. This method allows a greater surface area for growth. The production of the desirable substance and its recovery is generally easier and satisfactory. In the development of a fermentation process, the composition of the culture medium plays a major role and will determine to a very great extent the level of the end product. For example, a culture medium containing sucrose enables better production of citric acid than any other carbohydrate by *Aspergillus niger*. The pH, temperature of incubation, aeration, etc. are all important factors in fermentation, and these have to be optimized for each type of fermentation. Emphasis is generally placed on the use of cheap raw materials so that the cost of production is low.

13.3.2 Anaerobic Fermentation

In anaerobic fermentation, a provision for aeration is usually not needed. Nevertheless, in some cases, aeration may be needed initially for inoculum buildup. In most cases, a mixing device is also unnecessary, although in some cases initial mixing of the inoculum is necessary. Once fermentation begins, the gas produced in the process generates sufficient mixing. The air present in the headspace of the fermenter should be replaced by carbon dioxide (CO_2), hydrogen (H_2), nitrogen (N_2), or a suitable mixture of these. This is particularly important for obligate anaerobes like *Clostridium*. The fermentation usually liberates CO_2 and H_2, which are collected and used. For example, CO_2 is used for making dry ice and methanol and for bubbling into freshly inoculated fermenters. In case of acetogens and other gas-utilizing bacteria, O_2-free, sterile CO_2 or other gases are bubbled through the medium.

Acetogens have been cultured in 400 L fermenters by bubbling sterile CO_2 and 3 kg cells could be harvested in each run. Recovery of products from anaerobic fermenters does not require anaerobic conditions. However, many enzymes of such organisms are highly O_2 sensitive; therefore, when recovery of such enzymes is the objective, cells must be harvested under strictly anaerobic conditions. Under anaerobic conditions, the pyruvate molecule can follow other anaerobic pathways to regenerate the nicotinamide adenine dinucleotide (NAD^+) necessary for glycolysis to continue. These include alcoholic fermentation and lactate fermentation (Figure 13.3). In the absence of oxygen, the further reduction or addition of H_2 ions and electrons to the pyruvate molecules that were produced during glycolysis is termed fermentation. This process recycles the reduced or hydrogenated NAD (NADH), to the free NAD^+ coenzyme, which once again serves as the H_2 acceptor, enabling glycolysis to continue. Alcoholic fermentation, which is characteristic of some plants and

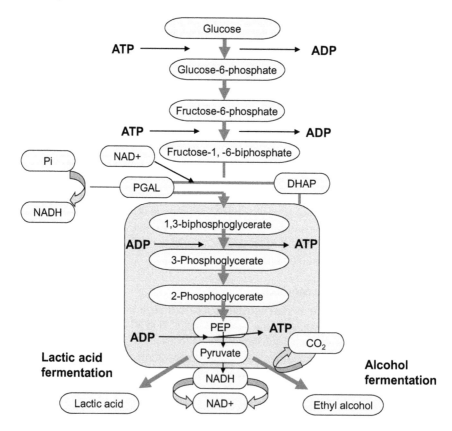

FIGURE 13.3 Typical bioreactor for producing lactic acid and ethyl alcohol.

many microorganisms, yields alcohol and CO_2 as its products. Yeast is used by the biotechnology industry to generate CO_2 gas necessary for bread making and in the fermentation of hops and grapes to produce alcoholic beverages. Reduction of pyruvate by NADH to release the NAD^+ necessary for the glycolytic pathway can also result in lactate fermentation, which takes place in some animal tissues and in some microorganisms. Lactic acid-producing bacterial cells are responsible for the souring of milk to produce yogurt. In working animal muscle cells, lactate fermentation follows the exhaustion of ATP stores. Fast-twitch muscle fibers store little energy and rely on quick spurts of anaerobic activity, but the lactic acid that accumulates within the cells eventually leads to muscle fatigue and cramp. The fermentation unit in industrial biotechnology is analogous to a chemical plant in the chemical industry. A fermentation process is a biological process and therefore has requirements of sterility and use of cellular enzymatic reactions instead of chemical reactions aided by inanimate catalysts, sometimes operating at elevated temperature and pressure. Industrial fermentation processes may be divided into two main types, with various combinations and modifications: *batch fermentation* and *continuous fermentation*.

13.3.2.1 Batch Fermentation Process

The tank of a fermenter is filled with the prepared mash of raw materials to be fermented. The temperature and pH for microbial fermentation are properly adjusted and nutritive supplements are occasionally added to the prepared mash. The mash is steam sterilized in a pure culture process. The inoculum of a pure culture is added to the fermenter from a separate pure culture vessel. Fermentation proceeds, and, after the proper time, the contents of the fermenter are taken out for further processing. The fermenter is cleaned and the process is repeated.

13.3.2.2 Continuous Fermentation Process

Growth of microorganisms during batch fermentation conforms to the characteristic growth curve, with a lag phase followed by a logarithmic phase. This, in turn, is terminated by progressive decrements in the rate of growth until the stationary phase is reached. This is because of the limitation of one or more of the essential nutrients. During continuous fermentation, the substrate is added to the fermenter continuously at a fixed rate, which maintains the organisms in the logarithmic growth phase. The fermentation products are taken out continuously. The design and arrangements for continuous fermentation are somewhat complex.

13.4 PRODUCTION OF BIOMOLECULES USING FERMENTER TECHNOLOGY

13.4.1 GLUCONIC ACID

Gluconic acid used in pharmaceutical industries is produced by the fermentation of glucose by strains of *A. niger*, *Penicillium* sp., or selected bacteria. In the commercial process, a nutrient solution containing 24%–38% glucose, corn steep liquor, a N_2 source and salts, with pH 4.5 is used to culture a selected strain of fungus in shallow pans or in submerged culture conditions to convert glucose into gluconic acid. The pH of the medium is controlled by the addition of a strong solution of sodium hydroxide. Fermentation is carried out at 33°C or 34°C. The medium composition and fermentation conditions determine the production of acids other than gluconic acid (such as citric acid and oxalic acid), and it is therefore important to select a mold strain and fermentation conditions that will avoid the formation of unwanted organic acid. After the fermentation, the cell-free broth is centrifuged and processed to recover gluconic acid.

13.4.2 CITRIC ACID

Citric acid, which is a key intermediate of the tricarboxylic acid (TCA) cycle, is produced by fungi, yeast, and bacteria as an overflow product owing to a faulty operation of the citric acid cycle. The ability of fungi to produce citric acid was first discovered by Carl Wehmer in 1893, and today all the citric acid commercially produced comes from mold fermentation. Among the organisms used for citric acid production, *A. niger* has been the mold of choice for several decades. A variety of carbohydrate sources such as beet molasses, cane molasses, sucrose, commercial glucose, and starch hydrolysates have been used for citric acid production. Among these, sucrose cane and beet molasses have been found to be the best. For citric acid production, the raw material is diluted to a 20%–25% sugar concentration and mixed with a N_2 source and other salts. The pH of the medium is maintained around 5 when molasses is used and at a lower level (pH 3.0) when sucrose is used. The fermentations are carried out either under the surface, submerged, or in solid-state conditions. In the surface culture method, shallow aluminum or stainless-steel pans are filled with the growth medium, inoculated with the fungal spores, and allowed to ferment. In the submerged culture method, the mold is cultured in fermenters under vigorous stirring and mixing, whereas in solid-state fermentation, the mold is grown over carrier material, such as bagasse, which is impregnated with the fermentation medium.

13.4.3 ACETONE BUTANOL

Acetone butanol fermentation is one of the oldest fermentation methods known. The fermentation is based on culturing various strains of *Clostridia* in a carbohydrate-rich media under anaerobic conditions to yield butanol and acetone. *Clostridium acetobutylicum* is the organism of choice in the production of these organic solvents. Until very recently, these fermentation methods were out of favor because of the availability of acetone and butanol from the petroleum industry. Today, there is considerable interest in these fermentations. However, the concentration of the end products in these fermentations is quite small, and the fermentations are a type of mixed fermentation yielding a mixture of compounds such as butyric acid, butanol, and acetone. Attempts to increase yields by use of genetically altered strains or changes in fermentation conditions have been partially successful.

13.4.4 ITACONIC ACID

Itaconic acid is used as a resin in detergents. The transformation of citric acid by *Aspergillus terreus* can be used for commercial production of itaconic acid. The fermentation process involves a well-aerated molasses—mineral salts medium at a pH below 2.2. At higher pH, this microbe degrades into itaconic acid. Like citric acid, low levels of trace metals must be used to achieve acceptable product yields.

13.4.5 GIBBERELLIC ACID

Gibberellic acid and related gibberellins are important growth regulators of plants. Commercial production of these acids helps in boosting agriculture. This acid is formed by the fungus *Gibberella fujikuroi* (imperfect state, *Fusarium moniliforme*) and can be produced commercially using aerated submerged cultures. A glucose–mineral salt medium, incubation at 25°C, and slightly acidic pH are used for fermentation. This process normally takes 2–3 days.

13.4.6 LACTIC ACID

Lactic acid is produced from various carbohydrates such as cornstarch, potato starch, molasses, and whey. Starchy materials when used are first hydrolyzed to simple sugars. The medium is then supplemented with an N_2 source and calcium carbonate. Arid fermentation is carried out by the inoculation with homofermentative lactobacilli such as *Lactobacillus bulgaricus* or *Lactobacillus delbruckli*. During fermentation, the temperature is controlled at 43°C–50°C, depending on the organism, and the medium is kept in constant agitation to keep the calcium carbonate in suspension. After completion of the fermentation (in 4–6 days), the fermented liquor is heated to 82°C and filtered. The filtrate containing calcium lactate is spray dried after treating with sodium sulfide. To obtain lactic acid, the calcium lactate is treated with sulfuric acid and the lactic acid thus obtained is further purified (Figure 13.3).

13.4.7 AMINO ACIDS

In recent years, there has been rapid development in the production of particular amino acids by fermentation. Microorganisms can synthesize amino acids from inorganic nitrogen compounds. The rate and the amount of synthesis of some amino acids may exceed the cells needed for protein synthesis, whereupon the amino acids are excreted into the medium. Some microorganisms are capable of producing sufficient amounts of certain amino acids to justify their commercial production. The amino acids can be obtained from hydrolyzing protein or from chemical synthesis, but in several instances the microbial process is more economical. In addition, the microbiological method yields the naturally occurring L-amino acids. The demand for amino acids for use in foods, feeds, and in the pharmaceutical industry is expanding. When production costs decrease, new usage is anticipated as raw material for amino acid polymers. Microbial production of amino acids shows outstanding features that are not usually encountered in the development of other microbiological processes; one is the importance of auxotrophic microorganisms.

Monosodium glutamate (sodium glutamate) is the sodium salt of glutamic acid, one of the most abundant naturally-occurring nonessential amino acids. MSG is found in tomatoes, Parmesan, potatoes, mushrooms, and other vegetables and fruits. Monosodium glutamate is used in the food industry as a flavor enhancer with an umami taste that intensifies the meaty, savory flavor of food, as naturally occurring glutamate does in foods such as stews and meat soups. The production of monosodium glutamate is generally made by bacterial fermentation as shown in Figure 13.4.

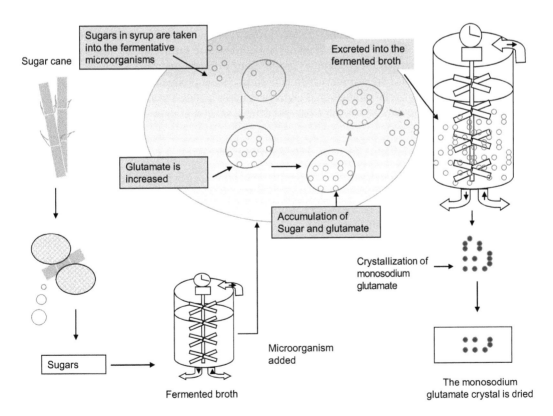

FIGURE 13.4 Production of monosodium glutamate by fermentation.

13.4.8 ENZYMES

Enzymes have important applications in industry and these are produced by different microbes, mostly by fungi. Some enzymes produced for industry are proteases, amylases, alpha-amylase, glucamylase, glucose isomerases, glucose oxidases, rennin, pectinases, and lipases. Of these, proteases, glucamylase, alpha-amylase, and glucose isomerase are produced extensively.

13.4.9 PROTEASES

Microbial proteases have been extensively used in modern detergent formulations and also in laundries, where they are used to remove stains caused by milk, eggs, or blood. Proteases are heat stable and are largely produced by *Bacillus licheniformis*. Other alkaline proteases are being developed using recombinant DNA technology to function over a wide range of pH and temperature. One such recombinant strain is *Bacillus* sp. *Gx 6644*, active for milk casein. Another strain, *Bacillus* sp. *Gx 6638*, produces several alkaline proteases. Through this technology, a recombinant strain of a bacterium has also been developed. It produces the enzyme kerazyme, which is used for dissolving hair and opening hair clogged drains. These enzymes also have important applications in the baking industry. Microbial proteases reduce mixing time and improve the quality of bread loaves. Fungal proteases are mainly obtained from *Aspergillus* spp., while bacterial proteases are obtained from *Bacillus* spp. These enzymes are used as meat tenderizers and for the bating of hides in the leather industry.

13.4.10 AMYLASES

Amylases are used for the preparation of sizing agents in the textile industry; for the preparation of starch sizing pastes for use in the paper industry; in the production of bread, chocolate syrup, and corn syrup; and for stain removal in laundries. There are various types of amylases: α-amylases, β-amylases, and glucamylases. *Aspergillus oryzae*, *A. Niger*, *Bacillus subtilis*, and *Bacillus diastaticus* are principally used. The conversion of starch to high-fructose syrup utilizes amylases (i.e., in producing sweeteners). Various other enzymes produced by different microbes also have industrial applications. Some examples are rennin, which is used in cheese production, and *Mucor pussilus*, which is used for its commercial production. Fungal pectinases are used to clarify fruit juices, while glucose oxidase is used to remove oxygen from soft drinks, salad dressings, etc.

13.5 DEVELOPMENT PROCESS OF MICROBIAL PRODUCTS

After describing the various components of fermentation technology, we will next take a brief look at the development process of microbial products using fermentation technology so that readers have a better understanding of the various phases and stages of product development. The list of microbial products is shown in the Table 13.1.

TABLE 13.1
Industrial Products and Microbes

Supplier	Bacteria	Product
Novo Nordisk, Denmark	*Bacillus sp.*	Detergent
Genencor International, USA	*Bacillus sp.*	Detergent
Gist-Brocades, Netherland	*Bacillus sp.*	Detergent
Solvay Enzymes, Germany	*Bacillus sp.*	Detergent
Wuxi Snyder Bioproducts, China	*Bacillus sp.*	Detergent

13.5.1 ISOLATION AND SCREENING OF MICROORGANISMS

The success of an industrial fermentation process chiefly depends on the strain of the microorganism used. An ideal producer or economically important strain should have the following characteristics:

It should be pure and free from phage.

It should be genetically stable, but amenable to genetic modification.

It should produce both vegetative cells and spores; species producing only mycelium are rarely used.

It should grow vigorously after inoculation in seed stage vessels.

It should produce a single valuable product and no toxic by-products.

Product should be produced in a short time.

It should be amenable to long-term conservation.

The risk of contamination should be minimal under optimum performance conditions.

13.5.1.1 Isolation of Microorganisms

The first step in developing a producer strain is the isolation of the relevant microorganisms from their natural habitats. Alternatively, microorganisms can be obtained as pure cultures from organizations that maintain culture collections, such as the American Type Culture Collection (ATCC) in Rockville, Maryland, USA; the Commonwealth Mycological Institute (CMI) in Kew, Surrey, England; the Fermentation Research Institute (FERM) in Tokyo, Japan; and the Research Institute for Antibiotics, in Moscow, Russia. Microorganisms of industrial importance are usually bacteria, actinomycetes, fungi, and algae. These organisms occur virtually everywhere, such as in air, water, soil, surfaces of plants and animals, and in plant and animal tissues. However, the most common sources of industrial microorganisms are soils and lake and river mud. Often, the ecological habitat from which a desired microorganism is more likely to be isolated will depend on the characteristics of the product desired from it and on the development process; for example, if the objective is to isolate a source of enzymes that can withstand high temperatures, the obvious place to look would be hot springs.

A variety of complex isolation procedures have been developed, but no single method can reveal all the microorganisms present in a sample. Many different microorganisms can be isolated by using specialized enrichment techniques such as soil treatment (UV irradiation, air drying or heating at 70°C–120°C, filtration or continuous percolation, washings from root systems, treatment with detergents or alcohols, preinoculation with toxic agents), selective inhibitors (antimetabolites, antibiotics, etc.), nutritional (specifically C and N) sources, variations in pH, temperature, aeration, and so on. The enrichment techniques are designed for selective multiplication of only some of the microorganisms present in a sample. These approaches, however, take a long time (about 20–40 days) and require considerable labor and money. The main isolation methods used routinely soil samples which are sponging (soil directly), dilution, gradient plate, aerosol dilution, flotation, and differential centrifugation. Often, these methods are used in conjunction with an enrichment technique.

13.5.1.2 Screening of Microorganisms

After isolation of the microorganisms, the next step is their screening. A set of highly selective procedures, which allows the detection and isolation of microorganisms producing the desired metabolite, constitutes primary screening. Ideally, primary screening should be rapid, inexpensive, predictive, and specific, but effective for a broad range of compounds and applicable on a large scale. Primary screening is time-consuming and labor-intensive, as a large number of isolates have to be screened to identify a few potential ones. However, this is possibly the most critical step, because it eliminates the large bulk of unwanted useless isolates, which are either nonproducers or producers of known compounds. Computer-based databases play an important role by instantaneously providing detailed information about the already known microbial antibiotic compounds.

Rapid and effective screening techniques, which utilize either a property of the product or that of its biosynthetic pathway for detection of desirable isolates, have been devised for a variety of microbial products. Some of the screening techniques are relatively simple, such as that for extracellular enzymes and enzyme inhibitors. However, for most microbial products of high value, the screening is usually complex and tedious and often may involve two or more steps, such as for antimicrobials. In some cases, it may be desirable to concentrate on a group of organisms expected to yield new products. For example, the search for new antibiotics now focuses on rare actinomycetes, (i.e., actinomycetes other than those belonging to the genus *Streptomyces*). Suitably designed specialized screening techniques may be used to detect compounds having various pharmacological activities other than antibiotics.

13.5.2 Inoculum Development

The preparation of a population of microorganisms from a dormant stock culture to an active state of growth that is suitable for inoculation in the final production stage is called *inoculum development*. As a first step in inoculum development, the inoculum is taken from a working stock culture to initiate growth in a suitable liquid medium. Bacterial vegetative cells and spores are suspended, usually in sterile tap water, which is then added to the broth. In case of non-sporulating fungi and actinomycetes, the hyphae are fragmented and then transferred to the broth. Inoculum development is generally done using flask cultures. Flasks of 50 mL to 12 L may be used and their number can be increased as needed. In some cases, small fermenters may be used. Inoculum development is usually done in a stepwise sequence to increase the volume to the desired level. At each step, the inoculum is used at 0.5%–5% of the medium volume, which allows a 20–200-fold increase in inoculum volume at each step. Typically, the inoculum used for the production stage is about 5% of the medium volume.

13.5.3 Culture Media

Inoculum preparation media are quite different from production media. These media are designed for rapid microbial growth, and little or no product accumulation will normally occur. Many production processes depend on inducible enzymes. In all such cases, the appropriate inducers must be included either in all the stages or at least in the final stages of inoculum development. This will ensure the presence of the concerned inducible enzymes at high levels for the production to start immediately after inoculation.

13.5.4 Contamination

The inoculum used for production tanks must be free from contamination. However, as the risk of contamination is always present during inoculum development, every effort must be made to detect as well as prevent contamination. The most common contaminants of different industrial processes are considerably different. Some examples are as follows:

- In the canning industry, *Clostridium butylicum* is the chief concern. This obligate anaerobe can grow in sealed cans, producing heat-resistant spores and a deadly toxin. However, this is not a problem for ketchup (too acidic), jam and jellies (too high a sugar concentration), and milk (stored at too low a temperature).
- Organisms like *Lactobacillus* are a problem in wine production.
- In the antibiotic industry, some potential contaminants are molds, yeast, and bacteria such as *Bacillus*.
- The most dreaded contaminants in the fermentation industry are phages. The only effective protection against phages is to develop resistant strains.

13.5.5 STERILIZATION

Sterilization is the process of inactivating or removing all living organisms from a substance or surface. In theory, it is regarded as absolute, in that all living cells must be inactivated or removed, usually in a single step at the given time; in practice, however, the success of sterilization procedures is only a probability. Therefore, the probability of a cell escaping inactivation/filtration does exist, although it is usually very small. When a closed system is sterilized once, it remains so indefinitely, as it has no openings for the entry of microorganisms. Most fermentation vessels are open systems. Such systems are initially sterilized and must be kept sterilized by ensuring the removal of living cells at their entry points, such as the cotton plug of a culture flask. Some methods of sterilization are discussed in the following.

13.5.5.1 Heating

Heat is the most commonly used and the least expensive sterilizing agent. Dry heat is used in ovens and is suitable for sterilization of solids that can withstand the high temperatures needed for sterilization. Steam, which is moist or wet heat, is used for sterilization of media and fermenter vessels. An autoclave uses steam for sterilization (at temperature 121°C and pressure 15 psi). The period of time at this temperature and pressure depends on the medium volume; for example, 12–15 min for 200 mL, 17–22 min for 500 mL, 20–25 min for 1 L, and 30–35 min for 2 L. However, sterilization of oils will require a few hours, and concentrated media (10%–20% solid) must be agitated for effective sterilization.

Autoclaves can also be used to sterilize laboratory vessels, small volumes of media, and even small fermenters. Large fermenters are sterilized either by a direct injection of steam or by indirect heating by passing steam through heat exchange coils or a jacket. The steam should always be saturated. Media sterilization may be achieved in a continuous flow sterilization system, either by direct steam injection or by indirect steam heating, and then filled in a sterile fermenter. Alternatively, the medium may be filled in the fermenter and steam sterilized with the latter. Heat killing is largely due to protein inactivation. In general, moist heat is far superior to dry heat. Bacterial spores are the most heat resistant. For example, spores of thermophilic bacteria can survive steam at 30 psi at 134°C for 1–10 min and dry heat at 180°C for up to 15 min.

13.5.5.2 Radiation

High-energy x-rays are used for sterilization of a variety of labware and food. In general, vegetative cells are much more susceptible than bacterial spores (*Clostridium* spores can resist nearly 0.5 M rad). However, *Deinococcus radiodurans* vegetative cells can survive 6 M rad. Viruses are usually similar to bacterial spores, but some viruses (such as the encephalitis virus) require up to 4.5 M rad. In practice, 2.5 M rad is used for sterilizing pharmaceutical and medical products. X-rays cause inactivation by inducing single- and double-strand DNA breaks and by producing free radicals and peroxides, to which SH enzymes are particularly susceptible.

13.5.5.3 Chemicals

The chemicals used for sterilization, such as formaldehyde, hydrogen peroxide (H_2O_2), ethylene oxide, and propylene oxide, cause inactivation by oxidation or alkylation. H_2O_2 (10%–25% w/v) is being increasingly used in the sterilization of milk and of containers for food products.

H_2O_2 is a powerful oxidizing agent, kills both vegetative cells and spores, and is very safe. Ethylene oxide is used to sterilize equipment that is likely to be damaged by heat and is very effective, but highly toxic and violently explosive if mixed with air.

13.5.5.4 Filtration

Aerobic fermentation requires a very high rate of air supply, often equaling one volume of air (equal to medium volume) every minute. Air contains both fungal spores and bacteria, which are

TABLE 13.2
Industrial Products and Microbes

End-Product	Starting Materials	Microorganisms
Ethanol	*Malt extract*	*Saccharomyces cerevisiae* (yeast and fungus)
Acetic acid	*Grape and other fruits*	*Saccharomyces cerevisiae* (yeast and fungus)
Lactic acid	*Milk*	Acetobacter

ordinarily removed by filtration using either a depth filter or a screen filter. Depth filters are made from fibrous or powdered materials pressed or bonded together in a relatively thick layer. Some materials used are fiberglass, cotton, mineral wool, and cellulose fibers in the form of mats, wads, or cylinders.

13.5.6 STRAIN IMPROVEMENT

After an organism producing a valuable product is identified, it becomes necessary to increase the product yield from fermentation to minimize production costs. Product yields can be increased by developing a suitable medium for fermentation, refining the fermentation process, and by improving the productivity of the strain. Generally, major improvements arise from the last approach, and all fermentation enterprises therefore place considerable emphasis on this activity. The techniques and approaches used to genetically modify strains and to increase the production of the desired product are called *strain improvement* or *strain development*. Strain improvement can be done by using mutant selection, recombination, and rDNA technology. Table 13.2 shows different approaches to improve strains.

13.6 UPSTREAM BIOPROCESS

The upstream part of a bioprocess refers to the first step in which biomolecules are grown, usually by bacterial or mammalian cell lines in bioreactors (Figure 13.5). When they reach the desired density (for batch and fed-batch cultures), they are harvested and moved to the downstream section of the bioprocess. The following discusses the different kinds of culture methods in biotechnology.

13.6.1 INDUSTRIAL MICROBIAL CULTURE

The use of microbes to produce natural products and processes that benefit and improve human healthcare has been a part of human history since early civilization. The production of secondary metabolites (such as antibiotics) often occurs during the stationary growth phase (or idiophase) of a bacterial culture. Secondary metabolites are nongrowth related and seem to play no obvious role in cell maintenance. The production of secondary metabolites largely concerns the first growth phase of a bacterial culture. This production is considered a two-stage process, with a trophophase (growth phase) and an idiophase (production phase), which usually occurs when culture growth has slowed or ceased. The major aim for mass production of secondary metabolites is to maximize the biomass in a short trophophase, while optimizing the conditions for high, sustained idiophase production. The growth of a bacterial culture can be represented by a curve that consists of the following four stages or phases:

1. *Lag phase*—growth and reproduction are just beginning
2. *Log phase*—reproduction is occurring at an exponential rate

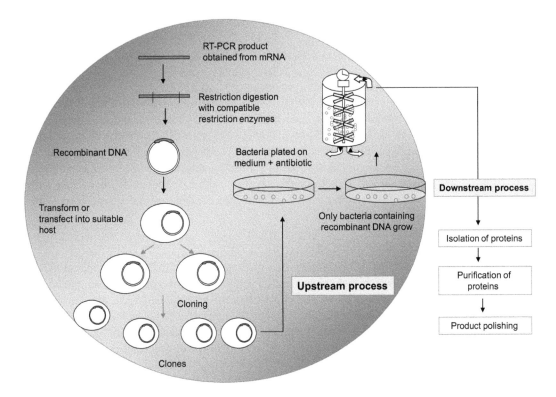

FIGURE 13.5 Production of biotechnology product: Upstream and downstream processes.

3. *Stationary phase*—the environmental surroundings and food supply cannot support any more exponential growth
4. *Death phase*—when all the nutrients have been exhausted and the population dies off

A variety of parameters are important to influence the production of desired secondary metabolites in bioreactor cultures, which include substrate, slow or fast utilization, N_2 requirements, amino acid supplements, temperature, pH, oxygen concentration, and pressure.

13.6.2 Mammalian Cell Culture

Cell culture is the process by which cells are grown under controlled conditions. In practice, the term "cell culture" has come to refer to the culturing of cells derived from multicellular eukaryotes, especially animal cells. The historical development and methods of cell culture are closely interrelated to those of tissue culture and organ culture. Animal cell culture became a common laboratory technique in the mid-1900s, but the concept of maintaining live cell lines separated from their original tissue source was discovered in the nineteenth century. Cells are grown and maintained at an appropriate temperature and gas mixture (typically, 37°C, 5% CO_2 for mammalian cells) in a cell incubator. Culture conditions vary widely for each cell type, and variation of conditions for a particular cell type can result in different phenotypes being expressed. Aside from temperature and gas mixture, the most commonly varied factor in culture systems is the growth medium. Recipes for growth media can vary in pH, glucose concentration, growth factors, and the presence of other nutrients. The growth factors used to supplement media are often derived from animal blood, such as calf serum. One complication of these blood-derived ingredients is the potential for contamination of the culture with viruses or prions, particularly in biotechnology medical applications.

Current practice is to minimize or eliminate the use of these ingredients wherever possible, but this cannot always be accomplished.

Cells can be grown in *suspension* or as *adherent* cultures. Some cells naturally live in suspension without being attached to a surface, such as cells that exist in the bloodstream. There are also cell lines that have been modified to be able to survive in suspension cultures so that they can be grown to a higher density than adherent conditions would allow. Adherent cells require a surface, such as tissue culture plastic or microcarrier, which may be coated with extracellular matrix components to increase adhesion properties and provide other signals needed for growth and differentiation. Most cells derived from solid tissues are adherent. Another type of adherent culture is the *organotypic culture*, which involves growing cells in a three-dimensional environment as opposed to two-dimensional culture dishes. This three-dimensional culture system is biochemically and physiologically more similar to *in vivo* tissue, but is technically challenging to maintain because of many factors (e.g., diffusion).

13.6.2.1 Manipulation of Cultured Cells

Generally, as cells continue to divide in a culture, they grow to fill the available area or volume. This can generate several issues, which include nutrient depletion in the growth media; accumulation of apoptotic/necrotic (dead) cells; cell-to-cell contact that can stimulate cell cycle arrest, causing cells to stop dividing, known as *contact inhibition* or *senescence*; and cell-to-cell contact that can stimulate cellular differentiation. Among the common manipulations carried out on cultured cells are media changes, passaging cells, and transfecting cells. These are generally performed using tissue culture methods that rely on sterile technique. Sterile technique aims to avoid contamination with bacteria, yeast, or other cell lines. Manipulations are typically carried out in a biosafety hood or laminar flow cabinet to exclude contaminating microorganisms. Antibiotics (e.g., penicillin and streptomycin) and antifungals (e.g., amphotericin B) can also be added to the growth medium. As cells undergo metabolic processes, acid is produced and the pH decreases. Often, a pH indicator is added to the medium in order to measure nutrient depletion.

13.6.2.2 Generation of Hybridomas

It is possible to fuse normal cells with an immortalized cell line. This method is used to produce monoclonal antibodies. In brief, lymphocytes isolated from the spleen (or possibly from the blood) of an immunized animal are combined with an immortal myeloma cell line (B cell lineage) to produce a hybridoma that has the antibody specificity of the primary lymphocyte and the immortality of the myeloma. Selective growth hypoxanthine-aminopterin-thymidine medium (HAT medium) is used to select against unfused myeloma cells. Primary lymphocytes die quickly in culture and only the fused cells survive. To start with, these are screened for production of the required antibody, generally in pools, and then single cloning is done afterwards.

13.6.3 Nonmammalian Culture Methods

13.6.3.1 Industrial Plant Cell Culture

Plant secondary metabolites are important for plant interactions with their environments. These metabolites give plants important properties such as antibiotic, antifungal, antiviral, UV protection, and pest deterrence. These chemicals have also been used as medicines by humans for thousands of years. Today, plants are responsible for the production of over 30,000 types of chemicals, including pharmaceuticals, aromas, pigments, cosmetics, nutraceutics, and other fine chemicals. Early on, it was discovered that metabolites could be accumulated in higher quantities using plant cell cultures as compared with whole plant cultivation. However, since plant cell cultures have low yield compared to their bacterial counterparts, they are primarily used only for high-value metabolites. The biosynthetic routes and mechanisms describing the production of plant secondary metabolites are often very complex and can involve dozens of enzymes. Through the study of plant metabolism, a new discipline called *metabolic engineering* has emerged. The goal of metabolic engineering is to improve cellular activity by manipulating enzymatic, transport, and regulatory functions of a cell. This is accomplished through the use of rDNA technology. Plant cell cultures are typically grown as cell suspension cultures in liquid medium or as callus cultures on solid medium. The culturing of undifferentiated plant cells and calli requires the proper balance of the plant growth hormones auxin and cytokinin.

13.6.3.2 Bacterial/Yeast Culture Methods

For bacteria and yeast, small quantities of cells are usually grown on a solid support that contains nutrients embedded in it, usually a gel such as agar, while large-scale cultures are grown with the cells suspended in a nutrient broth.

13.6.3.3 Viral Culture Methods

The culture of viruses requires the culture of cells of mammalian, plant, fungal, or bacterial origin as hosts for the growth and replication of the virus. Whole wild type viruses, recombinant viruses, or viral products may be generated in cell types other than their natural hosts under the right conditions. Depending on the species of the virus, infection and viral replication may result in host cell lysis and formation of a viral plaque.

13.7 DOWNSTREAM BIOPROCESS

The various processes used for the actual recovery of useful products from fermentation or any other industrial process are called *downstream processing* (DSP) (Figure 13.5). As the cost of DSP is often more than 50% of the manufacturing cost, and there is product loss at each step, DSP should be efficient and cost-effective and should involve as few steps as possible to avoid product loss. DSP refers to the recovery and purification of biosynthetic products, particularly pharmaceuticals, from natural sources such as animal or plant tissue or fermentation broth, including the recycling of salvageable components and the proper treatment and disposal of waste. It is an essential step in the manufacture of pharmaceuticals such as antibiotics, hormones (e.g., insulin and human growth hormone), antibodies (e.g., infliximab and abciximab), and vaccines; antibodies and enzymes used in diagnostics; industrial enzymes; and natural fragrance and flavor compounds. DSP is usually considered a specialized field in biochemical engineering, itself a specialization within chemical engineering, though many of the key technologies were developed by chemists and biologists for laboratory-scale separation of biological products. DSP and analytical bioseparation both refer to the separation or purification of biological products, but at different scales of operation and for different purposes. DSP implies manufacture of a purified product fit for a specific use, generally in marketable quantities, whereas analytical bioseparation refers to purification for the sole purpose of measuring a component or components of a mixture and may deal with sample sizes as small as a single cell.

13.7.1 Stages in DSP

A widely recognized heuristic for categorizing DSP operations divides them into four groups that are applied in order to bring a product from its natural state as a component of a tissue, cell, or fermentation broth through progressive improvements in purity and concentration.

13.7.1.1 Removal of Insolubles

This first step involves the capture of the product as a solute in a particulate-free liquid, for example the separation of cells, cell debris, or other particulate matter from fermentation broth containing an antibiotic. Typical operations to achieve this are filtration, centrifugation, sedimentation, flocculation, electroprecipitation, and gravity settling. Additional operations such as grinding, homogenization, or leaching, which are required to recover products from solid sources such as plant and animal tissues, are usually included in this group.

13.7.1.2 Product Isolation

Product isolation is the removal of those components whose properties vary markedly from that of the desired product. For most products, water is the chief impurity and isolation steps are designed to remove most of it, reducing the volume of material to be handled and concentrating the product. Solvent extraction, adsorption, ultrafiltration, and precipitation are some of the unit operations involved.

13.7.1.3 Product Purification

Product purification is done to separate those contaminants that resemble the product very closely in physical and chemical properties. Steps in this stage are expensive to carry out and require sensitive and sophisticated equipment. This stage contributes a significant fraction of the entire DSP expenditure. Examples of operations include affinity, size exclusion, reversed phase chromatography, crystallization, and fractional precipitation.

13.7.1.4 Product Polishing

Product polishing describes the final processing steps which end with packaging of the product in a form that is stable, easily transportable, and convenient. Crystallization, desiccation, lyophilization, and spray drying are typical unit operations. Depending on the product and its intended use, polishing may also include operations to sterilize the product and remove or deactivate trace contaminants that might compromise product safety. Such operations might include the removal of viruses or depyrogenation. A few product recovery methods may be considered to combine two or more stages; for example, expanded bed adsorption accomplishes removal of insoluble and product isolation in a single step. Affinity chromatography often isolates and purifies in a single step.

13.8 INDUSTRIAL PRODUCTION OF HEALTHCARE PRODUCTS

13.8.1 Antibiotic Manufacturing

The manufacturers of antibiotics maintain pilot plant facilities whose purpose is to upgrade yields and to bring about improvements in processing procedures. Studies are continually being made on strain improvement, inoculum conditions, fermentation conditions, and various combinations of these factors. For example, improved mutant strains almost always require adjustments in fermentation conditions in order to achieve higher yields in fermenters that are obtainable in shaken flasks. Because all of the antibiotics are made by aerobic fermentations, there are a number of similarities in the processes used for their production. Many of the specific details regarding the plant scale production of a given antibiotic remain trade secrets, with certain exceptions. However, the general outline of these methods is fairly well known to those connected with the industry.

13.8.1.1 Penicillin Production

The mold from which Alexander Fleming isolated penicillin was later identified as *Penicillium notatum*. A variety of molds belonging to other species and genera were later found to yield greater amounts of the antibiotic and a series of closely related penicillins. The naturally occurring penicillins differ from each other in the side chain (R group). Penicillin was produced by a surface culture method early in World War II. Submerged culture methods were introduced by 1943 and are now almost exclusively employed. Penicillin production needs strict aseptic conditions. Contamination by other microorganisms reduces the yield of penicillin. This is caused by the widespread occurrence of penicillinase-producing bacteria that inactivate the antibiotic. Penicillin production also needs a tremendous amount of air. In all methods, deep tanks with a capacity of several thousand gallons are filled with a culture medium. The medium consists of corn steep liquor, lactose, glucose, nutrient, salts, phenylacetic acid or a derivative, and calcium carbonate as a buffer. The medium is inoculated with a suspension of conidia of *P. chrysogenum*. The medium is constantly aerated and agitated, and the mold grows throughout as pellets. After about seven days, growth is complete, the pH rises to 8.0 or above, and penicillin production ceases. When the fermentation is complete, the masses of mold growth are separated from the culture medium by centrifugation and filtration.

13.8.1.2 Cephalosporins Production

Similar semisynthetic approaches can be used for the manufacture of other antibiotics. Cephalosporin C is made as the fermentation product of *Cephalosporium acremonium*. However, this form is not potent for clinical use. Its molecule can be transformed by removal of an aminoadipic acid side chain to form 7-α aminocephalosporanic acid (7-ACA), which can be further modified by adding side chains to form clinically useful broad-spectrum antimicrobials. Various side chains can be added to as well as removed from both 6-APA and 7-ACA to produce antibiotics with varying spectra of activities and varying degrees of resistance to inactivation by enzymes produced by pathogenic microbes. Thus, we have entered into the so-called third-generation cephalosporins, such as *moxalactam*, developed for control of bacteria that produce enzymes capable of degrading penicillins and cephalosporins.

13.8.1.3 Streptomycin Production

Streptomycin and various other antibiotics are produced using strains of *Streptomyces griseus*. Spores of this actinomycete are inoculated into a medium to establish a culture with a high mycelial biomass for introduction into an inoculum tank, with subsequent use of the mycelial inoculum to initiate the fermentation process in the production tank. The medium contains soybean meal (N-source), glucose (C-source) and NaCl. The process is carried out at 28°C and the maximum production is achieved at a pH range of 7.6–8.0. High agitation and aeration are needed. The process lasts for about 10 days. The classic fermentation process involves three phases. During the first phase, there is a rapid growth of the microbe with production of mycelial biomass. Proteolytic activity of the microbe releases ammonia (NH_3) to the medium from the soybean meal, causing a rise in pH. During this initial fermentation phase, there is little production of streptomycin. During the second phase, there is little additional production of mycelium, but the secondary metabolite, streptomycin, accumulates in the medium. The glucose and NH_3 released are consumed during this phase. The pH remains fairly constant—between 7.6 and 8.0. In the third and final phase, when carbohydrates become depleted, streptomycin production ceases and the microbial cells begin to lyse, pH increases, and the process normally ends by this time. After the process is complete, mycelium is separated from the broth by filtration and the antibiotic is recovered. In one method of recovery and purification, streptomycin is adsorbed onto activated charcoal and eluted with acid alcohol. It is then precipitated with acetone and further purified by use of column chromatography.

MANUFACTURING PROCESS OF ANTIBIOTICS

Step 1: Before fermentation can begin, the desired antibiotic-producing organism must be isolated and its numbers must be increased by many times. To do this, a starter culture from a sample of previously isolated, cold-stored organisms is created in the lab. In order to grow the initial culture, a sample of the organism is transferred to an agar-containing plate. The initial culture is then put into shake flasks along with food and other nutrients necessary for growth. This creates a suspension, which can be transferred to seed tanks for further growth.

Step 2: The seed tanks are steel tanks designed to provide an ideal environment for growing microorganisms. They are filled with all things a specific microorganism would need to survive and thrive, including warm water and carbohydrates like lactose or glucose sugars.

Additionally, they contain other necessary carbon sources such as acetic acid, alcohols, or hydrocarbons, and N_2 sources like ammonia salts. Growth factors like vitamins, amino acids, and minor nutrients round out the composition of the seed tank contents. The seed tanks are equipped with mixers, which keep the growth medium moving, and a pump to deliver sterilized, filtered air. After approximately 24–28 h, the material in the seed tanks is transferred to the primary fermentation tanks.

Step 3: The fermentation tank is essentially a larger version of the steel seed tank, which is able to hold about 30,000 gal. The fermentation tank is filled with growth media that provide an environment suitable for growth. Here, the microorganisms are allowed to grow and multiply. During this process, they excrete large quantities of the desired antibiotic. The tank is cooled to keep the temperature between 73°F and 81°F (23°C and 27.2°C). It is constantly agitated, and a continuous stream of sterilized air is pumped into it. For this reason, anti-foaming agents are periodically added. Since pH control is vital for optimal growth, acids or bases are added to the tank as necessary.

Step 4: After 3–5 days, the maximum amount of antibiotic will have been produced and the isolation process can begin. Depending on the specific antibiotic produced, the fermentation broth is processed by various purification methods. For example, for antibiotic compounds that are water soluble, an ion-exchange method may be used for purification. In this method, the compound is first separated from the waste organic materials in the broth and then sent through equipment that separates the other water-soluble compounds from the desired one. To isolate an oil-soluble antibiotic such as penicillin, a solvent extraction method is used. In this method, the broth is treated with organic solvents such as butyl acetate or methyl isobutyl ketone, which can specifically dissolve the antibiotic. The dissolved antibiotic is then recovered using various organic chemical means. At the end of this step, the manufacturer is typically left with a purified powdered form of the antibiotic, which can be further refined into different product types.

Step 5: Antibiotic products can take on many different forms. They can be sold in solutions for intravenous bags or syringes, in pill or gel capsule form, or as powders that are incorporated into topical ointments. Depending on the final form of the antibiotic, various refining steps may be taken after the initial isolation. For intravenous bags, the crystalline antibiotic can be dissolved in a solution and put in the bag, which is then hermetically sealed. For gel capsules, the powdered antibiotic is physically filled into the bottom half of a capsule, after which the top half is mechanically put in place. When used in topical ointments, the antibiotic is mixed into the ointment.

Step 6: From this point, the antibiotic product is transported to the final packaging stations where the products are stacked and put in boxes. They are loaded up on trucks and transported to various distributors, hospitals, and pharmacies. The entire process of fermentation, recovery, and processing can take anywhere from 5 to 8 days.

Step 7: Quality control is of utmost importance in the production of antibiotics. Since it involves a fermentation process, steps must be taken to ensure that absolutely no contamination is introduced at any point during production. To this end, the medium and all of the processing equipment are thoroughly steam sterilized. During manufacturing, the quality of all the compounds is checked on a regular basis. Of particular importance are frequent checks of the condition of the microorganism culture during fermentation. These checks are accomplished using various chromatography techniques. Also, various physical and chemical properties of the finished product are checked, such as pH, melting point, and moisture content. In the United States, antibiotic production is highly regulated by the FDA. Depending on the application and type of antibiotic, more or less testing must be completed. For example, the FDA requires that for certain antibiotics each batch must be checked by them for effectiveness and purity; only after they have certified the batch can it be sold for general consumption.

13.8.2 STEROIDS PRODUCTION

Microbial biotransformation of steroids is very important in the pharmaceutical industry. Steroids are used in the treatment of various disorders and are also involved in regulation of sexuality. Chemical synthesis of steroids is very complex because of the requirement to achieve the necessary precision of substitute location. For example, cortisone can be synthesized chemically from deoxycholic acid, but the process requires 37 steps, many of which must be carried out under extreme conditions of temperature and pressure, with the resulting product costing over $200 per g. The most difficult is the introduction of an oxygen atom at number 11 position of the steroid ring, but this can be accomplished by some microorganisms. The fungus *Rhizopus nigricans*, for example, hydroxylates progesterone, forming another steroid with the introduction of oxygen at the number 11 position. Similarly, the fungus *Cunninghamella blakesleeana* can hydroxylate the steroid cortexolone to form hydrocortisone with the introduction of oxygen at the number 11 position. Other transformations of the steroid nucleus carried out by microbes include hydrogenations, dehydrogenations, epoxidations, and removal and addition of the side chains.

13.8.3 TEXTILE PRODUCTION

There are two principal aspects to the microbiology of textiles. One is the use of microorganisms in preparing fibers such as flax and hemp; the other is deterioration of textiles, including cordage and ropes, and the preservation of such materials. The long fibers of flax, hemp, jute, and sisal are loosened from the plant stem by felting. The fiber bundles are just held within the outer layers of cells and outside the central pithy and woody layers by an intercellular cement of pectin. A number of bacteria and molds can digest pectin. This permits the fiber bundles to be separated mechanically from the stems and from each other. Fibers can then be collected and woven into linen or used in the form of ropes and packaging. There are two basic processes of retting. In the first process, the flax or hemp is spread out on the ground, particularly under somewhat acid conditions such as would occur in peat moss, and is exposed to the air and rain for some period. Under these conditions, the molds and aerobic bacteria that hydrolyzed pectin can grow.

Retting is achieved, but the process is somewhat uncertain and subject to continuous temperature and moisture changes. Frequently, certain organisms are observed that produce pigment or some growth on cellulose fibers, and thus color or weaken it. The fibers obtained by this method are frequently of poor quality and the yield is small. In the second process, the plant stalks are immersed in a flowing stream, ponds, or tanks of water and are allowed to remain there for several days. During this period, the anaerobic organisms, especially *Clostridium felsineum*, digest the pectin. The process, when properly operated, yields a nicer fiber that can be made into linen of quite good quality. Jute and sisal are tropical plants that are subjected to retting, generally under higher temperature condition and under water in bays and lagoons, or sometimes in flowing streams. The product resulting from these operations is relatively crude but relatively cheap. Much of the rope fibers and packaging cloth used throughout the world originates from these plants.

13.8.4 MICROBIAL SYNTHESIS OF VITAMIN B12

It seems probable that the only primary source of vitamin B12 in nature is the metabolic activity of microorganisms. It is synthesized by a wide range of bacteria and *Streptomycetes*, though not to any extent by yeasts and fungi. Although over 100 fermentation processes have been described for the production of vitamin Bl2, only half a dozen have apparently been used on a commercial scale. Some of these are as follows: (1) recovery of vitamin B12 as a by-product of the streptomycin and aureomycin antibiotic fermentations, and (2) fermentation processes using *Bacillus megatherium*, *Streptomyces olivaceus*, *Propionibacterium freudemeichii*, and *Propionibacterium shermanii*. The processes using the *Propionibacterium* species are the most productive and are now widely used commercially. Both batch and continuous processes have been described.

It is important to select microbial species that make the 5, 6-dimethyl benzimidazolylcobamid exclusively. Several manufacturers have been led astray by organisms that gave high yields of the related cobamides including pseudo-vitamin (adeninylcobamide). The natural form of the vitamin is Barker's coenzyme, where a deoxyadenosyl residue replaces the cyano group found in the commercial vitamin.

Practically, all of the cobamides formed in the fermentation are retained in the cells, and the first step is the separation of the cells from the fermentation medium. Large high-speed centrifuges are used to concentrate the bacteria to a cream, while filters are used to remove *Streptomyces* cells. The vitamin B12 activity is released from the cells by acid, heating, cyanide, or other treatments. Addition of cyanide solutions decomposes the coenzyme form of the vitamin and results in the formation of the cyanocobalamin. The cyanocobalamin is adsorbed on ion exchange resin IRC-50 or charcoal and is eluted. It is then purified further by partition between phenolic solvents and water. The vitamin is finally crystallized from aqueous-acetone solutions. The crystalline product often contains some water of crystallization.

PROBLEMS

Section A: Descriptive Type

Q1. What is a fermenter? Describe its significance.
Q2. Differentiate between aerobic and anaerobic fermentation.
Q3. Describe the screening method for microorganisms.
Q4. What is an inoculum?
Q5. What is an upstream bioprocess?
Q6. Describe the various stages in DSP.
Q7. Write a short note on *Corynebacterium*.
Q8. Describe the process for microbial synthesis of vitamin B12.
Q9. Describe the significance of contract manufacturing in biotechnology.

Section B: Multiple Choice

Q1. This is a device in which a substrate of low value is utilized by living cells to generate a product of higher value.
   a. PCR
   b. Bioreactor
   c. Chemical reactor
   d. Thermal reactor
Q2. Fermentation is the conversion of carbohydrates such as sugar into an acid or an alcohol. True/False
Q3. This is the organism of choice for acetone butanol fermentation.
   a. *Lactobacillus*
   b. *Nocardia*
   c. *Clostridium*
   d. *Aspergillus*
Q4. Inoculum development refers to the . . .
   a. Dormant stage of microbial culture
   b. Active stage of microbial culture
Q5. Sterilization is the process of inactivating or removing all living organisms from a substance or surface. True/False
Q6. An autoclave uses a temperature of this for steam sterilization.
   a. 100°C
   b. 150°C
   c. 121°C

Q7. Fermentation technology is used to make penicillin. True/False

Q8. The upstream part of a bioprocess refers to . . .

a. Growth of bacterial or mammalian cells

b. Synthesis of protein

c. Recombinant DNA technology

d. None of the above

Q9. The ideal temperature for mammalian cells to grow in a bioreactor is . . .

a. 37°C

b. 38°C

c. 30°C

d. 40°C

Q10. The actual recovery of useful products from fermentation is known as . . .

a. Upstream bioprocess

b. Downstream bioprocess

Q11. Cloning of *E. coli* K-12 was exempted from NIH guidelines. True/False

Section C: Critical Thinking

Q1. Why do mammalian cells fail to manufacture rDNA products as microorganisms do?

Q2. What would you do to improve the efficiency of existing bioreactors?

LABORATORY ASSIGNMENT

Write a protocol to make a cheese in the laboratory and discuss it with your course instructor. Then make a cheese using microbes. You may use different methods of making cheese.

REFERENCES AND FURTHER READING

American Cheese and Cheeses. *Consumer Reports* 728–732, November 1990.

Brown, B. *The Complete Book of Cheese.* Gramercy Publishing, New York, 1995.

Cambrosio, A., and Keating, P. Between fact and technique: The beginnings of hybridoma technology. *J. Hist. Biol.* 25(2): 175–230, 1992.

Carr, S. *The Simon and Schuster Pocket Guide to Cheese.* Simon & Schuster, New York, 1981.

Crueger, W. *Biotechnology: A Textbook of Industrial Microbiology.* Sinauer Associates, Inc., Sunderland, MA, 1989.

Demain, A.L., and Davies, J.E. *Manual of Industrial Microbiology and Biotechnology*, 2nd edn. ASM Press, Washington, DC, 1999.

Dinsmoor, R.S. Insulin: A never-ending evolution. *Countdown*, 22(2), 46, Spring 2001.

Dulbecco, R., and Ginsberg, H.S. *Virology*, 2nd edn. J.B. Lippincott Company, Philadelphia, PA, 1988.

Eli Lilly Corporation. *Humulin and Humalog Development.* CD-ROM. Eli Lilly Diabetes Web Page, November 16, 2001. www.lillydiabetes.com.

Kirk-Othmer. *Kirk Othmer Encyclopedia of Chemical Technology.* John Wiley & Sons, Inc., New York, 1992.

Kohler, G., and Milstein, C. Continuous cultures of fused cells secreting antibody of predefined specificity. *Nature* 1256: 495–497, 1975.

Kosikowski, F. *Cheese and Fermented Milk Foods.* Cornell University, Ithaca, NY, 1966.

Ladisch, M.R. *Bioseparations Engineering: Principles, Practice, and Economics.* John Wiley & Sons, Inc., Hoboken, NJ, 2001.

Mills, S. *The World Guide to Cheese.* Gallery Books, 1988. www.madehow.com/volume-1/cheese.html; www.enotes.com/how-products-encyclopedia/cheese.

Morell, V. Antibiotic resistance: Road of no return. *Science* 278: 575–576, 1997.

Novo Nordisk Diabetes Web Page, November 15, 2001. www.novonet.co.nz.

Olson, W.P. *Separations Technology: Pharmaceutical and Biotechnology Applications.* Interpharm Press, Inc., Buffalo Grove, IL, 1995.

PhRMA Reports Identifies More Than 400 Biotech Drugs in Development, Pharmaceutical Technology, August 24, 2006. http://pharmtech.findpharma.com/pharmtech/News/article/detail/367489 (retrieved on September 4, 2006).

Rang, H.P. *Pharmacology.* Churchill Livingstone, Edinburgh, UK, 2003, p. 241, for the examples infliximab, basiliximab, abciximab, daclizumab, palivusamab, palivusamab, gemtuzumab, alemtuzumab, etanercept and rituximab, and mechanism and mode. ISBN: 0-443-07145-4.

Riechmann, L., Clark, M., Waldmann, H., and Winter, G. Reshaping human antibodies for therapy. *Nature* 332: 323–327, 1988.

Schmitz, U., Versmold, A., Kaufmann, P., and Frank, H.G. Phage display: A molecular tool for the generation of antibodies—A review. *Placenta* 21(Suppl A): S106–S112, 2000.

Schwaber, J., and Cohen, E.P. Human x mouse somatic cell hybrid clones secreting immunoglobulins of both parental types. *Nature* 244(5416): 444–447, 1973.

Shepherd, P., and Christopher, D. *Monoclonal Antibodies.* Oxford University Press, New York, 2000.

Siegel, D.L. Recombinant monoclonal antibody technology. *Transfus. Clin. Biol.* 9: 15–22, 2002.

Stinson, S. Drug firms restock antibacterial arsenal. *Chem. Eng. News* 75–100, 1997.

14 Ethics in Biotechnology

LEARNING OBJECTIVES

- Discuss the significance of ethics in biotechnology
- Discuss ethical issues related to genetically modified (GM) food, plants, and animals
- Discuss ethical issues associated with animal testing
- Discuss ethical issues associated with xenotransplantation, human cloning, genetic screening, DNA fingerprinting, biometrics, organ transplantation, and embryonic stem cell research

14.1 INTRODUCTION

Bioethicists are people who raise ethical questions about the use of any living organism for experimentation purposes. Bioethicists are not against the development of new drugs or new therapies; they just wanted judicious use of animals and human subjects in experimentation. The whole objective of bioethicists is to minimize trauma in living organisms during experimentation. Human experimentation was one of the first areas addressed by modern bioethicists. The National Commission for the Protection of Human Subjects of Biomedical and Behavioral Research was initially established in 1974 to identify the basic ethical principles that should underlie the conduct of biomedical and behavioral research involving human subjects.

As a student of biotechnology, it's very important to know the ethical issues related to biotechnology-based products and how these ethical issues affect research and product development. You may recall the huge media debate on the ethical issues related to human cloning and human embryonic stem cells. In this chapter, we will learn about bioethics and how ethical issues pose a challenge to the biotechnology field.

14.2 GENETICALLY MODIFIED FOODS AND PLANTS

Millions of people worldwide have consumed foods derived from GM plants (mainly maize, soybean, and oilseed rape) and no adverse effects have been observed so far. The lack of evidence of negative effects, however, does not mean that new GM foods are without risk. The possibility of long-term effects of GM plants cannot be excluded and must be examined on a case-by-case basis. The question of the safety of GM foods has been reviewed by the International Council for Science (ICSU), which based its opinion on 50 authoritative independent scientific assessments from around the world. Currently available GM crops and the foods derived from them have been judged safe to eat, and the methods used to test them have been deemed appropriate.

Allergens and toxins occur in some traditional foods and can adversely affect some people, leading to concerns that GM plant-derived foods may contain elevated levels of allergens and toxins. Extensive testing of GM foods currently on the market has not confirmed these concerns. The use of genes from plants with known allergens is discouraged, and if a transformed product is found to pose an increased risk of allergies it should be discontinued. All new foods, including those derived from GM crops, should be assessed with caution. One concern about food safety is the potential transfer of genes from consuming food into human cells or into microorganisms within the body (Figure 14.1).

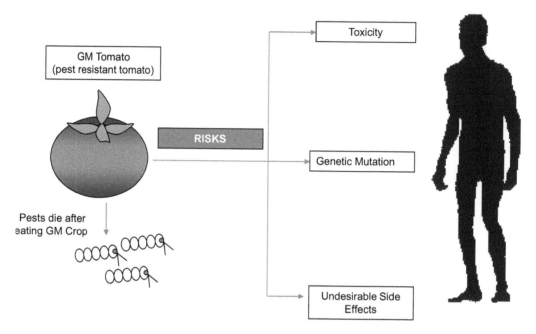

FIGURE 14.1   Concern raised by bioethical organizations with regard to use of GM crops.

Many GM crops were created using antibiotic-resistant genes as markers. Therefore, in addition to having the desired characteristics, these GM crops contain antibiotic-resistant genes. If these genes were to transfer in the digestive tract from a food product into human cells or to bacteria, this could lead to the development of antibiotic-resistant strains of bacteria. Although scientists believe the probability of such a transfer is extremely low, the use of antibiotic-resistant genes has been discouraged. Methods are now being developed whereby only the strict minimum of transgenic DNA is present in GM plants. Some of these techniques involve the complete elimination of the genetic marker once the selection process has been made. Scientists generally agree that genetic engineering can offer some health benefits to consumers. Direct benefits can come from improving the nutritional quality of food and from reducing the presence of toxic compounds and allergens in certain foods. Indirect health benefits can come from diminished pesticide use, less insect or disease damage to plants, increased availability of affordable food, and the removal of toxic compounds from soil. These direct and indirect benefits need to be better documented (Figure 14.2).

How do we identify a GM food or crop? The draft guidelines for the labeling of foods obtained through certain techniques of genetic modification/genetic engineering (FAO/WHO, 2003c) are still at an early stage of discussion and many sections are bracketed, indicating that the language has not yet been agreed upon. The guideline is proposed to apply to labeling of foods and food ingredients in three situations, when foods are (1) significantly different from their conventional counterparts; (2) composed of or contain GM and/or genetically engineered (GE) organisms or contain protein or DNA resulting from gene technology; and (3) produced from but do not contain GM/GE organisms, protein, or DNA from gene technology. According to the ICSU, scientists do not fully agree about the appropriate role of labeling. Although mandatory labeling is traditionally used to help consumers identify foods that may contain allergens or other potentially harmful substances, labels are also used to help consumers who wish to select certain foods on the basis of their mode of production, on environmental (e.g., organic), ethical (e.g., fair trade), or religious (e.g., kosher or halal) grounds. Countries differ in the type of labeling information that is mandatory or permitted. According to the ICSU,

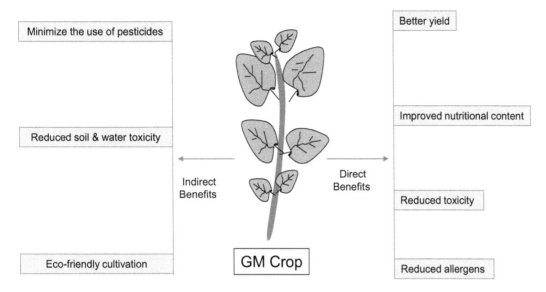

FIGURE 14.2   Benefits of GM crops.

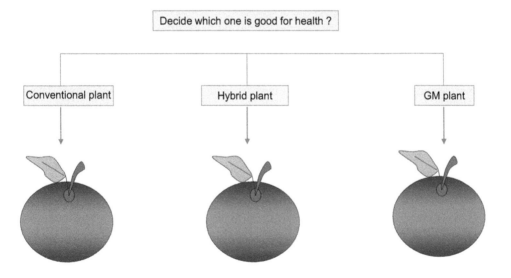

FIGURE 14.3   Can you decide which one is good for health?

labeling of foods as GM or non-GM may enable consumer choice as to the process by which the food is produced [but] it conveys no information as to the content of the foods, and whether there are any risks and/or benefits associated with particular foods.

The ICSU suggests that more informative food labeling that explains the type of transformation and any resulting compositional changes could enable consumers to assess the risks and benefits of foods (Figure 14.3).

14.3   USE OF ANIMALS AS EXPERIMENTAL MODELS

Animal testing is a phrase that most people have heard at some point in their lives, but they are perhaps still unsure of exactly what is involved in animal testing. Additionally, there is still a great deal of subjectivity around the meaning behind animal testing. Also, how animal testing is inter-preted is partly related to the person's personal views on animal testing. Whether it is called animal testing, animal experimentation, or animal research, it refers to the experimental use of non-human animals. This type of experimentation is not done directly for healing purposes, although the end result may involve medications used for healing both humans and other animals. Instead, healing an animal would be akin to veterinary medicine, which is entirely different from animal testing. It is also important to note that those who are opposed to animal testing may believe animal testing is nothing but torture and suffering of animals. Animal testing is conducted virtually everywhere, and its uses are broad. Global standards are quite strict and rigorous about animal testing and monitor-ing. Animal testing only occurs if there is no other viable alternative to the methods (Figure 14.4).

Animals are being tested at various centers located in universities, medical schools, pharmaceu-tical and biotechnology companies, and military defense establishments. Universities are often at the forefront of research that uses animal testing. Research students led by experienced scientists use the animals to better understand the anatomy and physiology of the body's function as well as to study the effects of pharmaceutical and drug treatments. Universities are quite often in the limelight, which means their treatment of animals is well scrutinized. Despite this, offenses have occurred at universities, such as previous primate experiments at the University of Cambridge. Medical research is considered integral to medical understanding and medical progress. For this reason, medical schools are generally avid supporters of animal testing for improving knowledge and health outcomes in disease prevention, treatment, and overall management. At the same time, medical schools are also strong supporters of animal testing alternatives whenever appropriate. In fact, new methods are being used as an alternative to animal testing even in the training of medi-cal students. Some of the leading, highly respected universities in the United States and Britain are using innovative, interactive clinical teaching methods to replace the more old-fashioned types of laboratories.

Pharmaceuticals comprise a billion-dollar industry, with enormous competition to find the next breakthrough drug. The new drug requires rigorous pharmaceutical testing to ensure its safety. Prior to performance of human clinical trials, drugs require preliminary testing on animals. Many drugs will never make it beyond this stage, which is precisely why they are first tested on animals rather than immediately on humans. This practice allows researchers to understand how the drugs

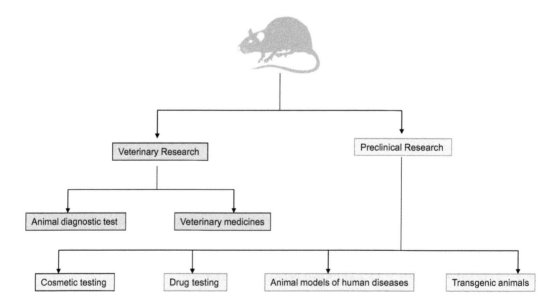

FIGURE 14.4 Application of animals in the development and testing of veterinary and therapeutic products.

are metabolized and how they affect the various body systems and allows researchers to determine their efficacy. Efficacy, however, must not carry an unacceptable level of risk, which means that a drug working for the intended purpose in an animal is not cause enough to move forward to human trials if the drug causes major side effects, harm, or death.

Many are not aware that animal testing occurs quite regularly in military establishments. One important reason for their use is to protect the military and the public from weapons and tactics that may be used by other countries. The military performs animal testing to determine how various biological weapons (such as viruses) will affect the intended recipients. By using animals, they can gauge the effect and develop the means to prevent and successfully handle the consequences. Vaccines are produced to prevent disease development, whereas the objective of medicine is to treat or cure disease, particularly on a wide scale, where an enormous number of people could be affected. Other testing uses include wound testing, where the military examines how various weapons create wounds and how those wounds heal.

14.3.1 Disadvantages of Animal Testing

In animal testing, countless animals are experimented on and then killed after their use. Others are injured and will still live the remainder of their lives in captivity. The unfortunate aspect is that many of these animals have undergone tests for substances that will never actually see approval for public consumption and use. It is this aspect of animal testing that many view as a major negative against the practice. This aspect seems to focus on the idea that the animal died in vain, because no direct benefit to humans occurred from the animal tests. Another disadvantage to the issue of animal testing is the sheer cost. Animal testing generally costs an enormous amount of money. Animals must be fed, housed, cared for, and treated with drugs or a similar experimental substance. The controlled environment is important, but it comes at a high cost. In addition, animal testing may occur more than once and over the course of months, which means that additional costs are incurred. The price of the animals themselves must also be factored into the equation. There are companies who breed animals, specifically for testing, and animals can be purchased through them. There is also the argument that the reaction of a drug in an animal's body is quite different from the reaction in humans. The main criticism here is that some believe animal testing is unreliable. Following on that criticism is the premise that because animals are in an unnatural environment, they will be under stress and therefore will not react to the drugs in the same way, compared to their potential reaction in a natural environment. This argument further weakens the validity of animal experimentation.

14.3.2 Attacks on Researchers

The increase in groups against animal testing led to a backlash against researchers who conduct these tests. There was less resistance in the earlier part of the twentieth century, but as experimental use of animals increased, the bonding of groups against animal testing began to take shape. The Internet also added glue to the network of those against animal testing, because it allowed them to more easily reach people around the world who would support their belief that animals should not be used in experimental testing. People for the Ethical Treatment of Animals (PETA) is one group that has led the way in animal rights campaigns as well as spurring on the creation of many other groups. PETA campaigns are not only directed at animal testing but also at areas such as promoting vegetarianism and ending the killing of animals for fur. Their campaigns regarding animal testing have, however, been well publicized, garnering a great deal of media attention and reaching people around the world. Threats to researchers and their families left some of the more prominent ones requiring bodyguards for protection. Arson attacks and violence became a reality for those who performed animal testing, particularly those who used primates. In fact, in 2006, a primate researcher at a university in the United States received numerous threats for his research. After regular demonstrations and threats to his well-being, he halted the research. In a similar incident in 2007, a bomb was placed under a researcher's car, but the bomb fortunately had a faulty fuse, and nobody was injured. In the United Kingdom, millions in estimated damages to property are caused each year by animal rights activists. Millions are also spent each year on policing and preventing damage.

14.3.3 REGULATIONS FOR ANIMAL TESTING IN THE UNITED STATES

A large part of the regulation process is governed by the Animal Welfare Act, which is supported and maintained through several sectioned agencies within the United States Department of Agriculture. The goals are to ensure that animals used for research receive a high standard of care and are treated humanely and with respect. Suffering should be minimized, but the stance is that it should not be minimized at the expense of the research goals and aims. This means that animal suffering is secondary to the goals of research and is only minimized if it does not interfere with the success of the experiment. The guidelines and regulations themselves are also only relevant to mammals, excluding birds, rats and mice, whether caught in the wild or purpose-bred for the experiment. Given the enormous numbers of rats and mice used in the United States each year, a great deal of legislation does not cover these animals, much to the criticism of animal welfare groups. The actual care of the animals and their subsequent use in experiments, however, is mostly monitored and handled by Institutional Animal Care and Use Committees.

14.3.4 ROLE OF ANIMAL WELFARE GROUPS

It is perhaps in the United States that animal welfare groups are the most prominent and play the largest roles. Members of PETA sometimes attempt to obtain "hidden" videos of abuse within government, private, or university laboratories in the hopes of exposing cruelty to animals and making an example of the offending research organization and scientists. PETA is also known to target researchers by threatening them or actively holding protests in front of their property. Indeed, in some instances, their actions have resulted in researchers withdrawing from their research. Although PETA targets animal testing across numerous species, they do tend to focus on the use of non-human primates as well as animals such as rabbits that are used for cosmetics and toxicology testing. The United States is competitive in terms of scientific advancement, which means that research into drugs and medical knowledge will continue to rise. The role of animal testing is a large one, but with aggressive groups like PETA that constantly challenge the laws and regulations within the country, it is anticipated that pressure to develop better alternatives to animal testing will prevail.

14.4 USE OF HUMANS AS EXPERIMENTAL MODELS

Biotechnology or pharmaceutical companies test their drug molecules in both animals and humans. The testing phase in animals is called the *preclinical phase*, whereas testing in humans is called the *clinical phase*. Clinical trials were first introduced in Avicenna's *The Canon of Medicine* in 1025 AD, in which he laid down rules for the experimental use and testing of drugs and wrote a precise guide for practical experimentation in the process of discovering and proving the effectiveness of medical drugs and substances. Avicenna laid out several rules and principles for testing the effectiveness of new drugs and medications, which still form the basis of modern clinical trials. One of the most famous clinical trials was a James Lind's demonstration in 1747 that citrus fruits cure scurvy. He compared the effects of various acidic substances, ranging from vinegar to cider, on groups of afflicted sailors, and found that the group who were given oranges and lemons had largely recovered from scurvy after 6 days. Frederick Akbar Mahomed (d. 1884), who worked at Guy's Hospital in London, made substantial contributions to the process of clinical trials during his detailed clinical studies, where he "separated chronic nephritis with secondary hypertension from what we now term essential hypertension." He also founded the Collective Investigation Record for the British Medical Association. This organization collected data from physicians practicing outside the hospital setting and was the precursor of modern collaborative clinical trials.

In planning a clinical trial, the sponsor or investigator first identifies the medication or device to be tested. Usually, one or more pilot experiments are conducted to gain insights for the design of the clinical trial to follow. In the United States, the elderly comprise only 14% of the population, but they consume over one-third of the pharmaceutical drugs that are manufactured. Women, children, and people with common medical conditions are also frequently excluded. In coordination with a panel of expert investigators, the sponsor decides what to compare the new agent with (one or more existing treatments or a placebo) and what kind of patients might benefit from the medication/device. If the sponsor cannot obtain enough patients with this specific disease or condition at one location, then investigators at other locations who can obtain the same kind of patients to receive the treatment would be recruited into the study. During the clinical trial, the investigators recruit patients with the predetermined characteristics, administer the treatment(s), and collect data on the patients' health for a defined time. These data include measurements like vital signs, the amount of the study drug in the blood, and whether the patient's health improves. The researchers send the data to the trial sponsor, who then analyzes the pooled data using statistical tests.

Note that while most clinical trials compare two medications or devices, some trials compare three or four medications, doses of medications, or devices against each other.

Beginning in the 1980s, harmonization of clinical trial protocols was shown as feasible across countries of the European Union. At the same time, coordination between Europe, Japan, and the United States led to a joint regulatory–industry initiative on international harmonization, named after 1990 as the International Conference for Harmonisation (ICH). Currently, most clinical trial programs follow ICH guidelines, aimed at

> ensuring that good quality, safe and effective medicines are developed and registered in the most efficient and cost-effective manner. These activities are pursued in the interest of the consumer and public health, to prevent unnecessary duplication of clinical trials in humans, and to minimize the use of animal testing without compromising the regulatory obligations of safety and effectiveness.

14.5   XENOTRANSPLANTATION

The controversial concept of cross-species transplantation, also known as *xenotransplantation*, has existed in myths and science fiction stories for a long time. Today, the lack of human organ donors has prompted an intense research effort throughout the medical community on the potential for animal organ transplants. Building on the overwhelming success of human-to-human transplantation, xenotransplantation aims to reduce the demand–supply gap for organs. In this section, we explore whether an individual who might benefit from xenotransplantation should be denied these benefits because of the ethical implications that accompany this practice on a mass scale. There are four components of xenotransplantation: (1) the individual receiving the xenograft, (2) the society which must incur the benefits and damages of decisions regarding xenotransplantation, (3) the long-term effect on humanity, and (4) the concern for animals used for humans' benefits.

There is no positive flip side to this danger of xenotransplantation, according to the experts in the field. Recently, a world-renowned primate virologist named Jonathan Allan from the Southwest Foundation for Biomedical Research in San Antonio addressed the possibility of viruses being able to jump the species gap and transfer to human beings. He declared that "primates are known to carry undiscovered viruses. In 1994, eight or nine baboons at Southwest were infected with a mysterious nervous-system virus which was later identified as a new virus." Adopting a precautionary approach could turn out to be ethical in the worst-case scenario. The consequentialist examination of society's recourse if a new virus is discovered probes the issue of our morality if we isolate the infected human recipients and treat them as specimens until the mode of transmission of the diseases is understood. The widespread objection to xenotransplantation is issued from utilitarian ethical theory, which was embraced by the Nuttfield Council on Bioethics. They reemphasized that it is not ethical for an individual to negatively affect the whole human population because of a singular decision: "the risks associated with the possible transmission of infectious diseases because of xenotransplantation have not been adequately dealt with. It would be unethical, therefore, to use xenografts involving human beings." While new epidemics remain, for the present time, only caveats to animal organ use, a very real consideration is the macro-allocation of resources in healthcare which will affect the society considering human-to-animal transplantation. The latter is just a small part of the medical realm, and yet it is certainly a very expensive field because of the complex manipulations necessary for a successful xenotransplantation. The ethical dilemma arises because our global economic system is built on limited resources juxtaposed with unlimited wants. The idea of depriving an individual suffering from organ malfunction of care because of society's interests in treating other diseases is not unique to xenotransplantation. Having said that, the appropriateness and ethicality of society's decision is extremely relevant to whether an individual who may survive a xenograft is going to have this option in a hospital. If we do decide to spend money on this field, then we should be quick to examine the benefits lost through the investment into other domains (e.g., cancer treatment) to make an ethical decision. Even if we attempt to disregard the economic burden of xenotransplantation as well as the potential for radiation of new diseases, human society will have to come to terms with what it means to be human in the long term.

Many will object to this characterization of humanity because they view appearance as an integral part of being human. Opponents of the perseverance in this domain often use the expression "playing God" to refer to the unethical use of animals to save human beings. Their most meaningful objection is based on the "unnaturalness" of such practices. It is indeed very difficult to define what is natural and to be able to distinguish it from what is unnatural without running into some difficulties. An article in the *Journal of Medical Ethics* states that "on one account, everything that humans do is unnatural, because it constitutes an interference with the non-human natural order. On another account, nothing that humans do is unnatural, since humans are themselves a part of nature" (Hugues 1998). Another relevant point which deters from the "naturalness" argument is the fact that any alterations made to the patient are somatic (i.e., they will not be passed on to any offspring) and temporary, since the transplanted organ can at any time be removed or replaced with a human organ. Looking at xenotransplantation as a temporary transition phase that will help many patients survive until a better solution is uncovered is a rational method of evaluating the ethical dimensions of cross-species transplantation without becoming overwhelmed with horrifying images originating from science fiction.

On the other hand, our reckless use of animals may cause mayhem in the "natural order" in its effect on the animals, and this should lead us to question the morality of doing so. A hastened assessment will direct us to a false conclusion that may be phrased in the following way: since it is ethical to use animals for consumption, then their use as organ donors should be even more acceptable. Indeed, using animals for organs may well be looked upon as a more efficient use of animals, which will lead to an increase in their social value and worth. In addition, justifying the use of non-humans for organs as being ethical simply because of our membership in the human species is misguided and superficial. To make animal use for saving a human being's life ethical, one must justify this use based on the superiority of humans over animals in a certain aspect. Superior brain and mental capacities are obvious candidates, but that excludes mentally retarded individuals from the batch.

14.6   GENETIC SCREENING

Despite having already succeeded in mapping and identifying the whole genetic sequence of humans, scientists do not yet know the exact function of most genes. A gene contains information on hereditary characteristics such as hair color, eye color, and height, as well as susceptibility to certain diseases. The term *embryo* refers to the first 6 weeks of development after a sperm fertilizes an egg. Due to variation in fertility in few couples, some seek medical assistance in the form of *in vitro* fertilization (IVF), during which fertilization occurs outside of the mother's body, with the embryos then being placed in her womb. Although it is unknown to what extent genes can influence someone's skills, psychology, or overall health, it is possible to carry out tests on embryos created during IVF (before placement in the womb) to detect severe genetic defects. This technique is known as *preimplantation genetic diagnosis* (PGD). Concerns have been raised that this technique will result in a modern form of eugenics where people with certain disabilities will be screened out of existence, while others argue that PGD is a way of reducing pain and suffering. PGD is also used to help treat ill children. For example, in the United States, the parents of a girl named Molly Nash, who was born with a rare incurable disease called Fanconi's anemia, underwent a controversial procedure to conceive another child who could help cure Molly. Embryos created using IVF were tested for signs of the gene responsible for the disease. Embryos that did not carry the disease underwent further testing to find those that would be a tissue match for Molly. As a result, a boy, Adam, was born, who was free of the disease and who could donate cells from his umbilical cord to treat his sister. Many people argue that there are several ethical problems with this "savoir sibling" technology. The question that arises is whether Adam would have been born at all had Molly been a healthy child or whether he was born only to save her life. Adam and Molly's parents refute this and say that they would have had another child anyway, but that this technology ensured Adam would be healthy and that Molly's life would be saved. Critics also express concern over whether children like Adam will constantly be called upon to donate tissue or maybe even organs (e.g., kidneys) to their sick siblings. Concerns are also raised about the possibility of "savior siblings" viewing their birth merely as a means to an end and seeing themselves as an instrument for curing disease. Another argument against selecting "savior siblings" is that several healthy embryos that are not exact tissue matches will be discarded.

14.7 BIOMETRICS

Biometrics refers to the technique of identifying people using unique physical characteristics such as fingerprints, the iris of the eye, voice pattern, or facial pattern. Biometric data can serve many purposes, including civil and criminal identification, surveillance and screening, healthcare, e-commerce, and e-government. Given the post-9/11 political and social anxieties, biometric data is being used for security purposes as a means of identifying citizens. Indeed, biometrics is increasingly being encountered in air travel as passengers are required to undergo fingerprint and iris scans before being allowed to travel to certain destinations. Under the US Visa Waiver Program, participating countries must begin using biometric passports. In 2005, Irish citizens made approximately 500,000 visits to the United States. Therefore, the department of Foreign Affairs launched the new e-passports, which have an inbuilt microchip containing biographical information as well as a digital photograph of the holder. Although the use of biometric data promises to augment security and reduce the occurrence of specific crimes such as identity theft and the abuse of social funds, it evokes a range of social, legal, and ethical concerns. It raises questions about the effect it will have on the concept of identity and whether its use will result in discrimination against individuals or communities. As with DNA, there are fears that personal information contained in e-passports might be accessed by organizations not involved in security, such as insurance companies and employers, which might lead to discrimination against individuals. There are also concerns about the reliability of biometrics and whether people will be at risk of having their personal information copied. At a recent hacker's conference, a German security consultant showed that he could clone data contained in e-passports like those introduced. There are also concerns that biometrics will allow governments to intrude into citizens' privacy and that people might be monitored without their knowledge or consent.

14.8 DNA FINGERPRINTING

The discovery of the polymerase chain reaction (PCR) technique made it possible to map the entire human genome and has tremendous diagnostic potential, especially in the fields of forensic science and individual identification. Over the past decades, DNA fingerprinting has made an invaluable contribution to crime detection and crime prevention. The UK DNA database provides many positive statistics. For instance, it is estimated that in a typical month, the database links suspects to 15 murders, 31 rapes, and 770 car crimes. However, numerous ethical questions arise with respect to who should have samples taken, for what crimes samples should be taken, how long samples should be retained on file, and who shall have access to stored genetic information. Under what circumstances should people have their DNA taken is unclear—that is, should a database include samples from people who are cautioned, arrested, or charged but acquitted, or only those who are convicted of a crime? Should samples be taken only in the case of serious crimes such as murder or assault, or should samples also be taken for lesser crimes such as driving and public order offenses? There are those who argue that only people guilty of criminal behavior would fear having their profile stored in a database and that everybody should be willing to provide a sample. However, 32% of black men in England and Wales are DNA profiled, as opposed to only 8% of white men, which has led to concerns about discrimination. Concerns are also raised about how long DNA should be stored in a forensic database. There is no doubt that scene-of-crime samples should be retained, in case an individual convicted of an offense alleges that a miscarriage of justice has occurred. In fact, DNA evidence was used to vindicate several convicted individuals who were on death row in the United States. However, what of an individual's sample? For instance, should it be destroyed once a convicted person has completed his or her sentence or should it be retained indefinitely? Some say the retention of DNA samples will greatly benefit society, while others fear that even those who have repaid their debt to society would be continually under suspicion and discriminated against. DNA can provide sensitive information about individuals and their families, such as their susceptibility to a genetic disorder like Huntington's disease. This increases the potential for genetic discrimination by the government, insurance companies, employers, financial institutions, and other organizations. For instance, were genetic information about a person to be given to an insurance company, that person might be prohibited from taking out life insurance and in turn be prevented from acquiring a mortgage.

14.9 ORGAN DONATION AND TRANSPLANTATION

There is currently a severe global shortage in the number of donor organs available for transplantation. For example, despite Ireland having one of the highest rates of organ donation (per million population) in the world, the number of people on transplant waiting lists there is increasing. Presently, Ireland operates an opt-in system where people are asked to sign an organ donor card. However, to reduce transplant waiting lists, other options are being explored, each with their own social, ethical, and economic implications. Traditionally, human organs used in transplants were acquired from brain dead donors (i.e., where permanent stoppage of all brain activity occurs). Now, the expansion of posthumous donation criteria to include cardiac death (i.e., where stoppage of heart and lung function occurs) is being contemplated, as is the establishment of a live donor program where organs such as kidneys are donated by a relative. Another system being considered is that of "presumed consent," whereby if someone does not expressly opt-out (e.g., via their driving license form), they are presumed to be willing donors. Some international commentators have gone so far as to call for automatic posthumous donation, like other mandatory civil obligations like jury service or compulsory military service, where individual consent is not required. Due to the global shortage of donors, people have been using more sinister methods to procure organs for donation. In recent years, a black market in organs has arisen, where people from developed nations in desperate need of a transplant travel to developing countries and pay tens of thousands of dollars to receive a life-saving transplant. There are grave concerns that the commercialization of organ donation has led to the coercion and exploitation of the economically disadvantaged. People have also raised concerns regarding the practice in the Philippines where prisoners are offered the chance to have their death sentences commuted to life imprisonment in return for an organ, and in China, where organs are procured from executed prisoners.

14.10   EUTHANASIA

Euthanasia is the intentional and painless taking of the life of another person, by act or omission, for compassionate motives. The word *euthanasia* is derived from the ancient Greek language and can be literally interpreted as "good death." Despite its etymology, the question of whether euthanasia is, in fact, a "good death" is highly controversial. Correct terminology in debates about euthanasia is crucial. Euthanasia may be performed by act or omission, either by administering a legal drug or by withdrawing basic healthcare which normally sustains life (such as air, food, water, or antibiotics). The term euthanasia mostly refers to the taking of the life of a person on his or her request, a *voluntary euthanasia*. On the other hand, the taking of a person's life against that person's expressed wish/direction is called *involuntary euthanasia*.

Central to the discussion on euthanasia is the notion of intention. While death may be caused by an action or omission of medical staff during treatment in hospital, euthanasia only occurs if death was intended. For example, if a doctor provides a dying patient extra morphine with the intention of relieving pain but knowing that his actions may hasten death, he has not performed euthanasia unless his intention was to cause death (the principle of double effect). Euthanasia may be distinguished from a practice called *physician-assisted suicide*, which occurs when death is brought about by the person's own hand by means provided to him or her by another person. All practices of euthanasia and physician-assisted suicide are illegal in Australia.

14.11   NEUROETHICS

Neuroethics is a field of inquiry that is very broad in scope and is closely related to both cognitive neuroscience and bioethics, though it is now formally recognized as a discipline. Neuroethics can be roughly divided into two streams. One stream concerns the more direct or proximal implications of cognitive neuroscience, which can be referred to as the "ethics of neuroscience." It deals with the ethical implications of neuroscientific knowledge and technology, such as enhancing neurological function through novel neuropharmacological, neurostimulation, and neurogenetic engineering techniques. The implications of brain imaging technology, which is now commonly used in both research and medical practice, raises issues concerning mental privacy, diagnostics, and predicting behavior. Furthermore, knowledge gained through neuroscience, along with brain imaging technology, may one day allow us to probe the human mind to even observe a person's thoughts and predilections. The second stream of neuroethics can be referred to as the "neuroscience of ethics." This stream of neuroethics lies at the border between philosophy, metaethics, and normative ethics; one of the central issues concerns moral agency. How do we impute moral responsibility given that cognitive neuroscience may shed new light on the way humans make their decisions, as well as the nature of our underlying motivations to act in certain ways? How can we trust our moral beliefs if it turns out that one's belief was not the product of rational contemplation but a post hoc rationalization of an emotive judgment, an attitude of disapprobation, or a pre-reflective moral intuition that is distinct, impenetrable, and encapsulated from rational contemplation?

14.12   ASSISTED REPRODUCTIVE TECHNOLOGY

Assisted reproductive technology (ART) is a medical intervention developed to improve an "infertile" couple's chance of pregnancy. "Infertility" is clinically accepted as the inability to conceive after 12 months of actively trying to conceive. The means of ART involves separating procreation from sexual intercourse. The importance of this association is addressed in bioethics. Some techniques used in clinical ART include artificial insemination, IVF, gamete intrafallopian transfer (GIFT), gestational surrogate mothering, gamete donation, sex selection, and preimplantation genetic diagnosis. Issues addressed in bioethics are the appropriate use of these technologies and the techniques employed to carry out procedures for quality and ethical reviews. ART and its use directly impact the foundational unit of society—the family. ART enables children to be conceived who have no genetic relationship to one or both of their parents. Children can also be conceived who will never have a social relationship with one or both of their genetic parents, such as a child conceived using donor sperm. Non-infertile people in today's society—including both male and female homosexual couples, single men and women, and postmenopausal women—are seeking the assistance of ART. Concerns in all situations include the child and his or her welfare, including the right to have one biological mother and father. The fragmented family created by ART can disconnect genetic, gestational, and social child–parent relationships that have typically been one and the same in the traditional nuclear family. Other important bioethical issues include the appropriate use of PGD screening; use and storage; destruction of excess IVF embryos; and research involving embryos. ART research requires human participants, donors and donated embryos, oocytes and sperm. Ethical issues that arise in ART research surround the creation and destruction of embryos. One approach in bioethics involves preserving justice, beneficence, non-malfeasance, and the autonomous interests of all involved. Bioethicists contribute to ethical guidelines and moral evaluations of new technologies and techniques in ART as well as to the public discourse that lead to the development of national regulations and restrictions of unacceptable practices.

14.13   EMBRYONIC STEM CELLS

Ever since embryonic stem cell research came into existence, it has not only generated tremendous hopes for millions of patients who are suffering from cancer, Parkinson's diseases, and diabetes, but at the same time it has raised serious issues pertaining to how embryonic stem cells are being collected after killing human embryos. Debates over the ethics of embryonic stem cell research continue to divide scientists, politicians, and religious groups. However, promising developments in other areas of stem cell research might lead to solutions that bypass these ethical issues. These new developments tilt the pros and cons scale toward the former, as the main concern of those against this research seems to be the destruction of human blastocysts, considered by pro-life advocates to be an immoral act of disregard for human life. It all started in November 1998, with the publication of a research paper which reported that stem cells could be isolated from human living embryos. Subsequent research led to the ability to maintain undifferentiated stem cell lines (pluripotent cells) and techniques for differentiating them into cells specific to various tissues and organs. The debates over the ethics of stem cell research began almost immediately, in 1999, despite reports that stem cells cannot grow into complete organisms. In 2000–2001, governments worldwide were beginning to draft proposals and guidelines to control stem cell research and the handling of embryonic tissues, and to reach universal policies to prevent "brain-drain" (emigration of top scientists in the field) between countries. The Canadian Institute of Health Research (CIHR) drafted a list of recommendations for stem cell research in 2001. The Clinton administration drafted guidelines for stem cell research in 2000, but Clinton left office prior to their release. The Bush government had to deal with the issue throughout his administration. Australia, Germany, the United Kingdom, and other countries have also formulated their policies on this matter.

14.13.1   IN FAVOR OF EMBRYONIC STEM CELL RESEARCH

The excitement about stem cell research is primarily because of the medical benefits in areas of regenerative medicine and therapeutic cloning. Stem cells provide huge potential for finding treatments and cures for a vast array of diseases, including different cancers, diabetes, spinal cord injuries, Alzheimer's disease, multiple sclerosis, Huntington's disease, Parkinson's disease, and more. There is endless potential for scientists to learn about human growth and cell development from studying stem cells. Use of adult-derived stem cells from blood, skin, and other tissues has been demonstrated to be effective for treating different diseases in animal models. Umbilical cord-derived stem cells have also been isolated and utilized for various experimental treatments. Although cell lines derived from adult stem cells trigger no ethical issues, there are some disadvantages or shortcomings compared to embryonic cell lines, because they are difficult to isolate, poor in quantity, and unsuitable for long-term expansion.

14.13.2   AGAINST EMBRYONIC STEM CELL RESEARCH

The use of embryonic stem cells involves the destruction of human blastocysts. Many people believe that it is in the blastocyst that life begins, and they thus have the notion that it is a human life, which it is unacceptable and immoral to destroy. This seems to be the only controversial issue standing in the way of stem cell research in North America and elsewhere. In the summer of 2006, US President George W. Bush stood his ground on the issue of stem cell research and vetoed a bill passed by the Senate that would have expanded federal funding of embryonic stem cell research. Currently, American federal funding can only extend to research on stem cells from existing (already destroyed) embryos. Similarly, in Canada, as of 2002, scientists cannot create or clone embryos for research. They can only use existing embryos discarded by couples. On the other hand, the United Kingdom allows embryonic stem cell cloning. The use of stem cell lines from alternative non-embryonic sources has received more attention in recent years and has already been demonstrated as a successful option for treatment of certain diseases. For example, adult stem cells can be used to replace blood-cell-forming cells killed during chemotherapy in bone marrow transplant patients. Biotech companies such as Revivicor and ACT are researching techniques for cellular reprogramming of adult cells, use of amniotic fluid, or stem cell extraction techniques that do not damage the embryo, and that also provide alternatives for obtaining viable stem cell lines. Of necessity, the research on these alternatives is catching up with embryonic stem cell research and, with enough funding, other solutions might be found that are acceptable to everyone.

14.13.3   STATUS OF EMBRYONIC STEM CELL RESEARCH

On March 9, 2009, US President Barack Obama overturned Bush's ruling, allowing US federal funding to go to embryonic stem cell research. Despite the progress being made in other areas of stem cell research, such as using pluripotent cells from other sources, many American scientists were putting pressure on the government to allow their participation and competition with the Europeans. However, many people are still strongly opposed.

14.14   HUMAN CLONING

Cloning is a technology to create genetically identical organisms (such as bacteria, insects or plants) asexually. In biotechnology, cloning refers to processes used to create copies of DNA fragments (molecular cloning), cells (cell cloning), or organisms. The ethics of cloning refers to a variety of ethical positions regarding the practice and possibilities of cloning, especially human cloning. Although many of these views are religious in origin, the questions raised by cloning are faced with secular perspectives as well. As the science of cloning continues to advance, governments have dealt with ethical questions through legislation.

Advocates of human therapeutic cloning believe the practice could provide genetically identical cells for regenerative medicine and tissues and organs for transplantation. Such cells, tissues, and organs would neither trigger an immune response nor require the use of immunosuppressive drugs.

Both basic research and therapeutic development for serious diseases such as cancer, heart disease, and diabetes, as well as improvements in burn treatment and reconstructive and cosmetic surgery, are areas that might benefit from such new technology. One bioethicist, Jacob M. Appel of New York University, has gone so far as to argue that "children cloned for therapeutic purposes" such as "to donate bone marrow to a sibling with leukemia" may someday be viewed as heroes. Proponents claim that human reproductive cloning also would produce benefits. Severino Antinori and Panos Zavos hope to create a fertility treatment that allows parents who are both infertile to have children with at least some of their DNA in their offspring. Some scientists, including Dr. Richard Seed, suggest that human cloning might obviate the human aging process. How this could work is not entirely clear, since the brain or identity would have to be transferred to a cloned body. Dr. Preston Estep has suggested the terms "replacement cloning" to describe the generation of a clone of a previously living person and "persistence cloning" to describe the production of a cloned body for obviating aging, although he maintains that such procedures currently should be considered science fiction and current cloning techniques risk producing a prematurely aged child. In Aubrey de Grey's proposed Strategies for Engineered Negligible Senescence (SENS), one of the considered options to repair cell depletion related to cellular senescence is to grow replacement tissues from stem cells harvested from a cloned embryo.

At present, the main non-religious objection to human cloning is that cloned individuals are often biologically damaged owing to the inherent unreliability of their origin. For example, researchers are currently unable to safely and reliably clone non-human primates. Bioethicist Thomas Murray of the Hastings Center argues that "it is absolutely inevitable that groups are going to try to clone a human being. But they are going to create a lot of dead and dying babies along the way." UNESCO's Universal Declaration on Human Genome and Human Rights asserts that cloning contradicts human nature and dignity. Cloning is an asexual reproductive mode that could distort generation lines and family relationships and limit genetic differentiation, which ensures that human life is largely unique. Cloning can also imply an instrumental attitude toward humans, which risks turning them into manufactured objects and interferes with evolution, the implications of which we lack the insight or prescience to predict. Furthermore, proponents of animal rights argue that non-human animals possess certain moral rights as living entities and should therefore be afforded the same ethical considerations as human beings. This would negate the exploitation of animals in scientific research on cloning, cloning used in food production, or as other resources for human use or consumption. Rudolph Jaenisch, a professor at Harvard, has pointed out that we have become more efficient at producing clones that are still defective. Other arguments against cloning come from various religious orders (believing cloning violates God's will or the natural order of life) and a general discomfort some have with the idea of "meddling" with the creation and basic function of life. This unease often manifests itself in contemporary novels, movies, and popular culture, as it did with numerous prior scientific discoveries and inventions. Various fictional scenarios portray clones being unhappy, soulless, or unable to integrate into society.

QUESTIONS

Section A: Descriptive Type

Q1. Define bioethics.
Q2. Describe ethical issues associated with GM foods.
Q3. Explain why animal testing is important for research.
Q4. Discuss some major advantages of animal testing.
Q5. What are the different kinds of ethical issues associated with human clinical trials?
Q6. What is xenotransplantation? Discuss the ethical issues associated with it.
Q7. Define biometrics and cite suitable examples.
Q8. Describe organ donation and transplantation, with examples.

Section B: Multiple Choice

Q1. The safety of GM foods is supervised by the . . .
   a. International Council of Science
   b. National Science Foundation
   c. National Institutes of Health
Q2. Clinical trials were first introduced by . . .
   a. Avicenna (Ibne Sina)
   b. Alexander Fleming
   c. Robert Koch
Q3. Which of the following does not come under the ethical regulation?
   a. Human embryonic stem cell
   b. Human cloning
   c. Xenotransplantation
   d. Antibiotics
Q4. Biometrics refers to the technique of identifying people using unique physical properties. True/False

Section C: Critical Thinking

Q1. How does DNA fingerprinting violate human rights?
Q2. Do you think it is ethical to create a human embryo by ART? Why or why not?

ASSIGNMENTS

The following lists some questions raised, based on bioethics. Discuss these issues with your course instructor and submit your answers in the form of an essay. The information may be accessed through the Internet.

Research issues: Should scientists be held to some standard of integrity and honesty? Who should enforce this? Why is peer review so important? Should scientists be held responsible for creating (or discovering) technology that can be used to harm others or have unforeseen side effects (chemical, nuclear warfare)?

Reproductive technologies: *In vitro* fertilization, surrogacy, RU-486, preimplantation embryo screening, cloning: Is there a significant difference between cloning sheep for pharmaceutical production and cloning humans?

Human genome project: Should employers be able to screen job applicants for specific genetic conditions? Who should have access to this information: Family members, lawyers, insurance agencies?

Gene therapy: What are the potential ramifications of somatic and germ-line gene therapy? Should genes be tinkered with? If so, what limits should be placed on this type of technology?

Fetal rights: Does a fetus have rights? If so, what rights does a fetus have and who is responsible for representing its interests? Does a fetus have rights that supersede the mother's? Can the government step in to ensure the health of a fetus if the mother is not willing to do so? What about embryos?

Euthanasia: What is the right to die? How does withdrawing or withholding treatment differ from physician-assisted suicide? Who has the right to decide when and how a person dies? Should doctors be held legally responsible if they assist a patient's death? What laws should be passed to protect doctors and patients?

Environmental issues: How do we decide between conservation and economic interests? How much land should be allocated to other species and to parks? Should industries be responsible for damage done to the environment (pollution) by them?

Animal rights issues: Is animal testing acceptable when it benefits humans? Which animals should be tested for these tests and which should not? Is animal research justified by its benefits to mankind?

Population control: Who has the right to decide who should have children (and how many)? What measures should be taken to control the world population?

Human research: Should humans be used for medical and psychological studies? What guidelines should be used to protect subjects?

Minors and medicine: What medical procedures do minors have available to them without parental consent? Do doctors have an obligation to inform parents of conditions a teen has (such as pregnancy and AIDS) even if the teen does not wish it?

GM crops: What rights do consumers have? What rights do farmers have to grow GM crops? Who decides whether a food is safe?

REFERENCES AND FURTHER READING

Andre, J. *Bioethics as Practice*. University of North Carolina Press, Chapel Hill, NC, 2002.

Aulisio, M., Robert, A., and Stuart, Y. *Ethics Consultation: From Theory to Practice*. Johns Hopkins University Press, Baltimore, MD, 2003.

Beauchamp, T., and Childress, J. *Principles of Biomedical Ethics*. Oxford University Press, Oxford, 2001.

Bioethical issues. www.bioethics.ie/uploads/docs/IntroBioethics.pdf.

Callahan, D. Bioethics as a discipline. *Hastings Cent. Stud.* 1: 66–73, 1973.

Caplan, A.L. *Am I My Brother's Keeper? The Ethical Frontiers of Biomedicine*. Indiana University Press, Bloomington, IN, 1997.

A dozen questions (and answers) on human cloning. World Health Organization. http:/www.who.int/ethics/topics/cloning/en/ (retrieved on February 27, 2008).

Emanuel, E., Crouch, R., Arras, J. et al. *Ethical and Regulatory Aspects of Clinical Research*. Johns Hopkins University Press, Baltimore, MD, 2003.

FDA probes sect's human cloning. *Wired News*, 2002. www.wired.com.

Food Agriculture Organization World Health of the United Nations Organization (FAO/WHO). 2003. ftp://ftp.fao.org/es/esn/jecfa/jecfa61sc.pdf.

Friend, T. The real face of cloning. *USA Today*, 2003.

Glad, J. *Future Human Evolution: Eugenics in the Twenty-First Century*. Hermitage Press, Ewing, NJ, 2008.

Hugues, J. Xenografting: Ethical issues. *J. Med. Ethics* 24: 18–24, 1998.

Jonathan, B. *Against Bioethics*. The MIT Press, Cambridge, MA, 2006.

Jonsen, A.R. *The Birth of Bioethics*. Oxford University Press, Oxford, 1998.

Jonsen, A.R., Veatch, R., and Walters, L. *Source Book in Bioethics*. Georgetown University Press, Washington, DC, 1998.

Kaufmann, R.E. Clinical trials in children: Problems and pitfalls. *Pediatr. Drugs* 2: 411–418, 2000.

Khushf, T. *Handbook of Bioethics: Taking Stock of the Field from a Philosophical Perspective*. Kluwer Academic Publishers, Dordrecht, the Netherlands, 2004.

Korthals, M., and Bogers, R.J. *Ethics for Life Scientists*. Springer, Dordrecht, the Netherlands, 2004.

Kuczewski, M.G., and Polansky, R. *Bioethics: Ancient Themes in Contemporary Issues*. The MIT Press, Cambridge, MA, 2002.

McGee, G. *Pragmatic Bioethics*. Massachusetts Institute of Technology Press, Cambridge, MA, 2003.

Murphy, T. *Case Studies in Biomedical Research Ethics*. The MIT Press, Cambridge, MA, 2004.

Nuttfield Council on Bioethics. *Animal-to-Human Transplants: The Ethics of Xenotransplantation*. Nuttfield Council on Bioethics, London, UK, p. 27, 1996.

Politis, P. Transition from the carrot to the stick: The evolution of pharmaceutical regulations concerning pediatric drug testing. *Widener Law Rev.* 12: 271, 2005.

Reich, W.T. The word 'bioethics': Its birth and the legacies of those who shaped its meaning. *Kennedy Inst. Ethics J.* 4: 319–336, 1994.

Rhoden, N.K. Treating baby Doe: The ethics of uncertainty. *Hastings Cent. Rep.* 16: 34–43, 1986.

Rothman, D.J. *Strangers at the Bedside: A History of How Law and Bioethics Transformed Medical Decision Making*, 2nd edn. Aldine de Gruyter, New York, 2003.

Singer, P.A., and Viens, A.M. *Cambridge Textbook of Bioethics*. Cambridge University Press, Cambridge, UK, 2008.

Stevens, M.L.T. *Bioethics in America: Origins and Cultural Politics*. Johns Hopkins University Press, Baltimore, MD, 2000.

Sugarman, J., and Sulmasy, D. *Methods in Medical Ethics*. Georgetown University Press, Washington, DC, 2001.

Universal declaration on the human genome and human rights. UNESCO, 1997. http://portal.unesco.org.

Wikler, D. Ethics and rationing: Whether, how, or how much. *J. Am. Geriatr. Soc.* 40: 398–403, 1992.

Aetna Recommends Guidelines for Access to Genetic Testing—Company Advocates Education and Information Privacy. www.aetna.com.

American Academy of Actuaries—Issue Brief: Genetic Information and Voluntary Life Insurance. www.actuary.org.

American Academy of Actuaries—Issue Brief: The Use of Genetic Information in Bridging the Gap Between Life Insurer and Consumer in the Genetic Testing Era—Article from the *Indiana Law Journal*. www.ornl.gov.

Cancer Genetics—A Collection of Resources on Genetic Testing for Cancer. From the National Cancer Institute. www.nci.nih.gov.

Disability Income and Long-Term Care Insurance. www.actuary.org.

Ethical, Legal, and Social Issues of the Human Genome Project. www.ornl.gov.

Evaluating Gene Testing—Information on Evaluating Test Quality, the Potential Usefulness of Test Results. Available Preventive or Treatment Options, and Social Issues. From the Human Genome Project Information Website. www.ornl.gov.

Gene Tests Website. www.genetests.org.

Gene Tests—Find Reviews of Genetic Disorders, Genetics Glossary, Clinics, and Laboratories Conducting Gene Testing. www.genetests.org.

Genes, Dreams, and Reality: The Promises and Risks of the New Genetics—Article from Judicature. www.ornl.gov.

Genetic Testing—A Collection of News, Overviews, and Other Web Resources from MEDLINEplus. www.nlm.nih.gov.

Genetic Testing—An Introduction to Genetic Testing and the Different Types of Tests. Available from the Genetics and Public Policy Center. www.dnapolicy.org.

Genetic Testing for Inherited Predisposition to Breast Cancer—An Article from the National Breast Cancer Coalition Website, April 2003. www.natlbcc.org.

Genetic Testing of Newborn Infants—An Activity for Considering Government Policies to Guide a Program for Genetic Screening of Newborns. From the Genetic Science Learning Center. http://gslc.genetics.utah.edu.

Human Gene Testing 1996—An Article on Gene Testing and Science Behind Genetic Disorders from the National Academy Sciences Beyond Discovery Series That Traces the Origins of Important Recent Technological and Medical Advances. www.beyonddiscovery.com.

Human Genetic Testing and DNA Analysis—A Collection of Websites and Recorded Broadcasts of Radio Programs. From the DNA Files Provided by National Public Radio (NPR). www.dnafiles.org.

If Genetic Tests Were Available for Diseases Which Could Be Treated or Prevented, Many People Would Have Them. A Report on the Results of a Poll Conducted in 2002 by Harris Interactive. www.harrisinteractive.com.

Prenatal Testing: A Modern Eugenics?—From the DNA Files Provided by National Public Radio (NPR). www.ornl.gov.

Resources from the National Conference of State Legislatures.

Understanding Gene Testing—A Basic Introduction to Genes and Genetic Testing from the U.S. Department of Health and Human Services. www.accessexcellence.org.

Understanding Gene Testing—A Tutorial That Illustrates What Genes Are, Explains How Mutations Occur and Are Identified Within Genes, and Discusses the Benefits and Limitations of Gene Testing for Cancer and Other Disorders. From the National Cancer Institute. www.cancer.gov.

What Can the New Gene Tests Tell Us? Article from the Judges. *Journal of the American Bar Association*. www.ornl.gov.

15 Careers in Biotechnology

LEARNING OBJECTIVES

- Discuss various careers in biotechnology
- Explain how one can get a suitable position in the biotechnology industry
- Discuss the status of job openings in the academic and industrial sectors involving biotechnology
- Explain how jobs in academic institutes are different from those in companies

15.1 INTRODUCTION

Biotechnology is an emerging field that keeps on expanding with arrays of opportunities and challenges, especially in terms of finding out new methods and better treatment modalities for various human ailments, protecting the environment, reestablishment of extinct species, enhancing crops and food products, and finding out alternative and eco-friendly energy sources. One can imagine the impact of biotechnology on everyday life. It has now become obvious to think about how we can shape our career in various areas of biotechnology.

The biotechnology industry consists of six segments:

1. research and development (R&D);
2. laboratory-level synthesis and testing;
3. industrial-level synthesis/manufacturing;
4. quality control and quality assurance;
5. sales, marketing, and finance; and
6. legal and intellectual property rights (IPRs).

In these fields, people with diverse educational qualifications, backgrounds, and skill sets are in great demand. We will discuss each of these six components in greater detail to learn about challenging job opportunities.

15.2 CAREER OPPORTUNITIES IN BIOTECHNOLOGY

Biotechnology is a very exciting sector with tremendous job possibilities and career development potential. The biotechnology field offers promising jobs in universities, R&D centers, and corporate organizations such as drug manufacturing and drug marketing companies. The basic qualification for getting a job in any organized sector of biotechnology is a BS degree in biotechnology or any related field. However, a mere degree will not fetch jobs; candidates should also have at least 6 months to 1 year of work experience. Now you may well ask, "How can one get work experience?" Obviously, biotechnology companies do not provide jobs just to give work experience to fresh graduates. So, one option is to apply to university research labs and work as a trainee to get work experience. Of course, as a trainee, you would not earn much money, but it would positively help you to get ready for the next job. In case you do not want a job after completing a BS degree, then you can proceed to higher education by getting a MS degree in any of the various specialized domains of biotechnology, such as medical, agricultural, industrial, microbial and environmental biotechnology, nanobiotechnology, biotechnology business and marketing, bioinformatics, or IPRs and biotechnology regulations. An MS degree in biotechnology with specialization will open up tremendous job opportunities and you can choose to work in any organized sector of biotechnology. Even if you do not want a job after completing an MS degree, you can do research by doing a PhD in any specialized or super-specialized field of biotechnology such as stem cell research, cloning, or gene therapy. During PhD research work, you have an opportunity to conduct research by employing various laboratory techniques independently, and if the research finding is novel, it may lead to a very commendable research paper publication in scientific journals. A PhD degree in biotechnology with work experience and decent publication tracks brings tremendous job opportunities, both in academic institutes and in corporate organizations. If you do not want a job in R&D labs, you can either go in for a master's degree in business administration (MBA) or do short courses in bioinformatics, IPRs, and health care management (Figures 15.1, 15.2).

There is a tremendous job satisfaction working in the R&D sector of biotechnology. By publishing or patenting their research, scientists establish a name as well as large fortunes as fruits of their research. Researchers (mostly PhDs) not only perform experiments in labs, but also devote much of their time to attending seminars and conferences, teaching university students, and writing papers. A researcher has a very creative role working at the forefront of scientific discovery and can decide what to study and how to study, thus learning new things. Laboratory assistants (mostly BS degree holders) and lab technicians (mostly MS degree holders) work on a variety of sophisticated equipment, prepare chemicals, and maintain experimental microbial, plant, or tissue cultures in research labs. Lab technicians also help research scientists in their scientific discoveries. After doing an MBA, lab technicians can get management positions in biotechnology firms and can be involved in marketing the biotechnology industry products to doctors, farmers, and other industries, or sell and supply scientific equipment to biotechnology researchers. After a BS or MS degree, one can do courses in scientific instrumentation and can work as a quality assurance technician to test biotechnology products in labs or production facilities to ensure that they are safe and meet the required standards. Bioinformatics personnel work on designing and maintaining scientific data and analyzing the results of various research trials.

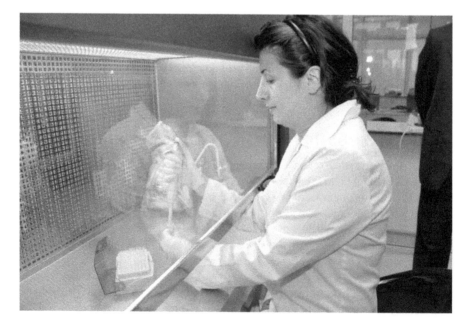

FIGURE 15.1 Career opportunities in research and development.

Source: Picture courtesy: Department of Biotechnology, Manipal University Dubai, UAE

FIGURE 15.2   Ready to take challenging opportunity in biotechnology

Source: Picture courtesy: Department of Biotechnology, Manipal University Dubai, UAE

As the biotechnology industry continues to grow and commercialize, the number of job vacancies continues to increase. Companies are hiring more and more sales force to push their products on the market. They are also employing trained people for clinical research to get through the clinical approval process. This has increased the demand for clinical trial managers, clinical research professionals, clinical research associates, and outsourcing managers. Because of the commercialization of the industry, the demand for finance and administration personnel is also increasing. Thus, biotechnology companies are recruiting people at all levels. A biotechnologist may also work in both the public and private sectors. Today, agriculture, dairy, horticulture, and drug companies are also employing large numbers of biotechnologists.

Compensation in biotechnology companies is competitive and includes incentives, such as stock option plans, 401K plans (in the United States), company-wide stock purchase plans, and cash bonus plans. The salary for any career in biotechnology varies from individual to individual and from organization to organization. Indeed, compensation is a big factor in choosing any career in biotechnology as a profession. There are some disparities in compensation between commercial biotechnology companies and academic or noncommercial organizations. Usually, the compensation package in commercial companies is higher than those in academic institutions. In addition, there is a greater flexibility in negotiating compensation packages in commercial companies than in academic institutions.

15.3   ENTRY-LEVEL JOB POSITIONS IN BIOTECHNOLOGY

This section discusses some typical entry-level biotechnology positions with their corresponding job descriptions.

15.3.1   RESEARCH AND DEVELOPMENT DIVISION

15.3.1.1 Glass Washer

A glass washer is responsible for washing and drying glassware and distributing it to appropriate locations within the laboratories. He or she maintains the glass washing facility, keeps it clean, and picks up dirty glassware. He or she also sterilizes glassware and other items using an autoclave. A glass washer performs routine maintenance of glass washing equipment and performs other related duties as required. This position requires a high school diploma or equivalent and a minimum of 0–2 years of laboratory experience.

15.3.1.2 Laboratory Assistant

A laboratory assistant is responsible for performing a wide variety of research laboratory tasks and experiments, making detailed observations, analyzing data, and interpreting results. He or she maintains laboratory equipment and inventory levels for laboratory supplies. He or she may also write reports, summaries, and protocols regarding experiments. A laboratory assistant also performs limited troubleshooting and calibration of instruments. An entry-level laboratory assistant position requires a basic associate degree in science and 0–2 years of laboratory experience.

15.3.1.3 Research Associate

A research associate is responsible for R&D in collaboration with the other staff involved in a project. He or she makes detailed observations, analyzes data, and interprets results. Research associates prepare technical reports, summaries, protocols, and quantitative analyses. An incumbent maintains familiarity with current scientific literature and contributes to the process of a project within his or her scientific discipline, along with investigating, creating, and developing new methods and technologies for project advancement. He or she may also be responsible for identifying patentable inventions and acting as principal investigator in conducting his or her own experiments. A research associate may also be asked to participate in scientific conferences and contribute to scientific journals.

15.3.1.4 Research Assistant

The job description for a research assistant is similar to that of a research associate. At the entry level, the job can require a BS or an advanced degree in a scientific discipline or an equivalent qualification, with little or no related experience.

15.3.1.5 Postdoctoral Fellow

A postdoctoral fellow has a PhD, but little or no job experience, and joins the staff for a maximum of 2–3 years to gain the necessary experience before moving on to a more senior scientist's position.

15.3.1.6 Media Preparation Technician

A media preparation technician, or media prep technician, is responsible for media preparation in the R&D area. He or she performs experiments as required and outlined and develops and maintains record keeping for procedures and experiments performed. An entry-level media prep technician position requires a basic associate degree in science and 0–2 years of related experience and/or completion of a company's on-the-job training program.

15.3.1.7 Greenhouse Assistant

A greenhouse assistant performs a variety of greenhouse research tasks and experiments. He or she may be required to make detailed observations, detecting horticultural or pest problems and instituting corrective action. Greenhouse assistants determine optimal cultural requirements and perform tasks related to disease and pest prevention. They are often required to collect, record, and analyze data and also to interpret results. In addition, a greenhouse assistant may perform troubleshooting and equipment maintenance. An entry-level greenhouse assistant position requires a high school diploma, an associate degree or equivalent, and a minimum of 0–2 years of relevant greenhouse/plant experience.

15.3.1.8 Plant Breeder

A plant breeder is responsible for the design, development, execution, and implementation of plant breeding research projects in collaboration with a larger research team. He or she may be responsible for project planning and personnel management within the project. Plant breeders may use exotic germ plasm, work with various mating systems, and integrate them with biotechnology, as needed, to enhance selection methods and accelerate product development. A plant breeder's diverse responsibilities can include contributing to and developing good public relations with scientific and other professional communities. He or she may participate in the development of patents or proposals and assist with the management and development of a plant breeding group. An entry-level plant breeder position requires a BS degree or equivalent. A minimum of 0–2 years of plant breeding or agronomical experience and/or training in plant breeding or plant science is also required.

15.3.2 QUALITY CONTROL

15.3.2.1 Quality Control Analyst

A quality control analyst is responsible for conducting routine and nonroutine analysis of raw materials. He or she compiles data for documentation of test procedures and reports abnormalities. A quality control analyst also reviews data obtained for compliance with specifications and reports abnormalities. He or she revises and updates standard operating procedures and may perform special projects on analytical and instrumental problem solving. An entry-level quality control analyst position requires a BS degree in a scientific discipline or equivalent and a minimum of 0–2 years of experience in quality control systems.

15.3.2.2 Quality Control Engineer

A quality control engineer is responsible for developing, applying, revising, and maintaining quality standards for processing materials into partially finished or finished products. He or she designs and implements methods and procedures for inspecting, testing, and evaluating the precision and accuracy of products and prepares documentation for inspection testing procedures. Depending on the job level, a quality control engineer is responsible for ensuring conformance to in-house specifications and good manufacturing practices and may conduct training programs. He or she may also be responsible for supervising the development and efforts of a quality control engineering group. An entry-level job requires a BS degree or equivalent and a minimum of 0–2 years of experience in quality control systems.

15.3.2.3 Environmental Health and Safety Specialist

An environmental health and safety specialist is responsible for developing, implementing, and monitoring industrial safety programs within the company. He or she inspects plant areas to ensure compliance with Occupational Safety and Health Administration (OSHA) regulations. He or she evaluates new equipment and raw materials for safety, and monitors employee exposure to chemicals and other toxic substances. Depending on the job level, a safety specialist may also conduct training programs in hazardous waste collection, disposal, and radiation safety regulations. An entry-level safety specialist job requires a BS degree or equivalent and a minimum of 0–2 years of related experience.

15.3.2.4 Quality Assurance Auditor

A quality assurance auditor is responsible for performing audits of production and quality control. He or she ensures compliance with in-house specifications, standards, and good manufacturing practices. The job requires a BS degree in a scientific discipline or equivalent and a minimum of 0–2 years of experience in biological or pharmaceutical manufacturing.

15.3.2.5 Validation Engineer

A validation engineer is responsible for the calibration and validation of equipment and systems and for assisting in the selection, specification, and negotiation of competitive pricing of equipment. He or she maintains all of the documentation pertaining to qualification and validation, and serves as an information resource for validation technicians, contractors, and vendors. At the entry level, the job requires a BS degree or equivalent, with 0–2 years of related experience.

15.3.2.6 Validation Technician

An entry-level validation technician would be responsible for developing, preparing the installation of, and revising test validation procedures/protocols to ensure that a product is manufactured in accordance with appropriate regulatory agency validation requirements, internal company standards, and current industry practices. A validation technician compiles and analyzes validation data, prepares reports, and makes recommendations for changes and/or improvements. He or she may also investigate and troubleshoot problems and determine solutions. He or she maintains appropriate validation documentation and files. An entry-level validation technician requires a high school diploma or equivalent and a minimum of 0–2 years of related experience.

15.3.3 CLINICAL RESEARCH

15.3.3.1 Clinical Research Administrator

A clinical research administrator is responsible for clinical data entry and validation to ensure legibility, completeness, and consistency of data. He or she assists users with requests for clinical documents and is responsible for working with physicians and/or their staff to clarify any questionable information. He or she may be responsible for auditing internal patient files and studies and for assisting with the development and evaluation of clinical record documents. At the entry level, a clinical research administrator typically develops an internal recordkeeping system, including maintaining, auditing data, and providing status and activity reports as required. The job requires a high school diploma or equivalent, with a minimum of 0–2 years of related experience.

15.3.3.2 Clinical Coordinator

A clinical coordinator must be familiar with the scientific/investigative process. Expertise may be limited to a specific functional area. A clinical coordinator must have good communication skills, both written and oral. He or she must also have project team experience and a familiarity with standard computer applications. Responsibilities include coordinating the clinical development plan as outlined by the company or clinical department and defining objectives, strategies, and studies. The clinical coordinator must provide support for planning, including detailed effort estimates, scheduling, and critical path analysis. He or she must monitor clinical activities to identify issues, variances, and conflicts, and analyze and recommend solutions. The clinical coordinator is also responsible for project staffing requirements and tracking drug supply to outside vendors, as well as providing ongoing objective updates on progress and problems with projects, tracking and following up on action items. The position requires a BS or a Bachelor of Arts (BA) degree in Health Science, Information Technology, or Business, and 3–5 years of experience in the healthcare industry.

15.3.3.3 Clinical Programmer

A novice clinical programmer is responsible for coordinating and monitoring the flow of clinical data into the computer database. He or she analyzes and evaluates clinical data, recognizes inconsistencies, and initiates the resolution of data problems. He or she implements data management plans designed to meet project and protocol deadlines and consults in the design and development of clinical trials, protocols, and case report forms. A clinical programmer also acts as a liaison between clinical management and subcommittees and project teams on an as-needed basis.

An entry-level position as a clinical programmer requires a BS degree or equivalent, although an MS degree is often preferred. A minimum of 0–2 years of experience in pharmaceutical programming in the clinical research area is also required.

15.3.3.4 Biostatistician

A biostatistician works with others to define and perform analyses of databases for publications, presentations to investigator meetings, and for meetings of professional societies. A position as a biostatistician requires at least an MS degree in biostatistics and 1–4 years of related experience.

15.3.3.5 Clinical Data Specialist

A clinical data specialist is responsible for collaborating with various departments on the design, documentation, testing, and implementation of clinical data studies. He or she develops systems for organizing data to analyze, identify, and report trends. A clinical data specialist also analyzes the interrelationships of data and defines logical aspects of data sets. A starting position as a clinical data specialist requires a BS degree or equivalent and a minimum of 0–2 years of related experience.

15.3.3.6 Drug Experience Coordinator

The major responsibility of a drug experience coordinator is to handle drug experience activities for marketed products. The candidate will also provide drug information on the products. He or she will oversee day-to-day processing of adverse event information for marketed products and will coordinate the receipt, classification, investigation, and processing of adverse experience reports. A position as a drug experience coordinator requires a Doctor of Pharmacy (Pharm D) degree or equivalent clinical training and 0–2 years of related experience.

15.3.3.7 Clinical Research Associate

A clinical research associate is responsible for the design, planning, implementation, and overall direction of clinical research projects. He or she evaluates and analyzes clinical data and coordinates activities of associates to ensure compliance with protocol and overall clinical objectives. He or she may also travel to field sites to supervise and coordinate clinical studies. An entry-level clinical research associate position typically requires a BS degree, a registered nurse degree, or an equivalent qualification, and a minimum of 0–2 years of clinical experience in medical research, nursing, or the pharmaceutical industry. Knowledge of FDA regulatory requirements is preferred.

15.3.3.8 Animal Handler

An animal handler is responsible for the daily care of research animals for experimental purposes. He or she cleans animal cages and racks, maintains records to comply with regulatory requirements and standard operating procedures, and performs preventive maintenance on facility equipment. The incumbent may also perform animal observation, grooming, and minor clinical tasks. An animal handler position requires a high school diploma or equivalent experience with a scientific background. A minimum of 0–2 years of relevant laboratory experience is also expected.

15.3.3.9 Animal Technician

An animal technician is responsible for the daily care of research animals for experimental purposes. He or she also coordinates with vendors and supervisors on operational, administrative, and technical responsibilities. The animal technician performs some surgery and postoperative care as directed and is responsible for overseeing procurement of animals and supplies, preventive maintenance of facility equipment, cleaning of animal cages and racks, daily rounds, and observation to check the animals' health status. He or she develops standard operating procedures and maintains records to comply with regulatory requirements. An entry-level animal technician job requires a high school diploma or equivalent experience with a scientific background and a minimum of 0–2 years of related laboratory experience.

15.3.3.10   Technical Writer

An entry-level technical writer is responsible for writing and editing standard operating procedures, clinical study protocols, laboratory procedure manuals, and other related documents. He or she edits and/or rewrites various sources of information into a uniform style and language for regulatory compliance, and assists in developing documentation for instructional, descriptive, reference, and/or informational purposes. An entry-level position requires a BS degree or equivalent and a minimum of 0–2 years of experience in technical documentation.

15.3.4   PRODUCT AND DEVELOPMENT

15.3.4.1   Product Development Engineer

An entry-level product development engineer is responsible for the design, development, modification, and enhancement of existing products and processes. The incumbent is involved in new product scale-up, process optimization, technology transfer, and process validation. He or she ensures that processes and design implementations are consistent with good labor and manufacturing practices. A product development engineer may also be responsible for contact with outside vendors and for the administration of contracts to accomplish goals. He or she works on problems of moderate scope, where analysis of a situation or data requires a review of identifiable factors. The job requires a BS degree or equivalent, and 0–2 years of related experience.

15.3.4.2   Production Planner/Scheduler

A production planner/scheduler is responsible for planning, scheduling, and coordinating the final approval of products through the production cycle. He or she coordinates production plans to ensure that materials are provided according to schedules to maintain production and provides input to management. When necessary, a production planner/scheduler works with the customer service, marketing, production, quality control, and sales departments to review backorder status, prioritize production orders, and deal with other potential schedule interruptions or rescheduling. An entry-level production planner/scheduler requires a bachelor's degree or equivalent and a minimum of 0–2 years of related experience.

15.3.4.3   Manufacturing Technician

Manufacturing technicians are responsible for the manufacture and packaging of potential and existing products. They operate and maintain small production equipment. They also weigh, measure, and check raw materials and ensure that manufactured batches contain the proper ingredients and quantities. They maintain records and clean production areas to comply with regulatory requirements, good manufacturing practices, and standard operating procedures. A manufacturing technician may also assist with in-process testing to make sure that batches meet product specifications. An entry-level position requires an associate degree in science and a minimum of 0–2 years of related experience in a manufacturing environment.

15.3.4.4   Packaging Operator

A packaging operator uses manual and/or automated packaging systems to label, inspect, and package final container products. He or she also enters data and imprints computer-generated labels, maintains records, and maintains the manufacturing/production area to comply with regulatory requirements, good manufacturing practices, and standard operating procedures. A packaging operator may also perform initial checks of completed documents for completeness and accuracy. The position requires a high school diploma or equivalent and a minimum of 0–2 years of experience in a manufacturing environment.

15.3.4.5   Manufacturing Research Associate

A manufacturing research associate is responsible for the implementation of production procedures to optimize manufacturing processes and regulatory requirements and has responsibilities in packaging and distribution processes. He or she may also help maintain production equipment. The position requires a BS degree in a scientific discipline or equivalent and a minimum of 0–2 years of experience in a manufacturing environment.

15.3.4.6 Instrument/Calibration Technician

An entry-level instrument/calibration technician is responsible for performing maintenance, testing, troubleshooting, calibration, and repair on a variety of circuits, components, analytical equipment, and instrumentations. He or she also calibrates instrumentation, performs validation studies, and specifies and requests purchase of components. He or she analyzes results, may develop, test specifications and electrical schematics, and maintains logs and required documentation. An instrument/calibration technician also maintains spare parts inventories and may prepare technical reports with recommendations for solutions to technical problems. An associate degree in electronics technology or equivalent is required, as is a minimum of 0–2 years of related experience.

15.3.4.7 Biochemical Development Engineer

A biochemical development engineer is responsible for the design and scale-up of processes, instruments, and equipment from the laboratory through the pilot plant and manufacturing process. He or she assists the manufacturing operations in problem solving with regards to equipment and systems and participates in the design and start-up of new manufacturing facilities and equipment. He or she develops and recommends new process formulas and technologies to achieve cost effectiveness and product quality. A biochemical development engineer also establishes operating equipment specifications and improves manufacturing techniques. He or she is involved in new product scale-up, process improvement, technology transfer, and process-validation activities. He or she works with various departments to ensure that processes and designs are compatible for new product technology transfer and to establish future process and equipment automation technology. The position requires a BS degree in biological, chemical, or pharmaceutical engineering, or a related discipline. A minimum of 0–2 years of experience is also required, preferably in the areas of pharmaceutical processes or research product development.

15.3.4.8 Process Development Associate

A process development associate is responsible for the implementation of production procedures to optimize manufacturing processes and regulatory requirements. He or she may also assist in process development, in creating scalable processes with improved product yield and reduced manufacturing systems costs. An entry-level process development associate may also be involved in packaging and distribution processes and in the maintenance of production equipment. He or she may research and implement new methods and technologies to enhance operations. The position requires a BS degree in a scientific discipline or equivalent and a minimum of 0–2 years of experience.

15.3.4.9 Assay Analyst

An assay analyst is responsible for doing cell cultures and performing assays and tests on tissue and cell cultures following standard protocols. He or she prepares glassware, reagents, and media for cell culture use. He or she also performs, prepares, and maintains tissues and cell cultures and maintains records required by good manufacturing procedures. An assay analyst also participates in the modification of assay procedures for routine implementation. The position requires a high school diploma or equivalent and a minimum of 0–2 years of related experience.

15.3.4.10 Manufacturing Engineer

A manufacturing engineer is responsible for developing, implementing, and maintaining methods, operation sequences, and processes in manufacturing. He or she works with the engineering department to coordinate the release of new products. He or she estimates manufacturing costs, determines time standards, and makes recommendations for process requirements of new or existing product lines. A manufacturing engineer also maintains records and reporting systems as required for the coordination of manufacturing operations. An entry-level job as a manufacturing engineer requires a BS degree in a scientific discipline or equivalent and a minimum of 0–2 years of related experience, preferably in research product development or in a manufacturing environment.

15.3.5 REGULATORY AFFAIRS

15.3.5.1 Regulatory Affairs Specialist

An entry-level regulatory affairs specialist coordinates and prepares document packages for submission to regulatory agencies, internal audits, and inspections. He or she compiles all materials required for submissions, license renewals, and annual registrations. An incumbent monitors and improves tracking and control systems and keeps abreast of regulatory procedures and changes. He or she may work with regulatory agencies and recommend strategies for earliest possible approvals of clinical trial applications. An entry-level position requires a BS degree or equivalent and a minimum of 0–2 years of related experience.

15.3.5.2 Documentation Coordinator

A documentation coordinator provides clerical and administrative support related to a company's documentation system requirements. He or she audits all documentation manuals to ensure that they are accurate, up-to-date, and available to appropriate personnel. A documentation coordinator also files and retrieves all master documents. The position requires a high school diploma or equivalent and a minimum of 0–2 years of related experience.

15.3.5.3 Documentation Specialist

An entry-level documentation specialist is responsible for coordinating all activities related to providing required documentation and implementing related documentation systems. He or she coordinates the review and revision of procedures, specifications, and forms. He or she also assists in compiling regulatory filing documents and in maintaining computerized files to support all documentation systems. The job requires a BS degree in a related field or equivalent and a minimum of 0–2 years of experience in documentation, quality assurance, technical writing, or an equivalent position.

15.3.6 INFORMATION SYSTEMS

15.3.6.1 Library Assistant

A library assistant maintains serial control and locates and orders journal articles and/or books on relevant subjects that are unavailable at local libraries. He or she performs special data-gathering projects as requested and is responsible for online computer searching of scientific databases. The position of entry-level library assistant requires an associate degree or equivalent and a minimum of 0–2 years of relevant library experience or the completion of an on-the-job training program.

15.3.6.2 Scientific Programmer/Analyst

A scientific programmer/analyst designs, develops, evaluates, and modifies computer programs for the solution of scientific or engineering problems and for the support of R&D efforts. He or she analyzes existing systems and formulates logic for new systems. A scientific programmer/analyst also devises logical procedures, prepares flowcharts, performs coding, tests, and debugs programs. He or she provides input for the documentation of new or existing programs and determines system specifications, input/output processes, and working parameters for hardware/software compatibility. An analyst also contributes to decisions on policies, procedures, expansion strategies, and product evaluations. An entry-level scientific programmer/analyst position requires a BS degree in a related discipline or equivalent and a minimum of 0–2 years of experience.

15.3.7 MARKETING AND SALES

15.3.7.1 Market Research Analyst

A market research analyst is responsible for researching and analyzing the company's markets, competition, and product mix. He or she performs literature research, analyzes and summarizes data, and makes presentations on new market and technical areas. He or she also analyzes the competitive environment as well as future marketing trends and makes appropriate recommendations. He or she conducts market surveys, summarizes results, and assists in the preparation, presentation, and follow-up of research proposals. An entry-level position as a market research analyst requires a bachelor's degree or equivalent and a minimum of 0–2 years of experience in market research, competitive analysis, and product planning, as well as excellent writing skills.

15.3.7.2 Systems Analyst

A systems analyst is responsible for system-level software and for maintenance of the operating system(s), various layered products, system tuning, and various levels of user assistance. The systems analyst is responsible for the operation of all system software, performing upgrades, and maintaining related system and user documentation. He or she assists in the implementation of system validation and documentation, in system capacity planning, and system configuration. In addition, the systems analyst troubleshoots system-related problems and interacts with vendors. An entry-level position as a systems analyst requires a degree in data processing and a minimum of 0–2 years of experience.

15.3.7.3 Sales Representative

A sales representative is responsible for direct sales of company products or services. He or she calls on prospective customers, provides product information and/or demonstrations, and quotes appropriate customer prices. A sales representative is also responsible for new account development and growth of existing accounts within an established geographic territory. A sales representative must meet assigned sales quotas and may handle key company accounts or act as an account manager for national or major accounts. Depending on the level of the position, an experienced incumbent may also assist in the training of other sales representatives. An entry-level sales representative position requires a bachelor's degree or equivalent, a minimum of 0–2 years of related sales experience, and some knowledge of the company's products.

15.3.7.4 Customer Service Representative

Customer service representatives are responsible for ensuring product delivery in accordance with customer requirements and manufacturing capabilities and for responding to customer product inquiries and satisfaction issues. He or she answers telephones, takes product orders, and inputs sales order data into consumer data systems. He or she also investigates problems related to the shipment of products, credits, and new orders. He or she may also be responsible for sale order administration and/or inside sales. The entry-level position requires a college education or equivalent, with preference toward a BS degree in a technical or scientific field. A minimum of 0–2 years of related experience in diagnosing and troubleshooting products in the pharmaceutical industry may also be required.

15.3.7.5 Technical Service Representative

A technical service representative provides technical direction and support to customers on the operation and maintenance of company products. He or she also serves as a contact for customers regarding technical and service-related problems. A technical service representative also demonstrates the uses and advantages of products. An entry-level technical service representative position requires an associate degree or equivalent and a minimum of 0–2 years of experience.

15.3.8 ADMINISTRATION

15.3.8.1 Technical Recruiter

A technical recruiter is responsible for recruiting, interviewing, and screening applicants for technical exempt and nonexempt positions. He or she coordinates preemployment physicals, travel, reporting dates, security clearances, and employment processing for new hires. He or she also conducts employee advertising and reviews employment agency placements. In addition, a technical recruiter maintains college recruiting, affirmative action, and career development programs. An entry-level position as a technical recruiter requires a BS degree or equivalent and a minimum of 0–2 years of experience.

15.3.8.2 Human Resource Representative

A human resources representative is responsible for a variety of activities in personnel administration, including employment, compensation and benefits, employee relations, equal employment opportunities, and training programs. He or she conducts job interviews, counsels' employees, maintains records, and conducts research and analyzes data on assigned projects. An entry-level position as a human resource representative requires a BS degree or equivalent and a minimum of 0–2 years of related experience.

15.3.8.3 Patent Administrator

A patent administrator is responsible for preparing and coordinating all procedural documentation for patent filings and applications. He or she tracks in-house research studies and recommends the need for and timing of patent filings. A patent administrator also assists attorneys with the drafting and editing of patent applications, collects, and evaluates supporting data. This position requires the maintenance of a tracking system to comply with trademark regulations. He or she may also be called upon to assist with determining the necessity of and approach to contracts to ensure protection of the company's proprietary technology. A patent administrator is also typically responsible for tracking and paying legal fees. An entry-level patent administrator position requires a BS degree or equivalent and a minimum of 0–2 years of experience.

15.3.8.4 Patent Agent

A patent agent is responsible for preparing, filing, and processing patent applications for the company. He or she negotiates and drafts patent licenses and other agreements. A patent agent also conducts state-of-the-art searches and may assist with the appeal and interference proceedings. A patent agent also performs other duties as required. The entry-level patent agent position requires a BS degree or equivalent with 0–2 years of related experience. It also requires registration to practice before the US Patent and Trademark Office.

15.4 WHICH JOB IS GOOD FOR ME?

Most jobs in academic institutes are usually available from BS graduates onward, starting with laboratory technicians or assistants to the laboratory head/director, who should have a PhD degree plus 10–15 years of laboratory experience. After getting a BS degree in biotechnology, students can either join any university or pursue an MS/MSc in a specialized branch of biotechnology, such as agricultural biotechnology, microbial biotechnology, industrial biotechnology, animal biotechnology, environmental biotechnology, or genetic and forensic fields. Although a laboratory assistant position may not be very lucrative, it will definitely give hands-on experience. Job prospects are high after 2–3 years of laboratory experience. After obtaining BS/MS degrees, candidates can also do specialized courses in patents and IPRs, healthcare management, and they can also do MBA courses. A BS degree with an MBA qualification from a reputed university/institute may also help candidates to enter into sales and marketing jobs in pharmaceuticals, biopharmaceuticals, biotechnology, or any other health care industries. Jobs in sales and marketing are very challenging and require numerous field visits, but they are also equally lucrative. Students who have obtained an MS/MSc degree but do not want to work in labs or in research projects can also choose to do an MBA and enter into the sales and marketing segment of biotechnology industries. Students with an MS/MSc degree in biotechnology can enroll for a PhD program in their respective field of interest and may pursue their careers as research scientists in various positions, from junior scientist to the president level in R&D. R&D labs offer the most challenging job opportunities in the biotechnology industry, because these are the people who bring about new discoveries and new inventions. If scientists work in universities and discover new inventions that cause the development of new fields, they may reap rewards in the form of a Nobel Prize in Physiology and Medicine, with a great deal of public recognition and appreciation. On the other hand, if scientists who work in industries discover new inventions, they do not get much public accolade and appreciation, but they get an exuberantly high salary from their company. It is up to the candidates to decide into which field he or she would like to take his or her career.

Another frequently asked question about careers in biotechnology is that of which field of biotechnology is the most challenging or career oriented. The answer would be that all fields of biotechnology provide challenging career opportunities. For example, researchers who work in agriculture, biology, genetics, and medicine are at the forefront of new biotechnology discoveries. These men and women are working to unravel the genetic codes that govern the biological processes of different forms of life so that they can be understood and, when appropriate, modified. Life sciences researchers may work in an academic environment such as a university, or for a company or a government agency. They can focus their work on animals, bacteria, humans, plants, viruses, or any other life forms in which they have a special interest. The discoveries made in government, university, or corporate laboratories are the first steps toward genetically engineered products or processes such as new vaccines, drugs, or plant varieties. There are many jobs related to biotechnology that are held by people without extensive science or engineering expertise. These individuals must understand the science of biotechnology, but their primary talent may be in communications or in some other area. Regulatory officials develop guidelines for biotechnology research and the development of new products and processes. They work with companies, government, or university researchers to review proposed research plans and assess the safety of resulting products. Regulatory officials must approve biotechnology research plans before they can be executed and biotechnology products before they can be marketed.

Individuals involved with the regulation of biotechnology research and product development generally work for a federal or state government agency. Public relations personnel provide comprehensible information to the general public about new biotechnology products and processes. They translate complex scientific information about new discoveries for nonscientists. Sales executives work with the dealers and distributors of biotechnology products. They have expertise in marketing skills and are knowledgeable about the products. Patent lawyers who specialize in biotechnology help scientists, companies, or universities protect their legal rights to new discoveries. They file patent applications for their clients and interact with the US Patent Office.

15.5 WHY ARE RESEARCH AND DEVELOPMENT JOBS THE MOST CHALLENGING?

Among all phases of biotechnology, the R&D phase is the most challenging one, as it allows scientists to discover something new, such as the discovery of a new drug molecule, better treatment modalities, or discovery of a new diagnostic test for the early detection of diseases. Challenging career opportunities are available in specialized fields of biotechnology where candidates can work in the domain of drug discovery to find new treatments for various dreaded diseases. We present a list of some of the major fields where one can work and be part of a research team.

15.5.1 MEDICAL AND DIAGNOSTICS SECTORS

15.5.1.1 Detecting and Treating Hereditary Diseases

Many diseases—including some types of anemia, cystic fibrosis, Huntington's disease, and some blood disorders—are the result of a defective gene that parents pass to their children. Biotechnologists are working to identify and locate where defects occur in genes that are related to hereditary diseases. Once the correct genetic code is known, healthcare professionals hope that in the future they will be able to replace the missing or defective genes to make the individual healthy. Currently, prospective parents can be screened for such genetic defects and counseled about the likelihood of their children being affected. Fetuses are being screened for genetic disorders before they are born and genetic counselors play an important role in informing parents concerning the test results. Genetic counselors prepare parents for the birth and early medical treatment of a child with a genetic disorder.

15.5.1.2 Heart Disease

Heart attacks occur when a blood clot enters one of the coronary arteries and cuts off blood flow to a portion of the heart. If the artery is not reopened quickly, severe damage to the heart can occur. Doctors can now prescribe a genetically engineered drug called *tissue plasminogen activator* (TPA) that travels to the blood clot and breaks it up within minutes, restoring blood flow to the heart and lessening the chance of permanent damage.

15.5.1.3 Cancer

Medical doctors are using biotechnology to treat cancer in several ways. Genetically engineered proteins called lymphokines seem to work with the body's immune system to attack cancer cells and growth inhibitor proteins seem to slow the reproduction of cancer cells. Highly specific and purified antibodies can be loaded with poisons that locate and destroy cancer cells.

15.5.1.4 AIDS

Genetic engineering has produced several substances that show promise in the treatment of AIDS. These substances stimulate the body's own immune system to fight the disease.

15.5.1.5 Veterinary Medicine

Veterinarians and professionals in animal science are using biotechnology discoveries to improve animal health and production. Genetically engineered vaccines, monoclonal antibody technology, and growth hormones are three developments that are making this possible. Questions concerning food safety, economic impacts, and animal health issues have been raised by those opposing the use of growth hormones and have made their use controversial.

15.5.1.6 Vaccines

Most vaccines are made from viruses or bacteria that have been weakened or killed. However, because live viruses or bacteria are often included in these vaccines, they are not without side effects. An animal could become sick from the vaccine. rDNA technology allows the production of synthetic vaccines that do not have this risk. rDNA vaccines have been developed for swine and cattle diarrhea, and research on other vaccines is continuing.

15.5.1.7 Monoclonal Antibodies

Antibodies are produced naturally by animals when invaded by a disease-causing organism. Each type of antibody is very specific in that it recognizes and attacks only one particular disease organism. Monoclonal antibody technology allows biotechnologists to produce large amounts of purified antibodies for use in the development of vaccines. Antibodies can also be used to diagnose illnesses and can detect drugs, viral and bacterial products, and other substances. For example, home pregnancy test kits use antibodies to detect the presence of a certain hormone in the urine.

15.5.1.8 Growth Hormones

Several biotechnology companies are seeking approval from the federal government for genetically engineered proteins that improve meat and milk production in cattle or pigs. Bovine somatotropin for cattle and porcine somatotropin for pigs could impact the life cycle of farm animals by increasing their rate of growth and milk production and producing leaner carcasses.

15.5.2 AGRICULTURAL SECTOR

Farmers and other agricultural professionals are being faced with decisions about the use of biotechnology products in their operations. In addition to animal health products and growth hormones that are available for livestock production, a host of crop-production products is (or soon will be) on the market. Scientists are exploring the genetic modification of food crops to achieve desirable characteristics such as high yield, increased protein or oil production, disease resistance, or pest resistance.

15.5.2.1 Crop Yield

Crop yields are controlled not by one gene but by many genes acting together. Scientists are working to identify these genes and their contribution to yield so that crops can be genetically modified for increased production.

15.5.2.2 Protein and Oil Content of Seeds

By modifying the genes that control the accumulation of protein and oil in seeds such as corn and soybeans, biotechnology researchers hope to develop more nutritious crops or crops that produce modified oils for food or industrial uses. For example, by changing the kinds or amounts of fatty acids stored in soybeans, new oils can be developed.

15.5.2.3 Environmental Conditions

Most crops do not grow well in dry, salty, or alkaline soils. Most cannot withstand heavy frosts or extreme temperature changes. Biotechnologists are trying to genetically engineer crops that will grow well in the poorest food-producing areas of the world, where these conditions are often present.

15.5.2.4 Disease and Pest Resistance

Genes for disease or pest resistance have been identified for several crops. If crops can be genetically modified to include a resistant gene that makes them undesirable to pests, the amount of chemical pesticide needed could be reduced, a less expensive and more environmentally friendly option. Crops that "tolerate" herbicides to which they are normally sensitive are now on the market. When used properly, the insertion of an herbicide-resistant gene into crops can give a farmer more choice in selecting an herbicide. There is also the opportunity to develop crops that are tolerant of herbicides that are less damaging to the environment. Many other diseases can be treated with genetically engineered products. Doctors can use a genetically engineered vaccine to treat human hepatitis B or a growth hormone to help children with dwarfism. Other treatments developed through genetic engineering techniques include a protein to control blood clotting in hemophiliacs, a hormone that stimulates red blood cell production to fight anemia, and antibodies that discourage organ rejection by transplant patients.

15.5.3 LAW ENFORCEMENT SECTOR

Biotechnology has provided law enforcement professionals with another way of placing a suspect at the scene of a crime. This area of study, called forensic biotechnology, uses a method called DNA fingerprinting. This method is based on the fact that each individual's DNA is highly unlikely to be identical to any other person's DNA (unless he or she has an identical twin). By examining traces of tissue, hair, tooth pulp, blood, or other body fluids left at the scene of a crime, a suspect can be linked to a crime location with great accuracy. Many states now accept DNA fingerprinting results as admissible evidence in criminal and civil trials.

15.5.4 PRODUCT MANUFACTURING SECTOR

After a biotechnology product has been approved for use, engineers with a BS, Bachelor of Technology (Btech), and/or Bachelor of Engineering (BE) degree in biotechnology engineering are needed to manufacture it. The engineers are trained to handle bioreactors to produce biotechnology products in large quantities while maintaining the quality as per FDA guidelines. A biotechnology company offers many of the same career opportunities as any other manufacturing business. However, in addition to specific skills in engineering, scale-up, quality control, and other manufacturing processes, individuals employed in the biotechnology industry will need a solid background in the biological sciences. Industrial chemists find that many natural biological products such as amino acids, enzymes, and vitamins can be manufactured more efficiently using biotechnology. The gene or genes that produce the natural biological product can be transferred into an organism, perhaps a bacterium that now starts producing it. Microorganisms are capable of producing many common organic chemicals, such as ethanol. They can also produce proteins for vaccines and for other uses through a fermentation process.

15.5.5 MICROBIAL ENGINEERING SECTOR

As the world's population grows, so does the problem of waste disposal. Biotechnology is helping waste management experts in several ways. Microorganisms such as bacteria and microbes can easily adapt to different environments and live off their surroundings. Biotechnologists have found bacteria in solid waste sites that can break down various kinds of waste for their own use. rDNA techniques can enhance these capabilities, and new strains of waste degraders could be developed. Biotechnology can also be used to improve microorganisms used in the treatment of wastewater to make the process cheaper and more efficient. Those involved in the energy industries are finding that living organisms modified by rDNA technologies can improve energy production and use. Ore-containing rock is often mixed with minerals that must be separated from the rock by a heat process called smelting. Some microorganisms can dissolve and absorb the minerals, lessening the need for smelting. Other bacteria can force oil out of rocks where conventional drilling is not possible. The production of energy from biomass, especially waste plant materials like wood chips or corn stalks, also benefits from biotechnology. Microorganisms can produce enzymes that degrade the plant materials, making them useful in energy production.

15.6 SALARY AND INCENTIVES IN BIOTECHNOLOGY

Biotechnology as a subject has grown rapidly and, as far as employment is concerned, it has become one of the fastest growing sectors. Employment records show that biotechnology will have great scope in the future. Biotechnologists can find careers with pharmaceutical companies, and in the chemical, agriculture, and allied industries. They can be employed in the areas of planning, production, and management of bioprocessing industries. There is large-scale employment in research laboratories run by the government as well as those in the corporate sector. Biotechnology students in India may find work in government-based entities such as state universities and research institutes or at private centers as research scientists/assistants. Alternatively, they may find employment in specialized biotechnology companies or biotechnology-related companies such as pharmaceutical firms, food manufacturers, or aquaculture and agricultural companies. Companies that are engaged in businesses related to life sciences (ranging from equipment, chemicals, pharmaceuticals, diagnostics, etc.) also consider a biotechnology degree relevant to their field. The work scope can range from research, sales, marketing, administration, quality control, animal breeders, and technical support. Armed with this powerful combination of fundamental cell and molecular biology and applied science, graduates are well placed to take up careers in plant, animal, or microbial biotechnology laboratories or in horticulture, food science, commerce, and teaching.

Biotechnology is a highly remunerative career option. Biotechnologists employed in public sector industries and research institutes have paid packages according to their qualifications and experiences. Those working in the private sector, corporate houses, and industries have very lucrative remuneration. Those employed in teaching have better pay packages, which are even very lucrative in private colleges and institutions. The average annual salary in the industrial sector is listed below for your information but note that a compensation package can always be negotiated with the employer.

- Lab technician (BS level): $15,000–20,000
- Lab manager (BS level plus experience): $20,000–30,000
- Research assistant (MS level plus experience): $30,000–50,000
- Postdoctoral fellow (PhD level): $35,000–50,000
- Research scientist (PhD level plus experience): $50,000–80,000
- Research director/vice president (PhD level plus experience): $80,000–120,000
- Company CEO/president: $150,000 or more

However, in academic institutions, the compensation package for all positions may vary depending on the university and location. An opportunity to negotiate a salary package is not a common phenomenon because most positions have a fixed compensation package.

PROBLEMS

Q1. Discuss how different countries invest money in the progress of biotechnology.
Q2. What career opportunities are available in R&D?
Q3. Describe the job situation in the US biotechnology sector.

ASSIGNMENTS

Make a chart showing various job opportunities available in the biotechnology field and discuss these with your course instructor and classmates. Next, organize a quiz contest based on who is who in biotechnology industries.

FIELD VISIT

With the help of your course instructor, organize a field visit to any biotechnology-based industry located in your own city or town. The field experience will help you to get real-time exposure to all segments of biotechnology—from R&D labs, manufacturing plants, animal testing facilities, to quality assurance/quality control labs—and may also help you to decide on the kind of field you might be keen to undertake.

REFERENCES AND FURTHER READING

Annual Report. 2008. UCB. www.ucb.com/_up/ucb_com_investors/documents/UCB_Annual_Report_2008_EN.pdf (retrieved on May 25, 2009).
Biogen Idec. Product Pipeline, in 2007 Annual Report, 2008.
Biogen Idec. SEC Form 10-K, pp. 1, 11, 2008. http://edgar.sec.gov/Archives/edgar/data/875045/000095013508001029/b68103bie10vk.htm (retrieved on June 7, 2008).
Calabro, S. Genzyme put patients first, and grew to become a multi-billion-dollar company: But empires don't survive on altruism. *Pharmaceutical Executive*, March 1, 2006. www.pharmexec.com/pharmexec/article/articleDetail.jsp?id=310976&&pageID=7&searchString=genzyme.

Coy, P. The search for tomorrow. *Business Week*, October 11, 2001. www.businessweek.com.

Dineen, J.K., and Leuty, R. Amgen slows its Bay Area expansion. *San Francisco Business Times* Website. www.bizjournals.com (retrieved on August 14, 2007).

Eugene, R. Special report: The birth of biotechnology. *Nature* 421: 456–457, 2003. www.nature.com.

Food and Drug Administration. FDA approves new orphan drug for treatment of pulmonary arterial hypertension. Press release, 2007. www.fda.gov/bbs/topics/NEWS/2007/NEW01653.html (retrieved on June 22, 2007).

Forbes. Henri Teemer profile, December 2007. www.forbes.com/finance/mktguideapps/personinfo/FromPersonIdPersonTearsheet.jhtml?passedPersonId=222120 (retrieved on July 10, 2008).

Genentech. Corporate Overview. www.gene.com.

Gilead Sciences. Donald H. Rumsfeld named chairman of Gilead Sciences. Press release, 1997. www.gilead.com/wt/sec/pr_933190157/ (retrieved on June 3, 2007).

Gilead Sciences. Gilead invests $25 million in Corus Pharma; establishes equity position in company with late-stage product candidate for Cystic Fibrosis. Press release, 2006. www.gilead.com/pr_842319 (retrieved on August 15, 2007).

Gilead Sciences. Gilead sciences completes acquisition of Raylo Chemicals Inc. Press release, 2006. www.gilead.com/wt/sec/pr_926645 (retrieved on June 7, 2007).

Gilead Sciences. Gilead sciences to Acquire Myogen, Inc. for $2.5 billion. Press release, 2006. www.gilead.com/pr_910704 (retrieved on August 15, 2007).

Gilead Sciences. Parion sciences and Gilead sciences sign agreement to advance drug candidates for pulmonary disease. Press release, 2007. www.gilead.com/pr_1040975 (retrieved on August 15, 2007).

Gilead Sciences. U.S. Food and Drug Administration approves Gilead's Letairis treatment of pulmonary arterial hypertension. Press release, 2007. www.gilead.com/wt/sec/pr_1016053 (retrieved on June 16, 2007).

Gilead Sciences and Bristol-Myers Squibb. U.S. Food and Drug Administration (FDA) approves Atripla. Press release, 2006. www.gilead.com/pr_881419 (retrieved on December 15, 2007).

Langer-Gould, A., Atlas, S.W., Green, A.J., Bollen, A.W., and Pelletier, D. Progressive multifocal leukoencephalopathy in a patient treated with natalizumab. *N. Engl. J. Med.* 353: 375–381, 2005.

The National Medal of Technology Recipients 2005 Laureates. The National Medal of Technology and Innovation. United States Patent and Trademark Office. www.uspto.gov/nmti/recipients_05.html (retrieved on February 26, 2009).

Pennsylvania Bio—Member Listings. *Pennsylvania Bio*. www.pennsylvaniabio.org.

Pollack, A. F.D.A. backs AIDS pill to be taken once a day. *New York Times*, 2006. www.nytimes.com/2006/07/13/business/13drug.html (retrieved on September 20, 2007).

Pollack, A. Gilead's drug is approved to treat a rare disease. *New York Times*, 2007. www.nytimes.com/2007/06/16/business/16gilead.html (retrieved on June 16, 2007).

Schouten, E. Genzyme's lifelong commitments. *NRC Handelsblad*, 2005. www.gaucher.org.uk.

Schwartz, N.D. Rumsfeld's growing stake in Tamiflu. *CNN*, 2005. http://money.cnn.com/2005/10/31/news/newsmakers/fortune_rumsfeld/?cnn=yes (retrieved on June 3, 2007).

Staff Writers. Roche makes $43.7B bid for Genentech. *Genetic Engineering and Biotechnology News*, July 21, 2008. ISSN: 1935472X. www.genengnews.com.

Who made America? Herbert Boyer. *PBS*. www.pbs.org.

Index